SCRAP TIRE DERIVED GEOMATERIALS
OPPORTUNITIES AND CHALLENGES

BALKEMA – Proceedings and Monographs
in Engineering, Water and Earth Sciences

PROCEEDINGS OF THE INTERNATIONAL WORKSHOP ON SCRAP TIRE DERIVED
GEOMATERIALS – OPPORTUNITIES AND CHALLENGES (IW-TDGM 2007),
23–24 MARCH 2007, YOKOSUKA, JAPAN

Scrap Tire Derived Geomaterials
Opportunities and Challenges

Editors

Hemanta Hazarika
Akita Prefectural University, Akita, Japan

Kazuya Yasuhara
Ibaraki University, Hitachi, Japan

Taylor & Francis
Taylor & Francis Group

LONDON / LEIDEN / NEW YORK / PHILADELPHIA / SINGAPORE

INTERNATIONAL WORKSHOP ON SCRAP TIRE DERIVED GEOMATERIALS
"OPPORTUNITIES AND CHALLENGES"

March 23-24, 2007   Yokosuka, Japan

Taylor & Francis is an imprint of the Taylor & Francis Group, an informa business

© 2008 Taylor & Francis Group, London, UK

Typeset by Charon Tec Ltd (A Macmillan Company), Chennai, India
Printed and bound in Great Britain by Antony Rowe Ltd (CPI-Group), Chippenham, Wiltshire

Published by: Taylor & Francis/Balkema
                P.O. Box 447, 2300 AK Leiden, The Netherlands
                e-mail: Pub.NL@tandf.co.uk
                www.balkema.nl, www.taylorandfrancis.co.uk, www.crcpress.com

ISBN 13: 978-0-415-46070-5 (hbk)
ISBN 13: 978-0-203-92932-2 (ebook)

Scrap Tire Derived Geomaterials – Opportunities and Challenges – Hazarika & Yasuhara (eds)
© 2008 Taylor & Francis Group, London, ISBN 978-0-415-46070-5

Table of Contents

Technical papers

Part 1 Mechanical properties, modeling and novel applications

Part 2   Case studies, design methods and field applications

Preface

Emerging geomaterials derived from scrap tires have been receiving increasing attention in recent years and gaining acceptance as suitable and alternative geomaterials worldwide, since they provide cost-effective solutions to many practical problems. Although, a considerable experience has accumulated in this specialized field over the past one decade, still a lot needs to be done and explored. SCRAP TIRE DERIVED GEOMATERIALS is a compilation of peer reviewed papers presented at the International Workshop on Scrap Tire Derived Geomaterials – Opportunities and Challenges – (IW-TDGM 2007), held at Yokosuka, Japan from 23–24 March, 2007.

This workshop was the first ever international forum to promote the opportunities for the exchange of knowledge and experiences, disseminating information about experiments, numerical modeling, design and practical applications related to scrap tire derived geomaterials (TDGM). The aim of this workshop was to: (1) bring together people from all spheres to exchange knowledge and discuss the new developments in this emerging field of geoenvironmental and geotechnical engineering. (2) exchange information to establish a consensus for common design and construction methods, and (3) explore novel applications of Scrap Tire Derived Geomaterials (TDGM) with ideas beneficial to our common future – a sustainable environment. The workshop, which was held under the auspices of the Port and Airport Research Institute (PARI), Japan and the Kanto Branch of the Japanese Geotechnical Society (JGS Kanto), Japan, brought together people from various discipline working in industry, academia and public sectors to discuss the latest ideas and development in the TDGM.

This book reflects the current research and advances made in the TDGM. It contains papers from across the world reflecting the international importance given to the areas such as: (1) use of whole tires, tire shreds and tire chips, (2) tire chips mixed with soils and/or treated materials, (3) mechanical properties and modeling, (4) geoenvironmental impact and assessment, (5) novel concept, case studies and field applications. The book contains thirty six contributions including two keynote lectures, three special lectures, four theme lectures and twenty seven technical papers. All the papers have been thoroughly reviewed by at least two reviewers selected from the international panel of experts in this field in order to check for relevance to the theme as well as the quality of technical contents and presentation.

The publication of this volume has been possible through the sustained efforts of some of the core members of the organizing committee of IW-TDGM 2007. The editors express their sincere thanks to the members (Dr. Ashoke K. Karmokar, Dr. Yoshio Mitarai, Dr. Kazuhiro Kaneda, and Mr. Takeshi Nagatome), who have worked tirelessly behind the scene for the successful publication of this book. Special thanks go to Dr. Takao Kishida of Toa Corporation, Japan for his timely and valuable advice and Ms. Yasuko Nishikawa of Kanto Branch of the Japanese Geotechnical Society for her tireless effort and dedication, which was instrumental in timely publication of this book.

SCRAP TIRE DERIVED GEOMATERIALS provides a wealth of knowledge and information and promotes the geotechnical approach for material recycling of tire derived geomaterials, which ultimately contributes towards a sustainable global environment. This book is aimed at people working in related industry, academia and public sector. It contains current thinking and observable facts that can impact on the development of those individuals and organizations concerned with geotechnical engineering, geoenvironmental engineering, waste reuse and management, sustainable construction and cost-performance.

The editors, therefore, hope that in the years to come, the developments and the state of arts on TDGM that have been compiled in this book will help in the advancement of research on TDGM, and offer possibilities of new ideas and applications. The dissemination of the latest information and technology that was possible through this manuscript, we can contribute to a better and safer environment by making use of this *smart geomaterial* aggressively in the geotechnical construction.

Hemanta Hazarika
Chairman, IW-TDGM 2007

Organization

INTERNATIONAL ADVISORY COMMITTEE

Patron: Mr. Makoto Owada, *President, Port and Airport Research Institute, Japan*
Advisors: Prof. Kenji Ishihara, *President, Kanto Branch of JGS, Japan*
(Past President, ISSMGE)
Mr. Kiyoyasu Mikanagi, *President, Recycle Solution, Japan*
(Past President, Japan Port and Harbour Association)

Promotional & advisory panel

Prof. Akira Asaoka, *Nagoya University, Japan*
Prof. Richard J. Bathurst, *Royal Military College of Canada, Canada*
Prof. Dennes T. Bergado, *Asian Institute of Technology, Thailand*
Prof. Malcolm Bolton, *Cambridge University, UK*
Prof. Tuncer B. Edil, *University of Wisconsin-Madison, USA*
Dr. Yasushi Hosokawa, *Port and Airport Research Institute, Japan*
Prof. Dana N. Humphrey, *University of Maine, USA*
Prof. Adrian F.L. Hyde, *University of Sheffield, UK*
Prof. Isao Ishibashi, *Old Dominion University, USA*
Prof. Masashi Kamon, *Kyoto University, Japan*
Prof. Madhira R. Madhav, *Vice President, ISSMGE, India*
Dr. Hiroshi Miki, *President, Japan Chapter of International Geosynthetic Society, Japan*
Mr. Shozo Naemura, *Public Work Research Center, Japan*
Prof. Pedro S. Seco e Pinto, *President, ISSMGE, Portugal*
Prof. G. Venkatappa Rao, *Indian Institute of Technology Delhi, India*
Prof. Eun Chul Shin, *University of Incheon, Korea*
Dr. Shigeo Takahashi, *Port and Airport Research Institute, Japan*
Prof. Fumio Tatsuoka, *President, Japanese Geotechnical Society, Japan*
Prof. Yeo Won Yoon, *Inha University, Korea*

ORGANIZING COMMITTEE

Chairman: Dr. Hemanta Hazarika, *Port and Airport Research Institute, Yokosuka*
Co-chairman: Prof. Kazuya Yasuhara, *Ibaraki University, Hitachi*
Executive Advisor: Dr. Takao Kishida, *Toa Corporation, Yokohama*

Scientific & advisory panel

Mr. Takashi Hasumi, *Recycle Solution, Tokyo*
Mr. Shigeru Higashiyama, *National Institute for Land and Infrastructure Management (NILIM), Yokosuka*
Prof. Masayuki Hyodo, *Yamaguchi University, Ube*
Dr. Sumio Horiuchi, *Shimizu Corporation, Tokyo*
Dr. Hajime Imanishi, *Samsung Corporation, Korea*
Mr. Minoru Kawaida, *Central Nippon Expressway Co., Ltd., Tokyo*

Dr. Yoshiaki Kikuchi, *Port and Airport Research Institute, Yokosuka*
Dr. Hideo Komine, *Ibaraki University, Hitachi*
Prof. Jun Otani, *Kumamoto University, Kumamoto*
Mr. Koichi Sahara, *Port and Airport Research Institute, Yokosuka*
Dr. Toru Sueoka, *Kanto Branch of Japanese Geotechnical Society, Tokyo*
Dr. Takahiro Sugano, *Port and Airport Research Institute, Yokosuka*
Mr. Kinya Suzuki, *Bridgestone Corporation, Tokyo*
Prof. Tomiya Takatani, *Maizuru National College of Technology, Kyoto*
Mr. Hideo Takeichi, *Bridgestone Corporation, Tokyo*
Dr. Norihisa Tatarazako, *National Institute for Environmental Studies, Tsukuba*
Prof. Ikuo Towhata, *University of Tokyo, Tokyo*
Mr. Ryuichiro Ushijima, *Port and Airport Research Institute, Yokosuka*

Secretary generals

Dr. Ashoke K. Karmokar, *Bridgestone Corporation, Tokyo*
Dr. Yoshio Mitarai, *Toa Corporation, Yokohama*

Scientific program & general affairs

Dr. Kazuhiro Kaneda, *Port and Airport Research Institute, Yokosuka*
Mr. Takeshi Nagatome, *Toa Corporation, Yokohama*

Executive members

Dr. Kiyoshi Fukutake, *Shimizu Corporation, Tokyo*
Dr. Yoshihisa Miyata, *National Defense Academy, Yokosuka*
Dr. Satoshi Murakami, *Ibaraki University, Hitachi*
Ms. Yasuko Nishikawa, *Kanto Branch of Japanese Geotechnical Society, Tokyo*
Mr. Shunsuke Nara, *Port and Airport Research Institute, Yokosuka*
Mr. Tomohiro Tanaka, *Port and Airport Research Institute, Yokosuka*
Dr. Juichi Yajima, *Meisei University, Tokyo*
Dr. Suguru Yamada, *Yamaguchi University, Ube*

Acknowledgements

FINANCIAL SUPPORT

The chairman of IW-TDGM 2007 and the editors of this book gratefully acknowledge the part of the financial support that was provided from the Grant-in-Aid for scientific research (Category A, Grant No. 18206052, Principal Investigator: Hemanta Hazarika) of the Japan Society of Promotion of Science (JSPS) and Ministry of Education, Culture, Sports, Science and Technology (MEXT), Japan.

SUPPORTING ORGANIZATIONS

The supports from the following organizations to the IW-TDGM 2007 and towards the publication of this book are also gratefully acknowledged.

Toa Corporation, Yokohama, Japan
Shimizu Corporation, Tokyo, Japan
Bridgestone Corporation, Tokyo, Japan

Ibaraki University, Hitachi, Japan
(NPO) Recycle Solution, Tokyo, Japan
Public Works Research Center, Tokyo, Japan

PANEL OF REVIEWERS

Each paper included in this book has been carefully reviewed for the quality of technical content and presentation by at least two members of a panel consisting of the following experts. The editors wish to express sincere gratitude to all the reviewers for their valuable time and contributions.

Prof. Adrian F.L. Hyde, UK
Prof. Arroyo Marcos, Spain
Prof. Ayşe Edinçliler, Turkey
Prof. Dennes T. Bergado, Thailand
Prof. Gökhan Baykal, Turkey
Prof. Hemanta Hazarika, Japan
Prof. Hirokazu Takemiya, Japan
Prof. Juichi Yajima, Japan
Dr. Ashoke Kumar Karmokar, Japan
Dr. Kazuhiro Kaneda, Japan
Prof. Kazuya Yasuhara, Japan
Dr. Kiyoshi Fukutake, Japan
Prof. Kiyoshi Hayakawa, Japan
Prof. Kudryavtsev Sergey, Russia

Prof. Noriaki Sako, Japan
Dr. Norihisa Tatarazako, Japan
Prof. Philippe Gotteland, France
Dr. Satoshi Murakami, Japan
Dr. Suguru Yamada, Japan
Dr. Sumio Horiuchi, Japan
Mr. Takeshi Nagatome, Japan
Prof. Taro Uchimura, Japan
Prof. Tomiya Takatani, Japan
Prof. Yeo Won Yoon, Korea
Dr. Yoshiaki Kikuchi, Japan
Dr. Yoshio Mitarai, Japan
Prof. Yukio Shimomura, Japan

Scrap Tire Derived Geomaterials – Opportunities and Challenges – Hazarika & Yasuhara (eds)
© 2008 Taylor & Francis Group, London, ISBN 978-0-415-46070-5

Photographs

Prof. Kenji Ishihara, President of JGS
Kanto Branch delivering the
welcome address

Prof. Hemanta Hazarika, Chairman of the
Workshop delivering the
opening address

Discussion from the floor

Discussion from the floor

Discussion from the floor

Discussion from the floor

Discussion during special lecture

Progress of the session

International color of the banquet

Banquet communication

Delegates on the first day

Welcome reception

Free discussion during tea break

Panel discussion and the panelist

Discussion Leader: Prof. A.F.L. Hyde,
University of Sheffield, UK
Panelist (from left):
Prof. Y.W. Yoon, Inha University, Korea
Dr. N. Tatarazako, NIES, Japan
Mr. H. Takeichi, Bridgestone Corporation, Japan
Dr. T. Sugano, PARI, Japan
Dr. H. Imanishi, Samsung Corporation, Korea
Prof. D.N. Humphrey, University of Maine, USA
Dr. S. Horiuchi, Shimizu Corporation, Japan

About the Editors

Dr. Hemanta Hazarika is Associate Professor in the department of Architectural and Environment System, Akita Prefectural University, Akita, Japan.

Dr. Hemanta Hazarika got his Bachelor of Technology degree in Civil Engineering from Indian Institute of Technology (IIT), Madras, India in 1990 and his Ph. D. in Geotechnical Engineering from Nagoya University, Nagoya, Japan in 1996. Before moving to the present position, he served few years as a practicing engineer in industry as well as several years in teaching and research in academia and public sector research institute in Japan.

Dr. Hazarika's present research activities include:

- Soil-structure interaction
- Seismic stability of soil-structure
- Recycled wastes and lightweight geomaterials
- Stability of cut slopes, landslides and their protection

He is one of the active and leading researchers in the field of tire derived geomaterials (TDGM) in Japan, especially on the dynamic characteristics and applications of such materials. He is the first to coin the term *Smart Geomaterials* for the tire derived geomaterials.

He has more than 100 publications in reputed international journals, international conferences, symposia and national conferences including contributed chapters in several books. He served and also has been serving presently as member of various research committees of the International Society of Soil Mechanics and Geotechnical Engineering (ISSMGE), The Japanese Geotechnical Society, The Kanto Branch of Japanese Geotechnical Society, The Kyushu Branch of Japanese Geotechnical Society, and the Japan Landslide Society.

Dr. Kazuya Yasuhara is Professor and currently Head of the Department of Urban and Civil Engineering, Ibaraki University, Ibaraki, Japan.

Dr. Kazuya Yasuhara got his Bachelor of Engineering degree in Civil Engineering from Kyushu University, Fukuoka, Japan in 1968 and his Ph. D. in Geotechnical Engineering from the same university in 1978. Before moving to the present position in 1991, he served for three years as Research Assistant at Kyushu University and thereafter served as Lecturer, Associate Professor and Professor for twenty years in Nishinippon Institute of Technology, Fukuoka, Japan.

Dr. Yasuhara's present research activities include:

- Cyclic behavior of soils
- Settlements of ground and structures
- Recycled waste and lightweight geomaterials
- Stability of coastal rocky cliffs
- Effects of global warming on geotechnical infrastructural instability and adaptation

He has published more than 150 papers in reputed international journals, international conferences, symposia and national conferences including contributed chapters in several books. He served and has also been serving presently as chairman and member of various research committees of Japanese Geotechnical Societies and Japan Society of Civil Engineers as well as chairman of committees of Governmental and Local Governmental Committees.

He got many academic prizes, such as Most Distinguished Young Researcher Award in 1980 from Japanese Geotechnical Society (JGS), Best Paper Award from ASCE in 1999 and Most Meritorious Achievement Award of JGS in 2005.

Keynote papers

Scrap Tire Derived Geomaterials – Opportunities and Challenges – Hazarika & Yasuhara (eds)
© 2008 Taylor & Francis Group, London, ISBN 978-0-415-46070-5

A review of environmental impacts and environmental applications of shredded scrap tires

Tuncer B. Edil
University of Wisconsin-Madison, Madison, Wisconsin, U.S.A.

ABSTRACT: Scrap tires shredded into small pieces (called "chips") alone or mixed with soil can have properties favorable to civil and environmental engineering applications. Although the reuse of scrap tires has become more common, questions regarding environmental suitability still persist, particularly the potential impact on ground and surface waters and aquatic life due to leaching. In contrast to this concern on contamination of the environment, the significant sorption capacity of tire material renders it a potential material for environmental protection and remediation when in contact with contaminated waters and leachate. This paper provides a review of the leaching characteristics of scrap tire chips as well as their sorption capacity that can be used in environmental applications based on the author's own research and the available literature.

1 INTRODUCTION

Scrap tires can be used in several ways either as whole or halved or shredded. They can be used alone as well as embedded or mixed with soils. Geotechnical applications of shredded tires include embankment fill, retaining wall and bridge abutment backfill, insulation layer to limit frost penetration, vibration damping layer, and drainage layer (Edil & Bosscher 1992; Humphrey 2003). Environmental applications of shredded tires include use as reactive drainage layer in landfills, septic tank leach field aggregate and nutrient barrier in golf courses and athletic fields (Lerner et al. 1993, Park et al. 1996, Edil et al. 2004, Aydilek et al. 2006, Lisi et al. 2004). Whole tires are used as retaining walls (Caltrans 1988), as a reinforcement layer in earthfill (O'Shaughnessy and Garga 2000a), in artificial reef construction (Candle and Prod 1983), and floating breakwater (Lee 1982). In these applications tire materials of varying sizes (from small chips to whole tires) will have contact with water in varying conditions (infiltrating, seeping, circulating, stagnant, etc.) resulting in interactions either releasing metals and organic chemicals or absorbing them depending on the chemical environment (i.e., pH and concentration of various chemicals in the contacting water). This paper presents a review of these tire-water chemical interactions based on the author's own research as well as the available literature.

2 TIRE COMPOSITION

Tires are principally composed of vulcanized rubber, rubberized fabric containing reinforcing textile cords, steel or fabric belts, and steelwire-reinforced rubber beads. The most commonly used tire rubber is styrene-butadiene copolymer (SBR) containing about 75% butadiene and 25% styrene by weight. Other elastomers such as natural rubber (cis-polyisoprene), synthetic cis-polyisoprene, and cis-polybutadiene are also used in tires in varying amounts. Carbon black is used to strengthen the rubber and to increase abrasion resistance. Extender oil, a mixture of aromatic hydrocarbons, serves to soften the rubber and improves workability. Sulfur, the vulcanizing agent, is used to cross-link the polymer chains within the rubber and to harden and prevent excessive deformation at

Table 1. Rubber compounding composition (Dodds et al. 1983).

Component	Weight percent
SBR	62.1
Carbon black	31.0
Extender oil	1.9
Zinc oxide	1.9
Stearic acid	1.2
Sulfur	1.1
Accelerator	0.7

elevated temperatures. The accelerator is typically an organosulfur compound that acts as catalyst for the vulcanization process and to enhance the physical properties of the rubber (Dodds et al. 1983). A typical composition of tire rubber is shown in Table 1.

3 POTENTIAL ENVIRONMENTAL IMPACTS

Although use of scrap tires above groundwater table is widely accepted, use of tires under the groundwater table or directly in surface water bodies is still not permitted in some states due to concerns regarding potential for groundwater contamination. A better understanding of the interaction between scrap tires and aqueous environments is essential for the further development of innovative uses for scrap tires. To comprehensively study the toxicity of tire chips, an extensive literature review of the leaching characteristics of tires and tire chips was performed.

A significant body of literature describes research with shredded automobile tires and their environmental suitability when used in civil engineering applications. The research consists of laboratory and field studies on the leaching behavior of tire chips. The following is a review of laboratory and field studies dealing with environmental suitability of tire chips. Since tires contain ingredients such as carbon black, vulcanizing agents, metallic reinforcement, antioxidants, pigments, accelerators, etc. (Miller and Chadik 1993) in addition to petroleum products that might be collected during use, laboratory and field leaching studies have addressed organic and inorganic compounds.

3.1 Leach testing of tire chips

Batch tests are typically used to determine the leaching characteristics of any by-product, such as tire chips. The mass of tire chips, the volume of the container i.e., solid-liquid ratio, and chemistry of the extraction liquid vary from study to another. The majority of studies dealing with the leachability of tire chips were performed under stagnant conditions with a fixed volume of liquid. This testing setup is not realistic since in the field the tire chips leach into a flowing groundwater and there is potential for dispersion and adsorption. The methods of analysis of these leachates usually consist of gas chromatography for organics and spectrometry for inorganics.

The Toxicity Characteristic Leaching Procedure (TCLP) is a popular test used to determine if a waste material or a by-product is hazardous. The volume of the container is 500 to 600 mL with zero head space. Liquids of pH of 4.93 and 2.88 are recommended for use in these tests, and time allowed for leaching is 16 hrs. The minimum amount of solids placed in the extractor is 100 g for the non-volatile analysis and 25 g for volatile organics. The EP-Toxicity test was developed before the TCLP and is used for the same purpose. The EP-Toxicity test specifies a pH of 5 for the liquid, a time period of 24 hrs, and no minimum size of container. Rubber Manufacturers Association (RMA) (1990) compared leaching results from tire chips using both tests and found no significant difference between the two tests.

Another batch leaching test is Water Leach Test (WLT) (*Standard Test Method for Shake Extraction of Solid Waste with Water* – ASTM D 3987) typically used for non-hazardous materials to

Table 2. Conversion of regulatory concentration limits.

| Compound | Concentration limits based on tire chip mass | | |
	TCLP[1] [mg/kg]	US EPA's MCLs[2] [mg/kg]	WI PALs[2] [mg/kg]
As	25.9	0.06	6.15×10^{-3}
Ba	518.0	2.27	0.25
Cd	5.18	6.15×10^{-3}	1.23×10^{-3}
Cr	25.9	0.15	6.15×10^{-3}
Pb	25.9	0.025	6.15×10^{-3}
Fe		0.37	0.18
Mn		0.06	0.03
Zn		6.15	3.1
Se	5.18	0.055	1.23×10^{-3}
Hg	1.03	3.7×10^{-3}	2.5×10^{-3}
NO_2^-/NO_3^-		1.23	2.46
Toluene	33.79	2.46	84.3×10^{-3}
Carbon disulfide	332.64		
Phenol	332.64		1.23×10^{-3}
Benzene	0.36	6.15×10^{-3}	0.08×10^{-3}

[1]TCLP are converted to mass of compound per kg of tire chips assuming that 100 g of tire chips with a specific gravity of 1.22 is used. (Volume of extractor = 600 mL; mass of solid = 100 g for inorganics and 25 g for volatile organics).
[2]U.S. EPA MCLs and Wisconsin PALs of 1.22 (1 mg/L = 1.22 mg/kg of tire chips).

assess the leaching potential. In this test, the solid material (70 g) is crushed to pass a US No. 4 Standard sieve (4.8 mm) and combined with 1400 ml of ASTM Type II deionized water in a 2-L sealed container. This mixture is agitated continuously in a tumbler for 18 hr. The mixture is allowed to settle and then the leachate is sampled and subjected to chemical analysis.

Converting leachate concentrations and the concentrations of different standards from the mass of compound per volume of liquid to the mass of compound per mass of tire chips helps to compare the findings of many studies. Such a comparison is presented in Tables 2 and 3 (Tatlisoz et al. 1996). Table 2 provides the conversion of several regulatory limits, i.e., TCLP, U.S. EPA recommended Maximum Concentration Levels (MCLs), and Wisconsin's Groundwater Preventive Action Limits (PALs).

Table 3 provides the conversions of concentrations from several batch tests reported in the literature. Zinc, iron, barium, manganese, selenium, lead, chromium, and cadmium were found at concentrations higher than the U.S. EPA's MCLs and Wisconsin's PALs in most studies. Zinc and iron (two compounds typically with the highest concentrations) are not classified as hazardous materials in TCLP limits. Concentration of arsenic exceeded Wisconsin PALs, but not MCLs (Miller & Chadik 1993) while arsenic was not detected in the Rubber Manufacturers Association (1990) study. Even though several inorganic and organic compounds have been detected in shredded tire leachates, concentrations were generally well below TCLP limits and were not significantly higher than recommended MCLs and Wisconsin PALs, regardless of differences in the test conditions. It is important to note that in many studies the extraction liquid or test setup was designed for worst case conditions that do not necessarily exist in the environment that tire chips are used. For example, laboratory tests performed with fixed volume of liquid or tire chips exposed to aggressive extractions do not simulate conditions likely to exist in most tire chips applications. Thus, in the field studies, lower concentrations are likely to exist.

For a more realistic assessment of leaching under flow-through conditions and to understand the temporal characteristics of leachate concentrations, Column Leaching Test (CLT) can be conducted. Specimens for the CLTs are prepared as for the WLTs and placed in the CLT cell in a density state

Table 3. Compounds leached from tire chips.

Reported concentrations based on tire chip mass

Grefe[1] (1988) [mg/kg]	RMA[1] (1990) [mg/kg]	TCTC[2] (1990) [mg/kg]	Miller & Chadik[3] (1993) [mg/kg]	Park et al.[4] (2003) [mg/kg]
	–		0.02	
0.55	0.1	1.08		0.37
	–	0.27		
	0.008	0.51		0.019
0.075	0.003	0.92		0.14
1.15		1081		
1.5				
3.15		50.3	5.02	1.13
		0.44		0.05
	7.2×10^{-5}			
1.85				
	0.034		0.28	
	0.012			
	0.01			
	–		0.63	

[1]The liquid concentrations are converted to mass of compound per kg of tire chips based on the given quantities of tire chips and liquid used in the study. (RMA = Rubber Manufacturers Association).
[2]Conducted by Twin City Testing Corporation (TCTC).
[3]Concentrations for compounds are taken from approximately 16 hour readings which are usually the highest concentrations achieved during the test. The liquid concentrations are converted to mass of compound per kg of tire chips based on the given quantities of tire chips and liquid used in the study.
[4]Concentrations for compounds are the final concentration reached in 800 days. The liquid concentrations are converted to mass of compound per kg of tire chips based on the given quantities of tire chips and liquid used in the study.

similar to that expected in the field conditions and water flow is initiated at a hydraulic gradient or flow rate as expected in the field. Typically, ASTM Type II water can be used as the influent liquid. The ionic strength of this solution is comparable to that in natural waters percolating through the pavement base layer. A 0.1 M LiBr solution prepared with LiBr salts (99.9% purity) can be used to trace breakthrough. Leachate (effluent) from the CLTs is collected in airtight sampling bags made of Teflon. Leachate is collected periodically, filtered, acidified, and stored using the same protocol employed for the WLTs for chemical analysis.

J & L Testing Co. (1989) performed column tests on tire chips with the liquid flowing at a rate of 5.8 cm/day. The container was 20.3 cm in diameter and had a length of 121.9 cm. Kim (1995) used steel containers having diameters of 61 cm and lengths of 91.4 cm. When comparing the results of these tests, the concentrations should also be normalized with respect to the weight of tire chips. O'Shaughnessy and Garga (2000b) performed column tests on tire chips embedded in sand and clay. The former indicated increase in effluent certain metal elements (aluminum, iron, zinc, and manganese) which in some cases exceeded their respective drinking water standards. All target elements were below detection limits or background levels for tire chips embedded in clay.

3.2 *Factors affecting tire chips leachate*

The main factors affecting the characteristics of leachates from tire chips include the aquatic environment in which the tire chips are exposed, the age of the tires, the size of the tires, and the time of exposure of the tire to the liquid. Understanding the relationships between these factors and the chemical characteristics of tire chips leachate is crucial to any leaching study.

The aquatic environment is typically described by the pH of the extraction liquid. Twin City Testing Corporation (1990) used extraction liquids having pH of 3.5, 5.0, 7.0, and 8.0. This study showed that higher metal concentrations were obtained in the leachate when the pH was 3.5. Higher hydrocarbons concentration was observed when the pH was 8.0. Miller & Chadik (1993) performed leaching tests with liquids having pHs of 5.4, 7.0, and 8.6 and reported that there was no strong correlation between pH and leachate characteristics.

Twin City Testing Corporation (1990) and Rubber Manufacturing Association (1990) performed leach tests on new and old tire chips. They reported that both new and old tire chips leached organic and inorganic compounds, but at concentrations lower than the TCLP limits to classify as hazardous material. Twin City Testing Corporation (TCTC) (1990) reported that newer tire chips leached slightly higher concentrations of polycyclic aromatic hydrocarbons (PAHs).

3.3 *Tire chips and groundwater quality*

3.3.1 *Inorganic compounds*

Several laboratory leaching studies used worst case conditions (acidic or alkaline) when assessing the leachability of tires and tire chips. In civil engineering construction, however, the tire chips are typically used as fill material and these extreme conditions are not likely to appear. The leaching process is typically caused by the infiltration of rain water or when tire chips are in contact with groundwater. Using solutions that better represent the chemistry of rainwater and/or groundwater is of interest.

To further study the leaching characteristics of tire chips in various aquatic environments, a laboratory study consisting of batch tests was designed to simulate various environmental conditions, such as aquifers and wetlands (Gunter et al. 1999). The tire chips used in the laboratory investigation were mechanically processed from steel-belted automobile tires. The particle size of the tire chips ranged approximately from 2 to 10 cm (longest dimension). Six 266-liter stainless steel tanks were used. Three different waters were used: groundwater, marsh water, and deionized water. Groundwater was taken from a 90-m deep supply well and the pH of the groundwater was 6.7 to 7.4. Groundwater at this location is sampled quarterly and tested for the presence of organic and inorganic contaminants of concern. Lead and iron are the only compounds in the groundwater from the well that have exceeded the Wisconsin PAL in one sampling event of this well. The marsh water was collected from a marsh. Marsh water was collected since tire chips are used as light-weight fill on soft materials such as marsh deposits. This water was then passed through a #200 sieve to filter out soil and plant remains. Deionized water does not have dissolved chemicals; therefore, it represents a reference state for the other two aquatic environments.

Three different stagnant waters were simulated. Three tanks contained tire chips and water, whereas the other three tanks contained only water as control. The tanks were filled with tire chips and/or water to the top to eliminate any headspace and then were sealed. Initial samples of the water were analyzed for volatile organic compounds (VOCs) using a purge and trap method (EPA method 8021) and a headspace scan (gas chromatogram/mass spectrometer method). The tanks remained stagnant until sampling began and continued periodically for the next 12 months. Samples were collected using a hypodermic needle inserted through a sampling port that was located at mid-height of the tanks.

Table 4 shows the concentration of the inorganic parameters analyzed in the first sampling event. None of the inorganic parameters exceeded the Wisconsin PALs (PALs are lower limits than drinking water standards) in any of the water environments. Subsequent sampling events (not shown) confirmed this finding. Hardness of the groundwater increased with time and the tanks with tire chips contained oil and grease at the end of the monitoring period.

Twin City Testing Corporation (1990) performed a field study to determine how groundwater characteristics changed when exposed to tires and tire chips. The study consisted of comparing ground water samples collected beneath a tire stockpile to background concentrations of a typical

Table 4. Summary of inorganic parameters analyzed in the last sampling (12 months later).

Parameter	Unit	PAL	Tank 1 (GW)	Tank 2 (GW+T) Total	From tires	Tank 3 (MW)	Tank 4 (MW+T) Total	From tires	Tank 5 (DI)	Tank 6 (DI+T) Total	From tires
As	µg/L	5	1			<1			<1		
Ba	µg/L	400	42	120	78	15	220	205	<1	120	119
Ca	µg/L			42000	42000				<100		
Cd	µg/L	0.5				<0.08					
Cr	µg/L	10	<2	<2			<2		<2	<2	0
Co	µg/L										
Cu	µg/L	130		<8	<8				13		
Fe	µg/L	150	<20	15000	14980	1200	71000	69800	20	18000	17980
Pb	µg/L	1.5	1	2	1	2	1	1		20	20
Mg	µg/L			37000	37000				<30		
Mn	µg/L	25	6	770	764	110	3200	3090	3	1400	1397
Mo	µg/L										
Ni	µg/L	20									
Se	µg/L	10	2			<1			<1		
Na	µg/L			11000	11000				<700		
Zn	µg/L	2500	750	<19	731	32	<19	13	30	42	12
TOC	mg/Kg		<0.5	220	220	5.6	270	264.4	<0.5	150	150

groundwater samples. This study indicated that the average zinc concentrations increased from 0.1 to 0.87 mg/L, iron concentrations from 5.8 to 298 mg/L, and the magnesium concentrations from 6.2 to 383 mg/L. The arsenic, barium, cadmium, chromium, and lead concentrations for only the groundwater sample collected underneath the tire stockpile exceeded the Minnesota recommended allowable limits. Soil samples were also collected from the area and the concentrations of arsenic, barium, calcium, and selenium were higher than those for the background samples. In contrast, aluminum, iron, magnesium, and zinc were found in lower concentrations when analyzing the soil sample.

In one of the earliest embankment fill application of shredded tires, Edil & Bosscher (1992) installed two lysimeters beneath the areas containing tire chips to collect the leachate. The pH was found to be stable around 7.5. Leaching of zinc, manganese, and iron was observed. However, these may have leached from the surrounding soil as there was a history of high manganese concentrations in the areas around the test fill. The Wisconsin groundwater PALs was exceeded in the leachate for several metals. However, it was concluded that the possibility of tire chips affecting groundwater quality was highly unlikely (Bosscher et al. 1993).

Humphrey & Katz (2000) describes the results of a five year field study on the effects of tire chips placed above the water table on the groundwater quality. The tire chips were placed beneath a secondary highway and the leachate was collected using geomembrane-lined basins located below the tire chips. One control section (constructed with typical granular materials) was also constructed adjacent to the tire chips sections. Samples of leachate were collected on a quarterly basis for constituent testing. Filtered and unfiltered samples were analyzed for barium, cadmium, chromium, lead, and selenium. Humphrey & Katz (2000) reported that there was no evidence that the presence of tire chips altered the concentrations of these substances from their naturally occurring background levels. In additions they reported that there was no evidence that the tire chips have increased the concentration levels of aluminum, zinc, chloride, or sulfate, which are secondary drinking water standards. Humphrey & Katz (2000) added that under some conditions, iron, and manganese levels might exceed secondary standards. In another study, O'Shaughnessy & Garga (2000b) reported that field monitoring of leachate from a prototype test embankment constructed

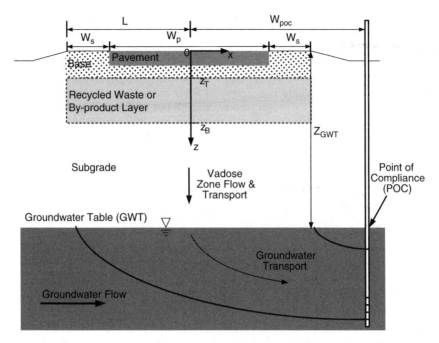

Figure 1. Conceptual model in WiscLEACH for predicting impacts to the vadose zone and groundwater caused by leaching from a pavement structure with a recycled waste or by-product layer.

with cut (one side wall removed) and whole tires above the groundwater table showed insignificant adverse effect on groundwater quality over a period of 2 years.

In a subsequent study, Humphrey & Katz (2001) constructed a field trial to evaluate the water quality effects of tire shreds placed below the water table. The study consisted of three sites, each with 1.4 metric tons of tire shreds buried in a trench below the water table. The tire shreds were made from a mixture of steel and glass belted tires and had a maximum size of about 75 mm. The soil types at the sites were marine clay, glacial till, and peat. At each site, one water sampling well was located upgradient to obtain the background water quality, one well was located in the tire shred filled trench, and two to four wells were located 0.6 m to 3 m downgradient of the trench. Samples were taken over a four-year period and analyzed for a range of metals, volatile organics, and semivolatile organics. Their results showed that tire shreds had a negligible effect on the concentration of metals with primary (health based) drinking water standards. For metals with secondary (aesthetic based) drinking water standards, samples from the tire shred filled trench had elevated levels of iron, manganese, and zinc. However, the concentrations of these metals decreased to near background levels for samples taken downgradient of the tire shred filled trench. Tire shreds placed below the water table appear to have a negligible off-site effect on water quality.

These investigations address the initial leachate concentrations but evaluating potential impacts on ground water quality caused by leaching of heavy metals from the tire chips is an important step in applying this technology. It can be done on a site-specific basis. WiscLEACH, a Windows-based computer application developed for assessing potential ground water impacts associated with use of recycled materials such as scrap tires in roadway construction provides a means of making such an evaluation (Li et al. 2006). The application is designed to be easy to use and to function rapidly so that a designer or regulator can quickly conduct a series of simulations without training in numerical modeling. WiscLEACH is freeware that is available at www.uwgeosoft.org. The conceptual model consists of a recycled material layer in a typical highway structure as shown in Figure 1. As water percolates down through the profile, heavy metals leach from this layer and then migrate downward through the subgrade soils until they reach the ground water table (GWT).

Flow in the recycled material layer and subgrade is assumed to occur only in the vertical direction and transport is assumed to follow the advection-dispersion-reaction equation (ADRE) with instantaneous and reversible sorption and a linear isotherm. Bin-Shafique et al. (2006) show that this assumption is valid for such systems and typical subgrades. Metals that reach the GWT are transported horizontally and vertically, although the flow of ground water is assumed to occur predominantly in the horizontal direction. Transport in ground water is also assumed to follow the ADRE with instantaneous and reversible sorption and a linear isotherm. In both layers, chemical and biological reactions that may consume or transform metals are assumed to be absent. The recycled material and all soils in the profile are assumed to be homogenous and isotropic.

An example problem has shown that maximum concentrations typically occur near the ground water table and the peak ground water concentration decreases as the depth to ground water increases (Li et al. 2006). Parametric studies also showed that the variables having the greatest influence on maximum concentrations in ground water are depth to the ground water table, thickness and width of the recycled material layer, hydraulic conductivity of the recycled material layer and the aquifer, and the initial metals concentration in the recycled material layer.

3.3.2 *Organic compounds*

A study conducted by Twin City Testing Corporation (1990) indicated that under alkaline conditions, the concentration of polycyclic aromatic hydrocarbons (PAHs) in tire chip leachate exceeded drinking water standards. Miller & Chadik (1993) suggested that these compounds may leach from carbon black, petroleum residues, and recipe extenders associated with the manufacturing of tires. Rubber Manufacturers Association (1990) reported that VOCs such as toluene, carbon disulfide, and methyl ethyl ketone might leach from tire chips. In addition, benzene was detected in tire chip leachates, and that only phenol was detected as a semi-volatile compounds. This study added that the concentrations of these organic compounds are below TCLP limits. Bosscher et al. (1993) used biological and chemical oxygen demand (BOD, COD) as indicators of organic compounds in leachates from tire chips and reported decreasing BOD and COD in their leachates. Miller & Chadik (1993) prepared 27 aquatic solutions in the laboratory at different pH levels and varying ionic strengths. Tire chips of various sizes were then submerged in these aquatic solutions and the liquids periodically analyzed. Miller & Chadik (1993) reported that organic compounds such as aromatic compound of gasoline, carboxylic acids, and aniline were leached from tire chips. Two of their solutions used were similar to groundwater. The leachates obtained using these liquids had benzene concentrations as high as 0.0115 mg/L, and toluene concentrations of 0.0112 mg/L. Humphrey & Katz (2001) found trace concentrations of a few organic compounds in the tire shred filled trenches in their study of tire shreds placed below the groundwater table, but concentrations were below method detection limits for virtually all the samples taken from the downgradient wells.

In the tank batch tests described above the organic chemicals were also analyzed. Table 5 shows the results for organics in the first sampling event. The steel tank that contained groundwater and tire chips (Tank 2) initially had concentrations above the Wisconsin PALs for benzene and trichloroethylene (TCE). This tank also exceeded the Wisconsin drinking water standard, which is 10 times higher than PAL, in methyl isobutyl ketone (MBIK). This compound is also known as hexanone, 2-methyl-4-pentanone. MBIK is a compound derived from the tire-manufacturing process. There were no organic compounds detected in Tank 1, the control tank that was filled with groundwater from the same source as Tank 2. In the second sampling for VOCs there were no contaminants found that exceeded any PAL. Initially, the steel tank that contained marsh water and tires chips (Tank 4) had a high concentration of MBIK. It exceeded the Wisconsin PAL. The control tank for marsh water (Tank 3) showed trace levels of toluene but nothing above the Wisconsin PALs. In the second testing for VOCs, neither Tank 3 nor 4 had detectable amounts of any organic compounds. Initially, the tank that contained deionized water and tire chips (Tank 6) exceeded the Wisconsin PAL in TCE. It also showed a small amount of toluene. MBIK was detected in high concentration. The control tank contained no detectable VOCs. In the second testing for VOCs, benzene was detected in a concentration above that of the PAL in Tank 6. Tank 5, the control tank was not sampled but was expected to be below the standard limits.

Table 5. Summary of organic compound analysis for laboratory tests.

Volatile organic compounds	Units	WI PALs	Tank 1 (GW)	Tank 2 (GW + Tire)	Tank 3 (MW)	Tank 4 (MW + Tire)	Tank 5 (DI)	Tank 6 (DI + Tire)
Acetone	mg/L	0.1479	ND	ND	ND	0.1149	ND	0.0014
Benzene	mg/L	0.0004	ND	0.0015	ND	ND	ND	ND
Ethylbenzene	mg/L	0.1035	ND	ND	ND	ND	ND	ND
Methyl ethyl ketone	mg/L	0.0666	ND	0.0147	ND	0.01724	ND	ND
Methyl isobutyl ketone (MIBK)	mg/L	0.0370	ND	1.1513	ND	0.9577	ND	1.0407
Toluene	mg/L	0.0509	ND	ND	0.0019	ND	ND	0.0012
Trichloroethylene	mg/L	0.0001	ND	0.0008	ND	ND	ND	0.0008

He overall conclusion of these studies is that tire shreds may increase levels of certain metals (e.g., iron and manganese) and some organic compounds; however, as concluded by Humphrey & Swett (2006) in a detailed literature review report, there appears to be limited effect on drinking water quality of groundwater from leachate derived from tires for a range of applications involving tires or tire shreds, so human health concerns are minimal.

3.4 Tires and aquatic life

B.A.R. Environmental Inc. (1992) reports that new and older tires are both toxic to rainbow trout, and therefore indicating that the toxicity is caused by materials associated with the manufacturing of tires and not by other materials accumulated during the life of tires. In addition they reported that the toxicity concentrations were slightly higher in older tires. The effects of tire size on the chemical characteristics were addressed by the Rubber Manufacturers Association (1990) study and Miller & Chadik (1993). Both studies show that, in general, the leachates obtained from ground and whole tires are comparable.

B.A.R. Environmental Inc. (1992) tested the toxicity of leachates generated from tires at 5, 10, 20, and 40 days of exposure. This study showed that toxicity concentrations reached a peak at 5 days and did not change then after. Miller & Chadik (1993) reported that benzene concentrations were highest at the beginning of the test and decreased exponentially with time, and that toluene concentrations were lowest at the start of the test but increased gradually with time. They also reported that the concentration of zinc increased with time for 63 days and then decreased thereafter. Edil & Bosscher (1992) also observed time-dependent composition of their leachate collected from field lysimeters underneath their shredded tire embankment. They attributed this change to the change in construction activities such as roadway dust treatment with calcium chloride and the asphalt paving operation performed on the roadway.

Abernethy (1994) reported tire leachate obtained by immersing an automobile tire in 300 L of water caused mortality in trout and other species. Abernethy (1996) also reported that tires placed in a tank of flowing water were non-lethal to trout. Hartwell et al. (2000) showed that toxicity decreased with increasing salinity and concluded that tire shreds are probably a greater threat to freshwater habitats than brakish or marine habitats. Sheehan et al. (2006) conducted aquatic toxicity tests on leachate samples collected from the below groundwater disposal study sites of Humphrey & Katz (2001). They found elevated levels of iron, manganese, and several other chemicals but no adverse effect to test organisms from the leachate of the tire chips placed above the groundwater table. However, leachate from the tire chips placed below the groundwater resulted in significant reduction to both survival and reproduction of a test organism. However, they expected these effects to decrease downstream from the tire chips. Sheehan et al. (2006) recommend a buffer zone of 3 to 11 m between the tire chips fill and the surface water to avoid impact on aquatic life.

Laboratory aquatic toxicity studies and field leachate samples indicate that tire shred leachate impacts test organisms. The implication of these tests is now confirmed. Tires placed in open water such as reef construction have proven to be a huge ecological blunder. Little sea life has formed on the tires. Some tires have broken loose from their bundles and scouring the ocean floor and washing up on beaches. Similar problems have been reported at tire reefs worldwide and expensive removal activities are underway as reported by The New York Times on February 18, 2007. However, the impact of leachate from tire shreds placed below the groundwater on the surface waters may be limited by allowing an appropriate separation zone between the tires and the surface water. More field evidence is needed.

4 ENVIRONMENTAL APPLICATIONS

The process of manufacturing tires combines raw materials into a special form that yields unique properties such as strength, resiliency, and high absorbency of chemicals (Edil & Bosscher 1994). These unique properties of tires can be exploited in a beneficial manner if scrap tires are used as a construction material. Scrap tires used in civil engineering applications have ten times better drainage than well graded soil, eight times better insulation than gravel, and 33 to 55% weight of soil (Rubber Manufacturers Association 2002). Tire chips and tire chips-sand mixture have high permeability and retain it even under high vertical loads (Edil et al. 1992). These properties make shredded tire chips a good lightweight substitute for aggregate in general construction.

Park et al. (1996) found that the tire chips have relatively high VOC sorption capacities based on batch isotherm tests on scrap tire chips. This suggests yet another innovative environmental application in which shredded tire chips could be used to eliminate VOCs from contaminated water or leachate thus alleviating the contamination problem. Kim et al. (1997) conducted a series of batch isotherm test with ground tire powder to investigate the effect of environmental conditions such as presence of other organic materials, ionic strength and pH of the solution, particle size of tire powder, and temperature. Since this study used only ground tire rubber not tire chips, it was possible to identify which part of tire chips played a major role in adsorbing organic compound. They found that organic compounds sorbed primarily onto tire rubber polymeric materials and partially other materials in tire rubber such as carbon black. Styrene butadiene rubber (SBR) is one of the major components of tires Table 1. Non polar organic compounds such as benzene, toluene, trichloroethylene, and tetrachloroethylene, are attracted by non polar materials such as SBR. Park et al. (1996) found that tire chips have 1.4 to 5.6% of the sorption capacity of granular activated carbon on a volume basis. In addition, Park et al. (1996) found that of the organic compounds sorbed in tire chips, only 3.5 to 7.9% were desorbed. Twin City Testing Corporation (1990) also indicated that the average concentrations of zinc, iron, and manganese increased in the groundwater samples under a tire stockpile but at the same time the concentration of petroleum hydrocarbons decreased from 11.8 to <0.5 mg/L compared to background groundwater samples indicating the potential of tires to absorb these compounds. Therefore, another application for scrap tires is to use them as a sorptive medium taking advantage of their sorptive quality to chemicals.

One such application is to use tire chips as a reactive drainage medium in landfills, i.e., in leachate collection layer. A wide range of VOCs have been detected in leachates from municipal solid waste and hazardous waste landfills (Plumb & Pitchford 1985; Gibbons et al. 1992; Klett et al. 2005). Research has also shown that diffusive transport (contaminant migration driven by the difference in concentration between the upper and lower sides of the liner) is often the dominant mode of contaminant transport in well-built liner systems (e.g., Edil et al. 1996, Kim et al. 2001, Foose et al. 2002), and that transport of VOCs generally is more critical than transport of inorganic compounds (e.g., toxic heavy metals), even though VOCs are often found at lower concentrations in leachates (Foose et al. 2002). VOCs generally are more critical for two reasons. First, VOCs are not as much retarded in clay liner and generally toxic at lower concentrations than many inorganic compounds. Second, geomembrane do little to inhibit the transport of VOCs in modern composite liner systems, because VOCs diffuse readily through geomembrane polymers (Park and Nibras 1993). A recent study showed that there is not a significant difference in concentration of VOCs in pan lysimeters placed under the liners in Wisconsin landfills at both compacted clay and composite (compacted clay plus geomembrane) lined sites (Klett et al. 2005). Furthermore, in many cases the

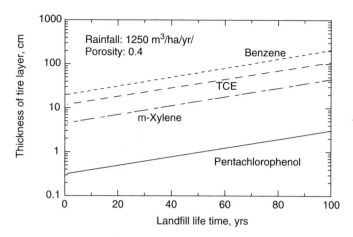

Figure 2.   Estimated tire layer thickness required for 90% removal over landfill life based on batch isotherm tests.

VOC concentrations in the lysimeters exceeded the Wisconsin PALs. Therefore, the effectiveness of modern landfill liner systems to minimize migration of VOCs is a concern.

Based on batch isotherm tests on tire chips in contact with various organic compounds, Park et al. (1996) proposed equations to determine the thickness of the tire chip layer needed to remove a given amount of organic compounds and the mass of tire chips required. When the leachate generation rate in a landfill is 1,500 m³/ha/yr and the porosity is 0.4, the required tire layer thicknesses for 90% organic compound removal over various design lives are shown in Figure 2. The mass density of tire chips used in the calculation was 1.22 g/cm³. If the design life of a landfill is 30 years, then the tire layer thickness required for 90% removal of benzene, trichloroethylene (TCE), m-xylene, and pentachlorophenol are approximately 50, 28, 15, and 1 cm, respectively. Polar organic compounds, such as methylene chloride and chloroform, that require large quantities of tire chips, tend to be more biodegradable in landfills. For general landfill application, the tire chip layer thickness of 30 to 45 cm is recommended. Since the rate of organic compound sorption is a function of the diffusion coefficient and not the surface area (Kim et al. 1997), tire chip size does not play a significant role. Therefore, the tire chip size ranging from 5 to 30 cm is recommended for use as a substitute for landfill leachate collection media. As a rule of thumb, a 1-ha landfill requires approximately 300,000 tires to fill 30 cm of a leachate collection layer. The hydraulic conductivity of tire chips is very high when they are not compressed under load and not mixed with soil (Edil et al. 1992). In tests for use of tire shreds in leachate collection layer, hydraulic conductivity decreased with increasing vertical pressure from 0.87 (no load) to 0.13 cm/s (at 478 kPa vertical stress) under 300 mm head (about a hydraulic gradient of 1) and higher values (down to 0.19 cm/s) under the lower hydraulic head of 150 mm (about a hydraulic gradient of 0.5). Having different hydraulic conductivities at different heads is not unusual because of large pore sizes the flow is more turbulent than laminar. The flow rate divided by the square root of hydraulic head was nearly constant at the two heads giving support to the observation that a more of a turbulent flow rather than laminar flow existed.

Although there are many advantages to using recycled tires in civil and environmental applications, concerns have been raised about their self-combustion potential and environmental suitability. However, guidelines have been developed to deal with this issue. It is recommended that the thickness of a tire shred layer be limited to 3 meters and that relatively large shreds with a minimum of rubber fines be used along with limiting the flow of air and water into the interior of tire-shred fills (ASTM D6270-98). Thus, it is possible to use tire chips alone and especially in mixture with granular soils (such mixtures are not known to self combust) as a sorptive drainage material taking advantage of their high permeability and high absorbency of chemicals without the fear of self-combustion.

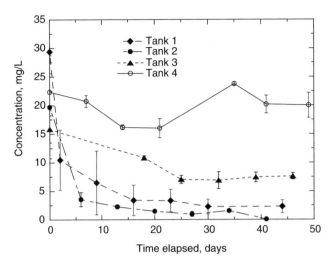

Figure 3. Toluene concentration in the upper reservoir in tank tests.

4.1 Large-scale laboratory tests

Edil et al. (2004) conducted large-scale laboratory tests simulating a clay liner with a tire-chips drainage layer over it to quantify the effect of tire chips in reducing the transport of VOCs through compacted clay liners. Tanks with diameters of 0.6 m were used in these tests. This experimental apparatus has been described in detail by Edil et al. (1994) and Kim et al. (2001). The influent and the effluent were supplied and collected using Teflon bags, which allow control of hydraulic heads. This system permits the water to flow evenly across the entire soil layer. All parts used were made of stainless steel and brass and tubing was made of Teflon. Tire chips (2 to 10 kg) were placed over the compacted clay in the upper influent reservoir. After the hydraulic conductivities of clay specimens stabilized, the upper reservoir and the influent bag fluid is replaced with the VOC containing solution. For the tank tests, methylene chloride, toluene, and trichloroethylene were the selected VOCs. The initial target concentration of each VOC was 16 mg/L.

The concentrations of toluene in the upper influent reservoir are shown in Figure 3. The organic compound concentrations in Tank 4 (T4), which did not contain tire chips, were relatively constant and higher than those in the three other tanks, which contained tire chips, because the sorption of organic compounds onto tire chips is much greater and faster than the mass transfer of organic compounds through a clay layer. However, in the tanks with tire chips (Tanks 1, 2, & 3), concentrations in the upper reservoirs decreased in a few days. Since no breakthrough of any of the VOCs was observed in the lower effluent reservoir of the tanks with tire chips (due to the sorption of VOCs by tire chips), the clay liner layers were cored to obtain the concentration-depth profiles. The cores were sectioned to obtain the concentration-depth data. The concentration-depth profile was obtained by analyzing the pore water of each section of the core. Figure 4 shows the concentration-depth profile for toluene. Toluene was not detected in any sectioned clay specimens of the tanks with 10-kg tire chips (T1 and T2) as well as in the upper compartment.

4.2 Field tests

To demonstrate the effectiveness of tire chips as a sorptive layer, two geomembrane lined collection cells were constructed in a landfill. Each basin was 7.32 m long by 7.32 m wide. Detailed schematics can be found in Park et al. (2003). A drainage layer was placed directly on top of the geomembrane. One of the test cells contained gravel (30-cm thick) and the other contained tire chips (30-cm thick). The gravel was the same gravel that was used in the drainage layer of the landfill. Municipal refuse was placed to a height of 4.57 m over the test cells and covered with wood chips. Pipes were

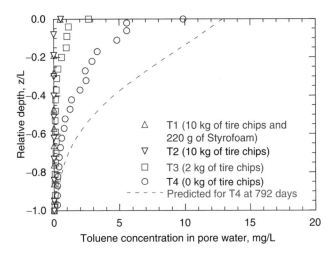

Figure 4.   Toluene concentration-depth profile in tank tests.

installed down-slope side of the collection cells in order to drain the leachate periodically. The base of the collection cells was sloped 4% to drain the leachate to the leachate collection pipe. The leachate collection pipe was constructed from a high density polyethylene (HDPE) pipe section welded with a boot to the HDPE liner using extrusion welding. A steel pipe with a valve was fitted and sealed to the HDPE pipe.

The leachate production volumes were estimated to be 3.9 and 5.2% over a 2-year period in gravel- and tire chip-containing cells, respectively. There was no indication of any retardation of leachate drainage due to use of tire chips instead of gravel in the drainage layer under 4.57-m of refuse and over 2 years. Leachate samples generated by rainfall and taken 1 and 6 months after construction showed a wide variety of metals and large amounts of oil and grease (i.e., 25,000 mg/L in gravel cell and 7,000 mg/L in tire chips cell). These two leachate samples were analyzed for VOCs following the EPA Method 8021. Methyl isobutyl ketone (MIBK) was detected only in the cell samples from the tire chip-containing cell, but not in those from the gravel-containing cell, indicating that tire chips might leach MIBK. In leaching tests performed on the shredded tires, MIBK was not reported (Gunter et al., 1999), indicating that the tire chips used in the collection cells might have been exposed to MIBK, or refuse containing MIBK might have been landfilled over the tire chip-containing cell. On the other hand, toluene was only detected in collection cell samples from the gravel-containing cell. The concentration increased from 1.4 to 32 μg/L, implying that toluene existed in the refuse and leached out. Since tire chips could sorb toluene (Park et al., 1996), collection cell samples from the tire chip-containing cell did not contain toluene. It can be said that the gravel-containing and tire chip-containing cells had different water quality that closely reflected the properties of tire chips and gravel.

The test cells were spiked with gasoline. First the leachate was drained entirely from the cells. Then, ten holes were dug in the refuse, five on each of the test cells. Approximately 3.79 L of gasoline was poured in each hole and the holes were covered with wood chips. A sample of leachate was taken from the drainage pipes two weeks later. In the gravel-containing cell, the leachate exceeded the U.S. EPA's MCLs in all of the BETX compounds while in the tire chip-containing cell, the leachate did not exceed the MCLs in its concentration of benzene. This supports the evidence from past laboratory batch isotherm tests in which tire chips were found to have beneficial sorption properties (Park et al., 1996). The leachate was sampled again on about one month after spiking. Results similar to the previous sampling were obtained further supporting the previous findings by Park et al. (1996) and provide evidence that tire chips have favorable sorptive characteristics. The gravel-containing cell exceeded the MCLs in benzene, toluene, and xylenes. The tire chip-containing cell did not exceed the MCLs, and had a much lower concentration than in the gravel-containing cell. It may not be possible to sorb all VOCs in leachate with tire chips. However, the concentration of VOCs in leachate will be lowered significantly with tire chips so that potential VOC migration from landfills will be minimized.

Another study conducted by Aydilek et al. (2006) to investigate the performance of tire chips as leachate collection material in municipal solid waste landfills indicated that the leachate flow rates and total leachate volumes generated by the two field test cells are comparable indicating no flow retardation of leachate drainage due to use of tire chips instead of gravel as the leachate collection layer material. The field temperatures inside the tire chips were between 28 and 61°C, which is comparable to the temperatures observed in solid waste landfills. Moreover, these temperatures were well below the approximate threshold temperature for potential combustion of tire chips. The leachate collected from the tire-chip layer had lower inorganic compound, dissolved metal and VOC concentrations than those collected from the gravel layer. Furthermore, the concentrations of the inorganics and VOCs of samples collected from the tire-chip cell were below the U.S. EPA's MCLs.

4.3 *Mitigating nutrient leaching*

Sand-based root zones, typically used for golf course putting green and athletic field construction, lack sufficient cation exchange capacity to restrict nitrogen and phosphorus migration through the root zone and into sub-surface drainage systems. The adsorptive properties of tire rubber for retaining nitrogen and phosphorus were studied by Lisi et al. (2004) for application as a distinct sub-surface drainage or intermediate layer in golf course putting greens. A statistically significant reduction in the concentration of nitrate in leachate was achieved by replacing traditional pea gravel with equally sized granulated tires for the drainage layer media, although the mechanism of nitrate mitigation remains unclear. The results indicate that using granulated tires as a drainage layer or fill material beneath sand-based root zones does not compromise the function of the profile or quality of the vegetation while creating an environmentally beneficial and value-added option for scrap tire reuse.

5 SUMMARY

Scrap tires shredded into small pieces (called "chips") alone or mixed with soil can have properties favorable to civil and environmental engineering applications. Although the reuse of scrap tires, either as whole tires but most commonly as tire chips (alone or mixed) has become more common, questions regarding environmental suitability still persist, particularly the potential impact on ground and surface waters and aquatic life due to leaching. This review indicates that a significant body of research has evolved over the last 25 years. Although the levels of certain metals (e.g., iron and manganese) increase, the impact of tire leachate on drinking water quality of groundwater is limited for a range of applications involving tires or tire shreds, so human health concerns are minimal. The compliance with the drinking water standards at a chosen point of compliance in the groundwater resulting from the initial increased tire leachate concentrations can be assessed through the advection-dispersion-reaction transport principles. Tires placed in surface water are shown to be toxic to aquatic life and this practice should be abandoned. Tires placed in groundwater also generate leachate that is toxic to aquatic life but again, if there is enough separation zone from the surface water, the toxic effects may decrease by the time the leachate reaches the surface water.

In contrast to this concern on contamination of the environment, the significant sorption capacity of tire material renders it a potential material for environmental protection and remediation when in contact with contaminated waters and leachate. This allows use of tires as a reactive drainage medium in landfills in reducing the strength of the landfill leachate but most importantly sorbing highly toxic volatile organic compounds. Similarly, the sorptive properties of tire rubber for retaining nitrogen and phosphorus can be beneficial beneath sand-based root zones in golf courses and athletic fields.

REFERENCES

Abernethy, S.G. 1994. The acute lethality to rainbow trout of water contaminated by an automobile tire. *Report* ISBN 0-7778-2381-0. Aquatic Toxicology Section. Standards Development Branch. Ontario Ministry of the Environment and Energy. Toronto.

Abernethy, S.G., Montemayor, B.P. & Penders, J.V. 1996. The aquatic toxicity of scrap automobile tires. *Report* ISBN 0-7778-4835-X. Aquatic Toxicology Section. Standards Development Branch. Ontario Ministry of the Environment and Energy. Toronto.

Aydilek, A.H., Madden, E.T. & Demirkan, M.M. 2006. Field evaluation of leachate collection system constructed with scrap tires. *Journal of Geotechnical and Geoenvironmental Engineering* 132(8): 990–1000.

B.A.R. Environmental Inc. 1992. Evaluation of the potential toxicity of automobile tires in the aquatic environment. *Report* to Environment Canada. National Water Institute. Burlington, Ontario. 15p.

Bin-Shafique, M.S., Benson, C.H., Edil, T.B.& Hwang, K. 2006. Leachate concentrations from water leach and column leach tests on fly-ash stabilized soil. *Environmental Engineering Science*, 23(1): 51–65.

Bosscher, P.J., Edil, T.B., & Eldin, N.N. (1993). Construction and Performance of a Shredded Waste Tire Test Embankment. *Transportation Research Record 1345*, Transportation Research Board, Washington, D.C., 44–52.

Caltrans. 1998. The use of discarded tires in highway maintenance. *Translab Design Information Brochure* No.TI/REC/1/88. California Department of Transportation. Sacramento, CA.

Dodds, J., Domenico, W. P., Evans. D. R., Fish, L. W., Lusahn, P. L. & Toch, W. J. 1983. Scrap tires: a resource and technology evaluation of tire pyrolysis and other selected alternative technologies. *Report*. EGG-2241, EG&G Idaho, Inc. Idaho Falls, Idaho.

Edil, T.B. & Bossher, P.J. 1992. "Development of Engineering Criteria for Shredded Waste Tires in Highway Applications," *Research Report* GT-92-9, University of Wisconsin, Madison, Wisconsin.

Edil, T.B. & Bosscher, P. J. 1994. Engineering Properties of Waste Tire Chips and Soil Mixtures. *Geotechnical Testing Journal*, ASTM, 17(4): 453–464.

Edil, T. B., Fox, P.J. & Ahl, S.W. 1992. "Hydraulic conductivity and compressibility of waste tire chips." *Proceedings of the Fifteenth Annual Madison Waste Conference*, 49–61.

Edil, T.B., Kim, J.Y. & Park, J.K. 1996. Reactive barriers for containment of organic compounds. *Proceedings* of the 3rd International Symposium on Environmental Geotechnology, San Diego, California, 523–532.

Edil, T.B., Park, J.K. & Kim, J.Y. 2004. Effectiveness of scrap tire chips as sorptive drainage material. *Journal of Environmental Engineering*, ASCE, 130(7): 824–831.

Edil, T.B., Park, J.K. & Heim, D.P. 1994. Large-size test for transport of organics through clay liners. ASTM *Special Technical Publication* 1142, 353–374.

Foose, G.J., Benson, C.H. & Edil, T.B. 2002. Comparison of solute transport in three composite liners. *Journal of Geotechnical and Geoenvironmental Engineering*, ASCE. 128(5): 391–403.

Gibbons, R.D., Dolan, D., Keough, H., O'Leary, K. & O'Hara, R. 1992. A comparison of chemical constituents in leachate from industrial hazardous waste and municipal solid waste landfills. *Proc. 15th Annual Madison Waste Conference, Municipal and Industrial Waste*, Madison, WI, 251–276.

Grefe, R 1992. Review of Waste Characterization of Shredded Tires. *Interdepartmental Memorandum*, Wisconsin Department of Natural Resources, Madison, Wisconsin.

Gunter, M., Edil, T R , Benson, C.II. & Park, J.K. 1999. The environmental suitability of scrap tire chips in environmental and civil engineering applications: a laboratory investigation. *Environmental Geotechnics Report*. Department of Civil and Environmental Engineering. University of Wisconsin, Madison, Wisconsin.

Hartwell, S.I., Jordahl, D.M. & Dawson, C.E.O. 2000. The effect of salinity on tire leachate toxicity. *Water, Air, and Soil Pollution* 121: 119–131.

Humphrey, D.N. 2003. Civil engineering applications of tire shreds. *Report* to California Integrated Waste Management Board. California Environmental Protection Agency.

Humphrey, D.N. and Katz, L.E. 2000. Water-quality effects of tire shreds placed above the water table: five-year field study. *Transportation Research Record* 1714, 18–24.

Humphrey, D.N., and Katz, L.E. 2001. Field study of the water quality effects of tire shreds placed below the water table. *Proceedings of the International Conference on Beneficial Use of Recycled Materials in Transportation Applications*, Arlington, VA, 699–708.

Humphrey, D.N. & Swett, M. 2006. Literature review of the water quality effects of tire derived aggregate and rubber modified asphalt pavement. *Draft Report* to U.S. EPA Resource Conservation Challenge. Department of Civil and Environmental Engineering University of Maine, Orono, Maine.

J & L Testing Co. 1989. Use of tire chips in liner protective cover. *Report* 89R414-01. Waste Management of Pennsylvania. Morrisville, Pennsylvania.

Kim, J. 1995. Soil and tire chip column testing, *Internal Report*. Department of Civil and Environmental Engineering University of Wisconsin, Madison, Wisconsin.

Kim, J.Y., Edil, T.B. & Park, J.K. 2001. Volatile organic compound (VOC) transport through compacted clay. *Journal of Geotechnical and Geoenvironmental Engineering*, ASCE, 127(2): 126–134.

Kim, J.Y., Park, J.K. & Edil, T.B. 1997. Sorption of organic compounds in the aqueous phase onto tire rubber. *Journal of Environmental Engineering*, ASCE. 123(9): 827–835.

Klett, N., Edil, T.B., Benson, C.H. & Connelly, J. 2005. Evaluation of volatile organic compounds in Wisconsin landfill leachate and lysimeter samples. *Final Report* to the University of Wisconsin System Groundwater Research Program, Department of Civil and Environmental Engineering, University of Wisconsin, Madison, Wisconsin.

Lee, D.T. 1982. Constructing a floating tire breakwater: the Lorain, Ohio experience. *Report* to the Lorain Port Authority, Ohio. 39p.

Lerner, A., Naugle, A., LaForest, J. & Loomis, W. 1993. Study of waste tire leachability in potential disposal and usage environments. *Amended Volume 1: Final Report*. Department of Environmental Engineering & Sciences. College of Engineering, University of Florida, Gainesville, Florida.

Li, L., Benson, C.H., Edil, T.B. & Hatipoglu, B. 2006. WiscLEACH: "A model for predicting groundwater impacts from fly-ash stabilized layers in roadways", *GeoCongress 2006*, ASCE, CD-ROM.

Lisi, R.D., Park, J.K. & Stier, J.C. Mitigating nutrient leaching with a sub-surface drainage layer of granulated tires. *Waste Management*. 24: 831–839.

Miller, W.L,. & Chadik, P.A. 1993. A study of waste tire leachability in potential disposal and usage environments: *Amended Volume 1: Final Report*. Department of Environmental Engineering & Sciences. College of Engineering, University of Florida, Gainesville, Florida.

O'Shaughnessy, V. & Garga, V.K. 2000a. Tire-reinforced earthfill, part 1: Constructionof a test fill, performance, and retaining wall design. *Canadian Geotechnical Journal*. 37(1): 75–96.

O'Shaughnessy, V. & Garga, V.K. 2000b. Tire-reinforced earthfill, part 3: Environmental assessment. *Canadian Geotechnical Journal*. 37(1): 117–131.

Park, J.K., Kim, J.Y & Edil, T.B. 1996. "Mitigation of organic compound movement in landfills by shredded tires." *Water Environment Research*, 68(1): 4–10.

Park, J.K., Kim, J.Y., Edil, T.B., Huh, M., Lee, S.H., and Lee, J.J. 2003. Suitability of shredded tires as a substitute for a landfill leachate collection medium. *Waste Management and Research*, 21(3): 278–289.

Park, J.K. & Nibras, M. (1993). "Mass Flux of Organic Chemicals through Polyethylene Geomembranes." *Water Environment Research*, Washington, D.C. 65(3): 227–237.

Plumb, R.H.& Pitchford, A.M. 1985. Volatile organic scans: implications for groundwater monitoring. *Proceedings. NWWA/API Conf. on Petroleum Hydrocarbons and Organic Chemicals in Ground Water-Prevention, Detection and Restoration*, Houston, TX, 207–222.

Rubber Manufacturers Association (1990) RMA leachate study. *Radian Corporation* Washington DC.

Rubber Manufacturers Association (2002). U.S. Scrap Tire Markets 2001, Washington DC.

Sheehan, P.J., Warmerdam, J.M., Ogle, S., Humphrey, D.N. & Patenaude, S.M. 2006. Evaluating the toxicity of leachate from tire shred fill in roads and the risk to aquatic ecosystems. *Journal of Environmental Toxicology and Chemistry*, 25(2): 400–411.

Tatlisoz, N., Edil, T.B., Benson, C.H., Park, J.K. & Kim, J.Y. 1996. Review of environmental suitability of scrap tires. *Environmental Geotechnics Report* No: 96-7. Department of Civil and Environmental Engineering. University of Wisconsin, Madison, Wisconsin.

Twin City Testing Corporation. 1990. Environmental study of the use of shredded waste tires for roadway sub-grade support. *Report* to Waste Tire Management Unit, Site Response Section, Groundwater and Solid Waste Division, Minnesota Pollution Control Agency, St. Paul, Minnesota.

Scrap Tire Derived Geomaterials – Opportunities and Challenges – Hazarika & Yasuhara (eds)
© 2008 Taylor & Francis Group, London, ISBN 978-0-415-46070-5

Recent Japanese experiences on scrapped tires for geotechnical applications

Kazuya Yasuhara
Department of Urban and Civil Engineering, Ibaraki University, Ibaraki, Japan

ABSTRACT: This paper introduces recent Japanese experiences related to scrapped tires for geotechnical applications. Applications of whole and shredded tires for construction of embankments and foundations have been proposed. Techniques for using tire shreds and chips are classifiable into two categories: tire chips mixed and not mixed with soil. The former techniques include cement-treated clay with tire chips, which possess high ductility and toughness, and non-cement-treated sand mixed with tire chips, which is intended to reduce liquefaction potential during earthquakes. In contrast, ongoing projects of the latter category include tire-chip drains that replace gravel drains as a countermeasure against liquefaction in sand. Furthermore, their feasibility has been confirmed not only for drainage materials used for accelerating compression and consolidation of volcanic-ash-based cohesive soils for applications on embankments for highways, but also as materials for protection against frost heave in subgrades and side ditches of highways in cold and mountainous sites.

1 INTRODUCTION

Scrapped tires provide numerous advantages from the viewpoint of civil engineering practices. They have light weight, high vibration-absorption, high elastic compressibility, high hydraulic conductivity, and temperature-isolation potential. New techniques have emerged to utilize these advantageous characteristics for practical purposes. Mainly, two types of scrapped tire materials are used for civil engineering applications: with and without shredding or cutting into small pieces of several tens of centimeters' or several centimeters' diameter. The former, without shredding, is useful for infrastructural retaining walls and foundations. The material is sometimes reinforced with geosynthetics. On the other hand, in cases with shredding or cutting into pieces of similar size, as with gravel and sand, they can be put into practical use because they have advantageous characteristics like those described above. This latter material type is more costly than unshredded or uncut material.

This paper describes attempts to introduce recent Japanese experiences related to scrapped tires for geotechnical applications. Possible applications of shredded tires for construction of embankments and foundations have been proposed. For utilization of whole used tires, application to foundations by themselves with geosynthetics is described.

Techniques for using tire shreds and chips are classifiable as cases of tire chips mixed and not mixed with soil. The former techniques include cement-treated clay with tire chips, which possess high ductility and toughness, and non-cement-treated sand mixed with tire chips, which is intended to reduce liquefaction potential during earthquakes. Ongoing projects of the latter category include tire-chip drains that can replace gravel drains as a countermeasure against liquefaction in sand. Their utility has been demonstrated not only for drainage materials used for facilitating compression and consolidation of volcanic-ash-based cohesive soils for applications on embankments for highways,

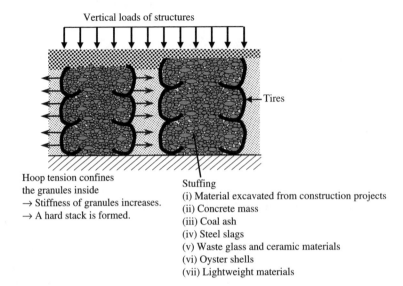

Vertical loads of structures

Tires

Hoop tension confines
the granules inside
→ Stiffness of granules increases.
→ A hard stack is formed.

Stuffing
(i) Material excavated from construction projects
(ii) Concrete mass
(iii) Coal ash
(iv) Steel slags
(v) Waste glass and ceramic materials
(vi) Oyster shells
(vii) Lightweight materials

Figure 1. Earth reinforcement using a stack of tires.

but also as materials to protect against frost heave in subgrades and side ditches of highways in cold regions or at mountainous sites.

The discussion presented in this paper pertains to the two important issues of economical and environmental aspects regarding both cases of chipped and shredded tires that are mixed and not mixed with soils. In particular, not only chemical impacts, but also biological impacts on the environment are described in this paper.

As a consequence, because the present study outlines the above-mentioned utilization techniques for geotechnical applications, this paper represents a state-of-the-art report on the situation of newly developed techniques in Japan, making use of chipped and shredded tires and whole tires, originating from waste tires, for geotechnical applications.

2   EFFECTIVE UTILIZATION OF WHOLE TIRES

Effective construction with high bearing capacity and good cost-performance will be attained if used tires are recycled to their original form. Fukutake et al. (2003, 2006) attempted construction of laminar structures using tires that had been packed with granular materials inside the hollow portion of tires. They validated the characteristics of the laminar structures through compression tests. The experimental results revealed that laminar bodies had high strength, especially in prestressed conditions.

2.1   Basic principles of method

Applying loads on a cylindrical stack of tires filled with granules causes the granules to expand the tires horizontally, as shown in Figure 1 and Photograph 1. Then, the hoop tension is exerted on the tires and confines the granules. The stiffness of granules increases with the confining force. As a result, a hard cylinder is formed for stabilizing soil foundations and structures. Tires contain wires and fibers and are highly resistant to hoop tension. The great confining stress works on the granules. Sufficiently hard granules in a stack of tires can resist large overburden loads.

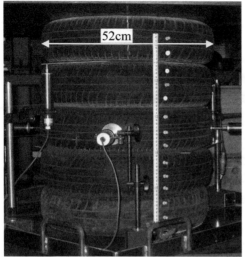

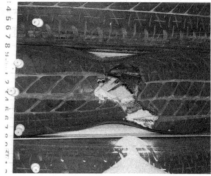

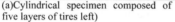

| (a)Cylindrical specimen composed of five layers of tires left) | (b) Fracture of a tire (ultimate state whereToyoura Sand was used) (below) |

Photograph 1.   Stuffed used tire stack compression test (for whole vehicle tires).

Stacks are filled with stuffing by (i) cutting part of the tire to facilitate stuffing (Fukutake et al. 2006) or (ii) placing stuffing into a stack of uncut whole tires (Fukutake et al. 2003, 2004). Method (i) uses bowl-shaped tires, like those shown in Photograph 2, to facilitate stuffing.

Recycled materials might be used as granules as long as their density is expected to increase through compaction. Lightweight aggregate, paper sludge ash, and shredded tires are useful for constructing lightweight fills.

2.2   Outline of model tests

Used vehicle tires were stacked in three layers. Tests were also conducted using five layers of tires (Photograph 1 (a)), and also with new tires for comparison. Recycled materials as well as real soil materials were used as stuffing from environmental considerations. Recycled aggregate is low-quality material with high adhesive mortar content. Paper sludge granules are obtained by mixing the ash produced from the incineration of paper sludge, cement, and lime. Vibratory filling was made easier by connecting the beads of tires with clips (Fukutake et al. 2004). Photograph 1b shows the fractured surface at the end of loading. Tread was fractured in numerous cases in a similar manner to that in this photograph, but beads were seldom fractured.

2.3   Test results

Figure 2 shows vertical stress-strain curves in all the cases of model tests. Loads were removed once each at a stress of 1200 kPa and 4800 kPa, and four times at 2400 kPa (depicted as black triangles along the vertical axis). Unloading took place at the other stress levels because of the severance of wires in the tread. Stress-strain curves at the time of loading tended to sag because small voids that remained in the tire at the initial stage were crushed, causing the stuffing to fit in with the tire. As the load increased, the hoop tension was fully exerted on tires, creating a highly confined state. The large confining force and the strength of tires were confirmed because stuffed tires had much higher stress than sandbags containing crushed stones (Fukutake et al. 2004). Several occurrences of severed wires in the tread all ended in fracture (Photograph 1b). No fracture occurred suddenly: the tires proved to be tough. In reality, the earth surrounding the tires confines them. Neither rapid

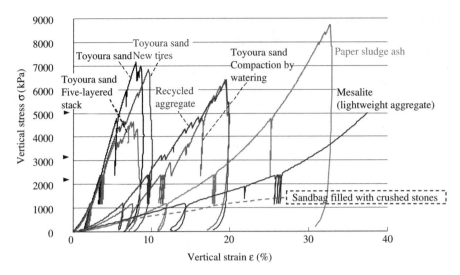

Figure 2. Stress-strain curves in compression and unloading tests of stuffed whole tire stacks.

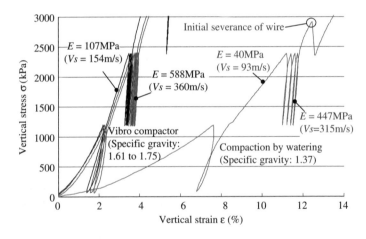

Figure 3. Enlarged diagram of stress-strain curves for stacks of tires filled with Toyoura sand.

fracture nor settlement is therefore unlikely to occur. The ultimate strength was lowest in the case of Toyoura sand filling five layers of tires. Fracture occurred at 4700 kPa in this case. Wires were severed earliest at 2900 kPa, or 600 kN, in the case of compaction of Toyoura sand by watering. Therefore, it was confirmed that whole vehicle tires provided a minimum strength of 600 kN. Used and new tires proved almost equally effective unless wires were severed.

Figure 3 shows an enlarged diagram of stress-strain curves for stacks of tires filled with Toyoura sand, and Young's modulus E. The curves obtained as a result of unloading and re-loading are sloped sharply. Under these conditions, tires can provide high stiffness. Prestressed stacks of tires can function as very hard members. Voids in the tire that are not filled with stuffing are crushed by prestressing. Stacks of tires are available for a wide range of applications such as preloaded prestressed bridge piers (Tatsuoka et al. 1998) (Figure 4) and retaining walls because of their high stiffness and strength.

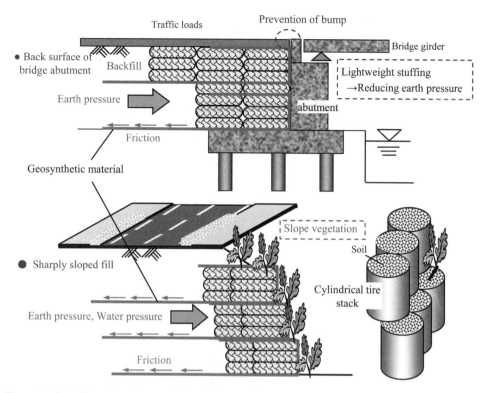

Figure 4.   Backfill, reclamation materials and slope vegetation.

2.4 *Various applications*

Stacks of tires are applicable to actual structures as tire piles. Combining tire stacks with geosynthetics is thought to enhance overall stability. Then, the friction between tires and the geotextile wrapping numerous tires increases the stability of earth reinforcement. Earth reinforcement can be formed by various shapes. Consequently, greater freedom of design is provided.

An image of tire stacks applied to back-filling and road fill is shown in Figure 5, which illustrates that geosynthetics integrate into tires to produce a unified structure, which ensures safety even in the case of failure of the fill. A vertical fill can be constructed with no inclination by applying geosynthetics densely.

Tires were cut by removing the top surface to use the bowl-shaped lower part in tests (Photograph 2). Tires were cut easily using existing cutting machinery. In land reclamation in the field, soil is spread over the cut tires that are laid and compacted by rollers on a layer-by-layer basis. In the compacted soil, land can be reclaimed easily on a large scale. Stacking tires vertically, or in the shape of a cylinder, is not necessary. Geotextile is inserted between layers of tires whenever required for ensuring safety, as portrayed in Figure 6. Securing traffic on at least one lane is sometimes required during a great earthquake with level-two ground motions. Stacks of tires might be used to provide one-lane traffic.

3 UTILIZATION OF SHREDDED TIRES

Two case studies have been carried out using used-tire shreds: one application was for thermal isolation to avoid frost heave and another was used for horizontal drainage material for embankments. Both case studies were executed in Hokkaido, in northernmost Japan.

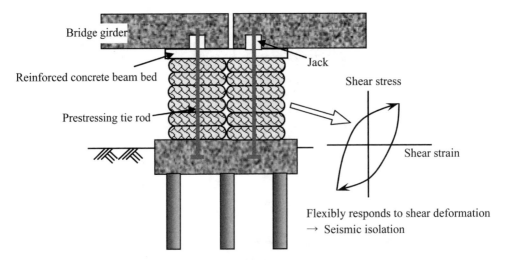

Figure 5.   Preloaded prestressed bridge pier.

Photograph 2.   Basin-shape tire cut upper section (→easy to filling).

3.1   *Thermal isolation utilization*

3.1.1   *Thermal insulation behavior of rubber materials*

For highlighting the differences, if any, in properties of thermal insulation of scrap-tire shreds and gravel, simple laboratory experiments to assess the thermal conductivity behavior of scrap-tire-derived rubber grains and sand are performed (Karmokar et al. 2006a). For this purpose, a Thermal Conductivity Meter is used, which is meant for measuring thermal conductivity of materials in sheet to blanket and plate forms with thickness of less than 10 mm. Because of such instrumental limitations, only finely grained materials, namely Toyoura and Sohma sand, and similarly distributed (see Figure 7) scrap-tire-derived fine grained rubber materials, are selected in

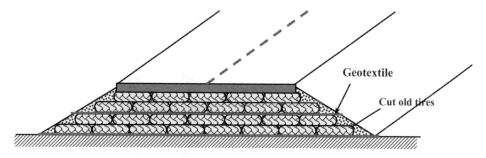

Figure 6. Easy embanking method using cut tires.

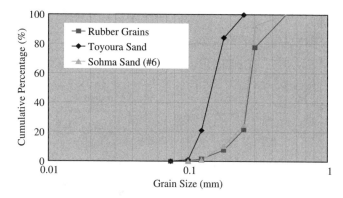

Figure 7. Grain size distribution curves.

(a) rubber grain blanket   (b) measurement in progress

Figure 8. Thermal conductivity test.

this study for evaluating relative thermal conductivity behavior. For the study, a rectangular frame is made to form a blanket-shaped test piece from selected material, which is then sandwiched between known-conductivity reference plates and a sensor, sequentially, for measuring thermal conductivity (Figure 8). Five observations are made for each kind of material.

Results of thermal conductivity of finely grained rubber material and sand are shown in Figure 9. As might be readily apparent, thermal conductivity of sand is 3–4 times higher than that of rubber materials. Therefore, it would be reasonable to say that the thermal insulation behavior of tire rubber material is superior to that of sand. Although results of the tests performed here are only relative in nature, and the selected materials are far from representative as compared to generally used gravel and/or other equivalent materials, one might argue at this point that the use of scrap-tire shreds is beneficial for mitigating frost penetration phenomena civil engineering structures in cold regions.

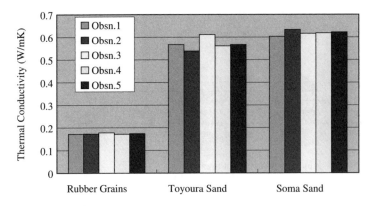

Figure 9. Relative values of thermal conductivity.

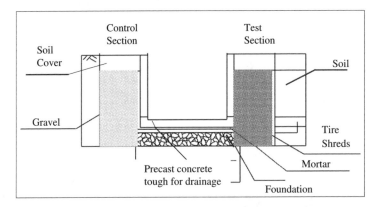

Figure 10. Cross-sectional view of control/test sections.

3.1.2 *Field study using scrap-tire shreds*

To investigate the *in-situ* performance of tire shreds in thermal insulation applications, e.g., frost-penetration phenomena in cold regions, an instrumented test section was constructed in Hokkaido. A highway drainage system trench extending more than 100 m was constructed using large pre-cast concrete troughs. One side of the concrete trough was backfilled with tire shreds; the other side was backfilled with gravel in a very similar fashion to serve as a control section in this study. The width of the backfill (insulation) layer is about 500 mm. A cross-sectional view of the constructed trench is shown in Figure 10. Scrap-tire shreds and gravel used in the backfills are shown in Figure 11. Tire shreds are uniformly graded (20–60 mm), in contrast to gravel, which is conventionally used for this purpose.

Numerous sensors are embedded within the backfill as well as in neighboring soil during construction for measuring temperature. Figure 12 (inset) shows the sensor positions, which are in fact mirror-imaged for this comparative study. A few of the typical temperature distributions in the tire shred test section are shown in Figure 10 along with their counterpart positions in the gravel section. Results showed that all the temperatures recorded in the backfill and neighboring soil are greater than 0°C despite a local minimum atmospheric temperature of −20°C. Temperature distributions in the tire-shred section were higher than those in the control section, which might indicate the effectiveness of scrap-tire shreds for mitigating frost penetration phenomena in civil engineering structures in cold regions. Scrap-tire shreds are demonstrably better as an insulation barrier over a control section; consequently, it might be said that scrap-tire shreds can be used for mitigating

Figure 11.   Materials used (a) tire shreds (b) gravel.

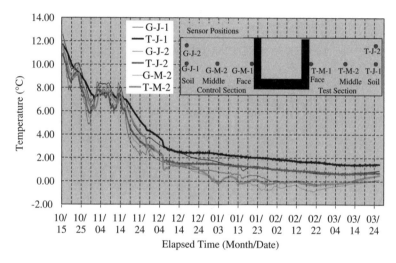

Figure 12.   Temperature distribution at control/test section.

frost penetration phenomena. Conservatively put, it might be said that performance of scrap-tire shreds is equivalent to, if not better than that of gravel.

3.2   *Embankment drainage*

3.2.1   *Drainage behavior of shredded used tire geomaterials*

At present, attempts are being undertaken to produce drainage/filtration applications of scrap tire materials. It might be argued that rubber chips/tire shreds are compressive in nature under vertical load, and that drainage quality would change according to their level of compression. Accordingly, drainage tests on scrap-tire-derived rubber chips and tire shreds were carried out using a compressible/expandable specimen. Some details of the test apparatus and experimental methods are described elsewhere in the literature (Karmokar et al. 2006b).

Two sample types are included in this study: scrap-tire-derived rubber chips (TC20) and tire shreds (TS20). Both samples comprise mainly non-spherical particles, and contain chips/shreds of size ranging 4.75–19 mm. The cumulative particle size distribution curves of these two samples are quite identical. Tire shreds contain steel and textile cords embedded inside the rubber in the tire, whereas tire chips sample are rubber-only materials; that is, steel and textile cords are separated out from rubber. Consequently, tire shreds behave less compressively than tire chips. The specific gravity of chips and shreds are 1.15 and 1.20, respectively. Tire rubber chip and shred specimens are compacted into the mold. The respective compacted densities maintained for TC20 and TS20

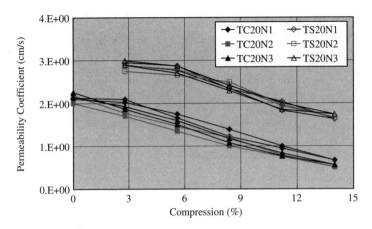

Figure 13.   Effect of compression on permeability coefficients.

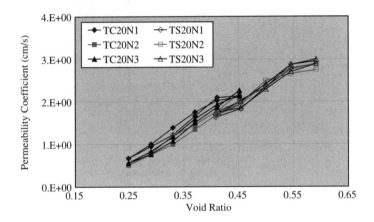

Figure 14.   Effect of void ratio on permeability coefficients.

samples are 0.66 and 0.71 kN/m³. To simulate the field consolidation of specimens by overlying soil, the specimen in the mold is then compressed for up to a level of normal pressure. The specimen so set is used for permeability testing.

The permeability coefficients of TC20 and TS20 samples for different compression and expansion are shown in Figure 13. The three repetitions for each test conducted also show identical permeability behavior. As the figure shows, the levels of permeability coefficients for tire-derived materials are of a high ($\times 10^{00}$) level in absolute terms: they are comparable to those of highly permeable geomaterials such as coarse grained sand and gravels. In spite of the high level of compression in rubber chips/shreds materials that might be incurred in field because of surcharge, the permeability coefficients remain at the higher level. Figure 14 shows the relationships between the void ratio and permeability of the samples. Similarly to permeable geomaterials, both the rubber chips and tire shreds samples show an overall linear response. In comparison to tire rubber chips, tire shreds show slightly higher permeability, which might be related to material compressibility. Chips of TC20 (rubber only) are deformable under normal stress. For that reason, the probability of better packing of chips engenders a lower void ratio than that of tire shreds (TS20 sample). In general, the permeability coefficients obtained in the compression stages and expansion stages are in the same order, which might indicate that the rubber chips and tire shred specimens are almost

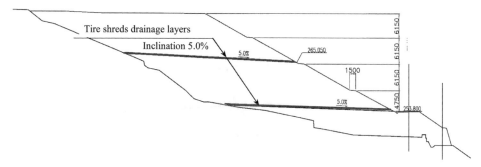

Figure 15. Cross-sectional view of test embankment.

Figure 16. View of test embankment after one year of construction.

recovered to their previous forms even after undergoing a considerable degree of compression. Such material properties might be very useful in many field conditions.

The high levels of permeability characteristics, along with the non-clogging phenomenon of tire shreds has led us to undertake a full-scale field trial in Hokkaido using tire shreds in a drainage and filtration layer under an embankment. The constructional process was almost entirely machine-controlled, and finished quite smoothly. The performance of tire shreds in the field was found to be highly satisfactory.

3.2.2 Field trial
Based on laboratory test results, a full-scale field trial was undertaken in Hobetsu, Hokkaido. Altogether, 240 tons of scrap-tire shreds were used in the form of a blanketed drainage/filtration layer under 20-m-high embankment. A cross-sectional diagram of the test embankment is presented in Figure 15. The view of completed embankment after one year of construction is shown in Figure 16. The construction process was almost entirely machine-controlled, and finished quite smoothly. The performance of tire shred was found to be superior in the field.

4 UTILIZATION OF USED-TIRE CHIPS

The tire chips were made from used tires by crushing and subsequent removal of the constituent textiles and metal fibers. The chips therefore have a rough or serrated surface, as illustrated in

29

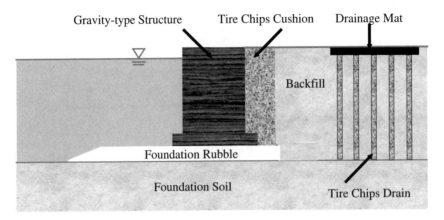

Figure 17. Sandwiched earthquake resistant technique.

Figure 17. The specific gravity of tire chips used in this study is 1.15 and the grain size is 2 mm on average. Tire chips are an elastically compressible material, of which Poisson's ratio is nearly 0.5 and the elastic modulus is approximately 4–6 MPa on average, corresponding to 0–15% strain.

4.1 Application of tire chips without soil mixing

4.1.1 Sandwiched backfilling technique

The environmentally friendly and the cost-effective disaster mitigation technique that has been developed involves placing a cushioning layer of tire chips as a vibration absorber immediately behind the structure. The beneficial effects of such a sandwiched cushioning technique have been described in Hazarika et al. (2006b). In addition, vertical drains made out of tire chips can be installed in the backfill to prevent soil liquefaction. Yasuhara et al. (2004) used tire chips in vertical drains for reducing liquefaction-induced deformation.

Figure 18 shows a typical cross section of the earthquake resistant reinforcement technique. One function of the cushion is to reduce the load against the structure caused by the energy absorption capacity of the cushion material. Another function is to curtail permanent displacement of the structure attributable to inherited flexibilities derived from using such an elastic and compressible material.

4.1.2 Underwater shaking-table testing

Large three-dimensional underwater shaking table assemblies of the Port and Airport Research Institute (PARI) were used in the testing program. The shaking table is circular with 5.65 m diameter; it is installed on a 15 m long ×15 m wide ×2.0 m deep water pool. The details of the shaking table are available in Iai and Sugano (2000).

A caisson-type quay wall (model to prototype ratio of 1/10) was used for testing. Figure 19 shows a cross section of the soil box, the model caisson and the locations of the various measuring devices (load cells, earth pressure cells, pore water pressure cells, accelerometers and displacement gauges). The model caisson (425 mm in breadth) was made of steel plates filled with dry sand and sinker to bring its center of gravity to a stable position. The caisson consists of three parts: the central part (width 500 mm) and two dummy parts (width 350 mm each). All the monitoring devices were installed at the central caisson to eliminate the effect of sidewall friction on the measurements. The soil box was made of a steel container 4.0 m long, 1.25 m wide, and 1.5 m deep. The foundation rubble beneath the caisson was prepared using Grade 4 crushed stone with 13–20 mm particle size. The backfill and the seabed layer were prepared using Sohma sand (No. 5).

30

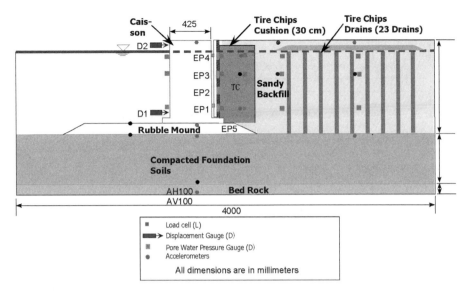

Figure 18.   Cross section of the experimental model.

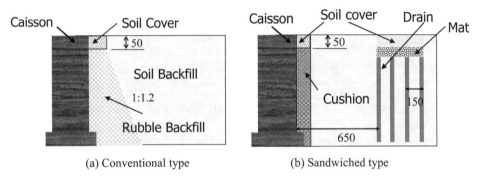

(a) Conventional type   (b) Sandwiched type

Figure 19.   Test cases of backfill.

The dense foundation sand representing the seabed layer was prepared in two layers. After preparing each layer, the entire assembly was shaken with 300 Gal of vibration starting with a frequency of 5 Hz and increasing up to 50 Hz. Backfill was also prepared in stages using free falling technique; it was then compacted using a manually operated vibrator. After constructing the foundation and the backfill, and setting up of the devices, the pool was filled with water, which gradually elevated the water depth to 1.3 m to saturate the backfill. This submerged condition was maintained for two days so that the backfill can attain a complete saturation stage.

4.1.2.1   Test cases
As was shown in Figure 19, two test cases were examined. In one case (Case A), a caisson with a rubble backfill with conventional sandy backfill behind it was used. In another case (Case B), behind the caisson, a cushion layer of tire chips (average grain size 20 mm) was placed vertically; its thickness was 0.4 times the wall height. In actual practice, the design thickness will depend upon many other factors such as the height and rigidity of the structure, in addition to compressibility and stiffness of the cushion material. In compressible buffer applications, there seems to be an optimum value for the cushion thickness, beyond which increased thickness will not engender

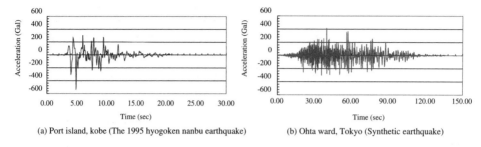

(a) Port island, kobe (The 1995 hyogoken nanbu earthquake)   (b) Ohta ward, Tokyo (Synthetic earthquake)

Figure 20.   Strong motion wave records.

a proportionate decrease of the load. The effect of the cushion thickness shaking table test was described using a small-scale model in Hazarika et al. (2006c).

4.1.2.2   Test procedures

The cushion layer was prepared by filling the tire chips inside a bag made from geotextile product. The geotextiles are required to wrap the tire chips so that they do not mix with the surrounding soil. Such confinement also simplifies the backfilling process. Also, the presence of geotextiles prevents flowing of sand particles into the chip structure; it therefore prevents clogging and mixing, which might affect the chips' compressibility and permeability. The average dry density of the tire chips achieved after filling and tamping was $0.675\,t/m^3$.

The relative densities that were achieved after the backfill preparation were about 50% to 60%, implying that the backfill soil is partly liquefiable. Liquefaction tends to increase the earth pressure. Therefore, the presence of a tire-chip cushion is expected to protect the structure from the adverse effect of liquefaction within a limited region surrounding the structures during an earthquake. For that reason, liquefiable backfill was deliberately selected. The foundation soils were compacted with a mechanical vibrator to achieve a relative density of about 80%, implying a non-liquefiable foundation deposit.

Vertical drains made out of tire chips (average grain size 7.0 mm) were installed in the backfill. Geotextile bags with the specific drain size were first prepared. They were then filled with tire chips having a pre-determined density. They were then installed with a spacing of 150 mm in triangular pattern. The drain diameter was chosen as 50 mm. The tops of all drains were covered with a 50-mm-thick gravel layer underlying a 50-mm-thick soil cover. The purpose of such a cover layer is twofold: one is to allow the free drainage of water and other is to prevent the likely uplifting of the tire chips during shaking because of its lightweight nature.

The similitude of various parameters in a 1g gravitational field for the soil-structure-fluid system was calculated using the relationship given in Iai (1989) for a model to prototype ratio of 1/10. It is worthwhile mentioning here that, the material particle size and compressibility of the material are assumed to remain unchanged for the model and the prototype.

Earthquake loadings of different magnitudes were imparted to the soil-structure system during the tests. The input motions selected were: (1) the Port Island (PI) wave – the N-S component of the strong motion acceleration record at the Port Island, Kobe, Japan during the 1995 Hyogo-ken Nanbu earthquake (M 7.2); and (2) the Ohta Ward (OW) wave – a scenario synthetic earthquake motion assuming an earthquake that is presumed to occur in the southern Kanto region with its epicenter at Ohta ward, Tokyo, Japan. It is noteworthy that the 1995 Hyogo-ken Nanbu earthquake is an intra-plate earthquake, while the scenario earthquake (synthetic) was constructed assuming an inter-plate earthquake. The wave records of the two input motions are shown in Figure 20.

The loading intensities were varied using the various maximum acceleration ratios (0.5, 1.0, 1.2, and 1.5) of the target acceleration to the actual acceleration. The loading steps in the test series are summarized in Table 1, which also shows the code names used according to the acceleration ratios used and the target maximum acceleration in each test series. Durations of the shaking in the

Table 1.   Loading sequences.

Series	Earthquakes types (Code names)	Acceleration ratio	(Maximum acceleration)
No. 1	PI (PI 0.5)	0.5	(339.39 Gal)
No. 2	OW (L2 0.5)	0.5	(243.47 Gal)
No. 3	PI (PI 1.0)	1.0	(678.78 Gal)
No. 4	PI (PI 1.2)	1.2	(814.54 Gal)
No. 5	OW (L2 1.0)	1.0	(486.94 Gal)
No. 6	PI (PI 1.5)	1.5	(1018.17 Gal)

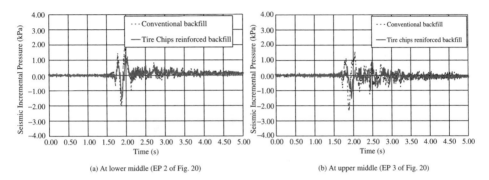

(a) At lower middle (EP 2 of Fig. 20)   (b) At upper middle (EP 3 of Fig. 20)

Figure 21.   Time history of incremental seismic thrust.

model testing were based on the time axes of these accelerograms, which were reduced by a factor of 5.62 according to the similitude relationship.

4.1.3   *Test result and discussion*

As displayed in Table 1, various types of earthquake motion with different magnitudes were adopted for this study. However, the discussion here will be mostly limited to series no. 3 (PI 1.0). The PI 1.0 data are the actual recorded data at Port Island, Kobe, with the time axes scaled to fit the model to prototype ratio of 1/10.

4.1.3.1   Incremental seismic thrust

Figure 21 shows the time history of the increment of the seismic earth pressure acting on the quay wall at the lower middle and the upper middle part of the caisson. It is readily apparent that, compared to conventional backfill condition (Case A in Figure 19), the seismic increment is decreased to a considerable extent in the tire-chip reinforced backfill (Case B in Figure 19). Considering the fact that the static earth pressure itself will also be reduced because of its low weight and compressible characteristics of the cushion materials, the total earth pressure acting on the structure will therefore be reduced to a greater extent. The result is the reduction of the total earth pressure, which will contribute to the stability of the structure during earthquakes.

4.1.3.2   Excess pore water pressures

To evaluate whether the developed method can minimize the liquefaction-related damage, the time histories of the excess pore water pressure during loading at two particular locations (A and B in Figure 18) are compared for the two test cases in Figures 22 and 23.

Comparisons reveal that the pore water pressure buildup is different for the two cases. In the case of conventional backfill, at a location closer to the gravel backfill (**A**), the pore water buildup

33

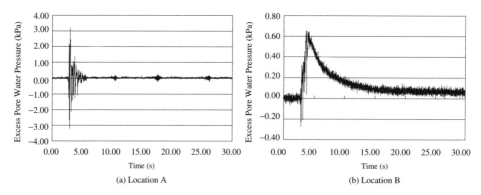

(a) Location A

(b) Location B

Figure 22. Excess pore water pressure (Conventional backfill).

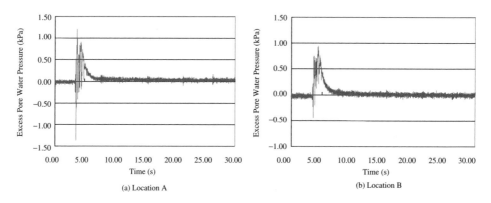

(a) Location A

(b) Location B

Figure 23. Excess pore water pressure (Sandwiched backfill).

is restricted because of dissipation of the permeable gravel backfill. However, at a distant location (**B**), the pore water pressure builds up and it takes considerable time (about 25 s) to dissipate. However, in the case of a tire-chip reinforced backfill, the built-up pore water pressure dissipates within a very short interval (2.5 s), which obviates any chance for the backfill to liquefy.

4.1.3.3 Seaward displacement of structures

The time histories of the horizontal displacements (D1 and D2 in Figure 18) during the loading for the two test cases are compared in Figure 24. Comparisons reveal that the maximum displacement experienced by the quay wall with a tire-chip reinforced caisson (thick continuous line) is toward the backfill, in contrast to the quay wall without any reinforcement (shown in dotted line), for which case it is seaward. The compressibility of the tire chips renders flexibility to the soil-structure system, which allows the quay wall to bounce back under its inertia force; this tendency ultimately (at the end of the loading cycles) aids in preventing the excessive seaward deformation of the wall. However, the wall with a conventional backfill experiences very high seaward displacements right from the beginning because of its inertia. As a consequence, the structure can not move back to the opposite side and ultimately suffers from a huge permanent seaward displacement.

4.2 Cement-treated marine clay mixed with tire-chips

4.2.1 Preparation of specimens

Dredged clay was used to make specimens of the cement-treated clay with addition of tire chips (CTCT). The dredged clay was taken from Tokyo Bay ($\rho_s = 2.72\,\text{g/cm}^3$, $w_L = 100\%$ and $Ip = 70$).

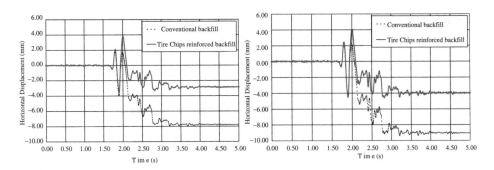

Figure 24. Time history of the caisson displacements.

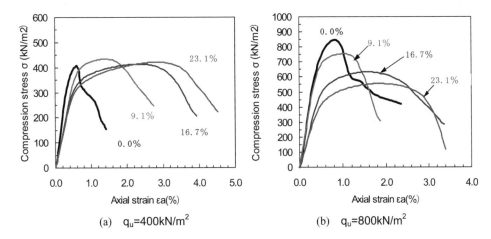

(a) $q_u = 400 \text{kN/m}^2$ (b) $q_u = 800 \text{kN/m}^2$

Figure 25. Stress vs. strain curves for unconfined-compression tests on CTCT specimens.

Its percentage of fine-grained fraction is about 90%. Slurry dredged clay with initial water contents of 250% was mixed with seawater to produce a 1.25 g/cm^3 for saturated-density and 330 mm for cylinder flow. The cement used was normal Portland cement, with particle density of 3.16 g/cm^3. The tire chips used for the mixture were made from used tires, with average grain size of 2 mm by screening, and soil particle density of 1.15 g/cm^3.

The specimens were prepared to provide two kinds of strength of cement-treated clay, and four kinds of tire chips' contents. The targeted unconfined compressive strength of cement-treated clays were $q_u = 400$ or 800 kN/m^2. Table 1 shows the mixing conditions of cement-treated clay and the percentage of added tire chips as 0%, 9.1%, 16.7%, and 23.1% within a whole volume of specimen under 28 days' curing. All specimens were cured under a high moisture content of greater than 95% and constant temperature at $20°C \pm 2°C$.

4.2.2 *Undrained behavior*
Figure 25 shows the stress-strain relation of unconfined compressive test (UC-test) on CTCT. These results imply the following.

a) The deformation property of cement-treated clay was changed from brittle into tough, merely by addition of tire chips,

b) The larger the percentage of tire chips in the mixture is, the larger the failure strain becomes. In the case of the cement-treated clay without adding tire chips, the stress-strain relation shows a marked strain-softening behavior. On the other hand, CTCT shows strain-hardening behavior.

35

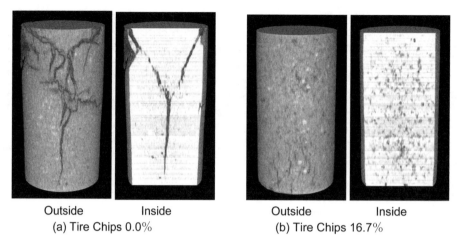

| Outside | Inside | Outside | Inside |
| (a) Tire Chips 0.0% | | (b) Tire Chips 16.7% | |

Figure 26.   3-dimentional images of UC-test specimens by X-ray CT scanner at axial strain $\varepsilon_a = 2.5\%$.

c) The larger the strength of cement-treated clays is, the smaller the tangent elastic modulus by adding tire chips becomes.

The three-dimensional images of unconfined compression test specimens by X-ray CT scanner (Kikuchi et al. 2006) are illustrated in Figure 26, in which (a) is the case of without added tire chips and (b) is the case with added tire chips of 16.7% by volume. The strength of cement-treated clay is the same in each specimen: $q_u = 400\,\text{kN/m}^2$. In addition, these figures show the same state of axial strain of nearly 2.5%. In the case without added tire chips (0%), it is apparent that some large vertical cracks are developed inside and outside of the specimen. On the other hand, the case with added tire chips presents no visible cracks. Results of X-ray images clarify that the characteristic of deformation of cement-treated clay was improved from brittle to tough by adding tire chips.

For comparison with the results from unconfined compression tests, undrained triaxial compression tests were conducted under the following conditions. Back pressure was 100 kPa and an isotropic effective consolidation stress was 200 kPa (2 h) and after consolidation, during undrained shear, the confining pressure was maintained constant with 300 kPa. The results from monotonic undrained triaxial tests were interpreted to investigate the effects of tire-chips contents and hardness of tire chips on toughness improvement.

Figures 27 (a) and (b) show the stress vs. strain curves for specimens with different contents of tire chips for the target undrained strengths of $400\,\text{kN/m}^2$ and $800\,\text{kN/m}^2$, respectively, although the strain at peak stress for specimens with $800\,\text{kN/m}^2$ is greater than that for specimens with $400\,\text{kN/m}^2$.

To confirm the effect of confinement, a comparison of the results from triaxial tests with those from unconfined compression tests is the most suitable way. Figure 28 illustrates a set of comparisons between results in which the target unconfined strength was $800\,\text{kN/m}^2$. Both tests were carried out on three sets of specimens with tire-chip contents of 9.1%, 16.7%, and 23.1%. This test series was carried out for verifying the effect of confinement on toughness characteristics, independently of the series of tests shown in Figures 28. The confining pressure was $200\,\text{kN/m}^2$ in triaxial compression tests, but no confining pressure was applied in unconfined compression tests. It is apparent from comparison between both test results that:

i) the stress and strain curves in triaxial tests exhibit ductile behavior, whereas those in unconfined compression test show brittle behavior, independently of tire-chip content;
ii) the ductile behavior in triaxial tests was improved according to the increase in the tire-chip content;

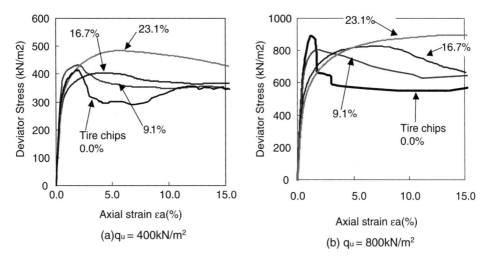

(a) $q_u = 400 kN/m^2$

(b) $q_u = 800 kN/m^2$

Figure 27.   Stress-strain curves (TC-test).

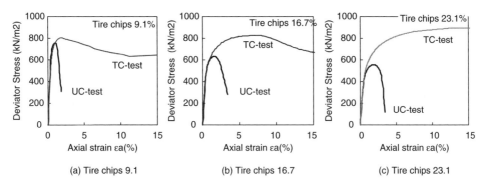

(a) Tire chips 9.1   (b) Tire chips 16.7   (c) Tire chips 23.1

Figure 28.   Effect of horizontal stress (Comparison UC-test with TC-test) (Strength of cement treated clay part: $q_u = 800 \, kN/m^2$).

iii) as in the ductile behavior improvement, less increase in triaxial undrained strength was observed, even with increasing the tire-chip content, while unconfined compressive strength rather decreases, even with increased tire-chip content

4.2.3   *Improvement of hydraulic conductivity*

As shown above, the cement-treated clay shows brittle deformation. Cracks develop with progress of shear deformation; then it is expected that the hydraulic conductivity increases by increasing deformation. But, in the case of adding tire chips, it can be expected that the hydraulic conductivity changes during deformation because the toughness was improved by the effect of the added tire chips. To verify that fact, the hydraulic conductivity was measured during shear deformation in the plate loading tests on CTCT.

To examine the hydraulic conductivity of a non-deformed specimen, we conducted a falling head permeability test. The authors adopted an acrylic fiber cell of 20 cm diameter and 20 cm, as shown in Figure 29. First, the saturated Toyoura-sand was poured at the bottom of the cell, and then the CTCT was poured onto the sand surface. The thickness of the sand tire and specimen was 5 cm. After the specimen was cured for 7 days in a high moisture condition of more than 95% under a constant temperature of $20 \circ C \pm 2°C$, the falling head permeability test was conducted under the water pressure of 20 kPa applied onto the specimen surface. Figure shows the result of

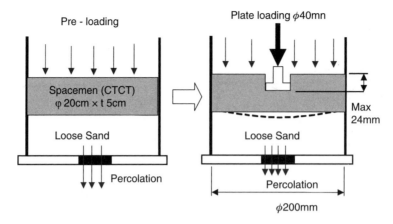

Figure 29.　Plate loading tests with measurement of hydraulic conductivity.

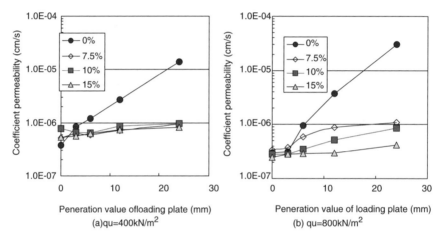

Figure 30.　Variations of hydraulic conductivity with penetration displacement.

the permeability tests of the cement-treated clay. It was obtained that the permeability was equal in the case of equivalent strength (mix proportion) of cement-treated clay, but the amounts of added tire chips were various, from 0% to 30%.

To examine the variations of the hydraulic conductivity of CTCT with deformation in the plate loading tests, we conducted the following examination: the specimen and devices were the same as above, and the specimen was loaded into the center using a steel loading plate, which had 4 cm diameter. The rate of loading was 15 mm/min; it was about 30 times the speed of the ordinary field plate-loading test. The loading to the specimen was applied gradually, and the permeability test was conducted after unloading. This process was repeated until cracks were readily apparent. In this examination, the loading plate was applied in the center of specimen at a rapid speed; then, the loading plate was penetrated into the specimen. Consequently, the bending deformation was developed in the specimen. Then, it was speculated that the coefficient of permeability might increase by developing some cracks, as compared to that before deformation.

Figure 30 shows the relation between the penetration displacement and the coefficients of permeability of CTCT specimen. In this figure, the specimen deformation is shown in the value of penetration by loading plate. As shown in this figure, the change of the coefficient of permeability with deformation was very small in the case of added tire chips (CTCT). On the other hand, the

(a) With no tire-chips addition

(b) With 10 tire-chips addition (at 24mm penetration)

Figure 31.   Observed appearance of specimens after penetration tests (left for surface and right for surface).

hydraulic conductivity of the cement-treated clay without added tire chips (0% tire-chips) increases with deformation.

Figure 31 shows a photograph of the specimen that was taken at 24 mm penetration of the loading plate. It is apparent that the specimen without added tire chips was cracked, but that the specimen of CTCT was not cracked. These observations indicate that the fewer cracks in CTCT are attributable to improving the ductility of cement-treated clay with the added tire chips.

4.2.4   *Application of CTCT to sealing materials of a coastal waste-disposal site*
A sealing material for use in coastal waste disposal sites must necessarily follow deformation of a revetment and ground to maintain its impermeable property. As described above, the authors found that CTCT has a low hydraulic conductivity and ductility, as clarified through triaxial tests and X-ray scanning, and plate loading tests. For that reason, we judged that the CTCT is applicable as a sealing material for coastal waste-disposal sites.

Figure 32 shows the cross-section of a revetment of coastal waste disposal site, which is the actual site of Shinkaimen disposal site in Tokyo Bay, Japan. The revetment was structured in double steel piles, and its foundation ground was stabilized by sand compaction piles (SCP). Figure 32 also shows that CTCT was applied between steel piles and cement deep mixing ground (CDM). The CTCT was implemented by mixing dredged clay and Portland cement with tire chips. The physical properties of dredged clay are shown in Table 2; the mixing proportion of CTCT is shown in Table 3.

In this construction work, the extension of the disposal yard was about 83 m and CTCT was executed in the amount of nearly 800 m^3. Therefore, tire chips were used in the amount of about

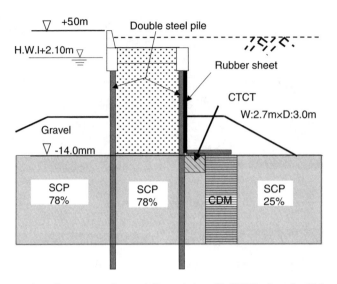

Figure 32. Cross-section of revetment of coastal disposal site with CTCT adopted at Tokyo bay.

Table 2. Index properties of dredged clay used for disposal yard.

Specific gravity g/cm^3	Grain size (%)			Consistency (%)	
	Sand	Silt	Clay	W_L	W_P
2691	1	26	73	120.3	50.3

Table 3. Mixing proportion of CTCT.

Table flow (adjusted dredged clay)	350 mm
Quantity of cement adding	57 kg/m^3
Taget strength (28 days)	qu = 400 kN/m^2
Quantity of tire chips adding	9.9% volume

80 tons, indicating that about 16 000 ordinary automobile tires were recycled in this construction work because about 5 kg tire chips can be made from each ordinary automobile tire.

5 CONCLUSION

This paper presented descriptions of some recent experiences of the utilization of used tires for geotechnical practices.

Some applications of whole used tires were attempted, without cutting them into such small pieces as shreds or chips. The feasibility of reinforced embankments and foundations using geosynthetics was clarified by carrying out model tests. This technique incorporates the idea of preloads and pre-stress: so-called PLPS. Using that technique, structures became very stable, even against cyclic loading.

Shredded tires, which are small pieces produced from cutting whole tires, have been widely used for embankments and back-filling materials of wall structures, particularly in the USA. This paper introduces case studies on the use of shredded tires for frost heave protection and drainage acceleration for construction of highway embankments in which layers of tire shreds were interbedded between volcanic-ash-origin cohesive soils.

In addition to applications using shredded tires, two techniques were proposed using tire chips that were processed by cutting them into smaller pieces of several centimeters' diameter on average. In one case, a series of model shaking table tests under a 1 g gravitational field was conducted to examine the performance of a newly developed sandwiched backfilling technique for earthquake disaster mitigation. In that technique, sandwiched cushions and vertical drains made out of an emerging and smart geomaterial known as tire chips were used as retaining materials behind massive rigid structures such as caisson quay walls.

To overcome a salient disadvantage of cement-treated clay (CTC), *brittleness*, an attempt was made to mix tire chips with CTC. This technique, abbreviated as CTCT, provided an additional characteristic, *toughness* or *ductility*, which was useful for resistance against occurrence of cracks in CTCT during development of shear displacement. This was already put into practice at a disposal yard where CTCT was used as a sealing material to protect leakage of contaminated materials.

Although this paper outlines some recent Japanese situations in which used tires were practically applied for civil engineering projects, some case studies were also described which were executed in the field, in addition to others that have been ongoing as research projects. In this regard, other important applications such as gravel-drain liquefaction mitigation (Yasuhara et al., 2005:2006) and traffic-induced vibration reduction (Yasuhara, 2002) were not fully described in this paper. Furthermore, discussions of associated environmental issues and cost-performance analyses were omitted from this paper, although they are also very important considerations for utilization of recycled materials.

ACKNOWLEDGEMENTS

The studies described in this paper have been financially supported by the Grants-in-Aid from the Ministry of Land, Transportation and Infrastructure and the Ministry of Education, Culture, Sports, Science and Education of the Japanese Government. The author is indebted to Dr. Takao Kishida of Toa Corp., Dr. Sumio Horiuchi of Shimizu Corp., Mr. Hideo Takeichi of Bridgestone Corp. for their strong support in carrying out the research project successfully. This paper was prepared by kind and capable assistance of Dr. Hemanta Hazarika of PARI, Dr. Yoshio Mitarai of Toa Corp., Dr. Takeshi Fukutake of Shimizu Corp., Dr. Ashoke Karmorkar of Bridgestone Corp. and Mr. Takeshi Nagatome of Toa Corp. The author would like to express sincere appreciation for their support.

REFERENCES

Dickenson, S. & Yang, D.S. 1998. Seismically-induced Deformations of Caisson Retaining Walls in Improved Soils. *Geotechnical Earthquake Engineering and Soil Dynamics III*, Geotechnical Special Publication 2(75): 1071–1082.

Fukutake, K., Horiuchi, S., Matsuoka, H. Liu, S. & Kawasaki, H. 2003. "Stable geostructure using recycled tires and properties of improved body using tires with granular materials." (in Japanese) *Proc. of the 5th Japan National Conference on Environmental Geotechnical Engineering*, 189 191.

Fukutake, K. & Horiuchi, S. 2004. "Forming method of geostructure using recycled tires and granular materials." (in Japanese) *Proc. of the 39th Japan National Conference on Geotechnical Engineering*, 653–654.

Fukutake, K. & Horiuchi, S. 2006. "Forming Method of Geostructure using Recycled Tires and Granular Materials." (in Japanese) *Proc. of the 41st Japan National Conference on Geotechnical Engineering*, 597–598.

Hashiguchi, K. 1996. "An elasto-plastic constitutive model based on the sub-loading and rotational hardening concepts." *J. Geotechnical Eng.* 547(36): 127–144 (in Japanese).

Hazarika, H., Kohama, E., Suzuki, H. & Sugano, T. 2006a. Enhancement of Earthquake Resistance of Structures using Tire chips as Compressible Inclusion. *Report of the Port and Airport Research Institute* 45(1): 1–28.

Hazarika, H., Sugano, T., Kikuchi, Y., Yasuhara, K., Murakami, S., Takeichi, H., Karmokar, A.K., Kishida, T. & Mitarai, Y. 2006b. Model Shaking Table Test on Seismic Performance of Caisson Quay Wall Reinforced with Protective Cushion. *International Society of Offshore and Polar Engineers (ISOPE) Transactions* 2: 309–315.

Hazarika, H., Sugano, T., Kikuchi, Y., Yasuhara, K., Murakami, S., Takeichi, H., Karmokar, A.K., Kishida, T. & Mitarai, Y. 2006c. Evaluation of Recycled Waste as Smart Geomaterial for Earthquake Resistant of Structures. 41st *Annual Conference of Japanese Geotechnical Society, Kagoshima,* pp. 591–592. http://www.bridgestone.co.jp/eco/report/

http://www.bousai.go.jp: "Cabinet Office, Government of Japan: Central Disaster Management Council," *Available: http://www.bousai.go.jp/jishin/chubou,* 2006 (in Japanese).

Humphrey D.N. Civil Engineering Applications of Tire Shreds, *Manuscript Prepared for Asphalt Rubber Technology Service,* SC, USA, 1998.

Humphrey, D. N., Whetten, N., Weaver, J., Recker, K. & Cosgrove, T. A. 1998. Tire shreds as lightweight fill for embankments and retaining walls. *Recycled Materials in Geotechnical Applications, ASCE, Geotechnical Special Publication* 79: 51–65.

Iai, S. 1989. Similitude for Shaking Table Tests on Soil-structure-fluid Model in 1 g Gravitational Field. *Soils and Foundations* 29(1): 105–118.

Iai, S. & Sugano, T. 2000. Shake Table testing on Seismic Performance of Gravity Quay Walls, *12th World Conference on Earthquake Engineering, WCEE,* Paper No. 2680.

Inagaki, H., Iai, S., Sugano, T., Yamazaki, H. & Inatomi, T. 1996. Performance of Caisson Type Quay Walls at Kobe Port. *Special Issue, Soils and Foundations* 1: 119–136.

Ishihara, K., Yasuda, S. & Nagase, H. 1996. Soil Characteristics and Ground Damage. *Special Issue of Soils and Foundations* 1: 101–118.

Ishihara, K. 1997. Geotechnical Aspects of the 1995 Kobe Earthquake. Terzaghi Orientation. 14th *Intl. Conf. of International Society of Soil Mechanics and Geotechnical Engineering, Hamburg, Germany* 4: 2047–2073.

Japanese Geotechnical Society (JGS) and Japan Society for Civil Engineers (JSCE). "Joint Report on the Hanshin-Awaji Earthquake Disaster," 1996 (in Japanese).

Kamon, M., Wako, T., Isemura, K., Sawa, K., Mimura, M., Tateyama, K. & Kobayashi, S. 1996. Geotechnical Disasters on the Waterfront. *Special Issue of Soils and Foundations* 1: 137–147.

Karmokar, K.A., Takeichi, H., Yasuhara, K. & Kawai, H. 2005. "Ageing and durability of used tire rubber materials embedded in cement treated soil for their use in civil engineering applications." *Proc. 60th Annual Conf. JSCE,* 779–800 (in Japanese).

Karmokar, A.K., Takeichi, H., Kawaida, M., Kato, Y., Mogi, H. & Yasuhara, K. 2006a. Study on thermal insulation behavior of scrap tire materials for their use in cold region civil engineering applications. *Proc. 60th JSCE Annual Meeting, JSCE, Japan,* pp. 851–852.

Karmokar, A.K., H., Takeichi, H. & Yasuhara, K. 2006b. Drainage behavior of shredded used tire geomaterials. *Proc. 41st Geotech. Res. Conf., JGS Japan,* pp. 587–588.

Kikuchi Y., Nagatome, T., Mitarai, Y. & Otani J. 2006. Engineering property evaluation of cement treated soil with added tire chips using X-ray CT Scanner. *Proc. of conference 5th International Congress on Environmental Geotechnics.*

Mitarai Y., Kawai H., Yasuhara K., Kikuchi Y. & Karmokar A.K. 2004. "The mechanical properties of composite geo-materials mixing with scraped tire-chips and application to harbor construction works." *Proc. of 49th Symposium on Geotechnical Engineering,* 77–84 (in Japanese).

Mitarai, Y. et al. 2006. Application of the cement treated clay with tire-chips added to the sealing materials of coastal waste disposal site. *Proc. 6th International Congress on Environmental Geotechnology, Cardiff, UK,* Vol. 1, 2006.

Murakami, S. et al. 2005. An elasto-plastic constitutive model with sub-loading surface concept and damage theory for cement-treated clay with tire-chips. *Geosynthetics Journal, Japan Chapter of IGS* 20: 129–136 (in Japanese).

Murakami, S., Yasuhara, K., Komine, H. & Mitsuyama, S. 2006. An elasto-plastic model for unsaturated soil based on sub-loading surface and rotational hardening concepts. *Proc. of the 11th International Symposium Plasticity 05: Plasticity and its Current Applications* 1: 280–282.

PIANC: *Seismic Design Guidelines for Port Structures,* Balkema Publishers, Rotterdam, 2001.

Tatsuoka, F., Uchimura, T., Tateyama, M. and Kojima, K. 1998. Performance of Preloaded-Prestressed Geogrid-Reinforced Soil Railway Bridge Pier. (in Japanese) *Tsuchi-to-Kiso* 46(8): 13–15.

Towhata, I., Ghalandarzadeh, A, Sundarraj, K.P. & Vargas-Monge, W. 1996. Dynamic Failures of Subsoils Observed in Waterfront Area. *Special Issue of Soils and Foundations* 1: 149–160.

Yasuhara, K., Unno, T., Komine, H. & Murakami, S. 2004. Gravel Drain Mitigation of Earthquake-induced Lateral Flow of Sand. 13th *WCEE,* No. 146 (CD-ROM).

Yasuhara, K., Komine, H., Murakami, S., Taoka, K., Ohtsuka, Y. & Masuda, T. 2005. Tire chips drain for mitigation of liquefaction and liquefaction-induced deformation in sand. *Proc. of Symposium on Technology of Using Artificial Geomaterials, Fukuoka, Japan,* pp. 115–118.

Yasuhara, K., Karmokar, A.K., Kato, Y., Mogi, H. & Fukutake, K. 2006. "Effective utilization technique of used tires for foundations and earth structures." *Kisoko* 34(2): 58–63 (in Japanese).

Special invited lectures

Scrap Tire Derived Geomaterials – Opportunities and Challenges – Hazarika & Yasuhara (eds)
© 2008 Taylor & Francis Group, London, ISBN 978-0-415-46070-5

Reinforced lightweight tire chips-sand mixtures for bridge approach utilization

D.T. Bergado & T. Tanchaisawat
Geotechnical and Geoenvironmental Program, School of Engineering and Technology, AIT, Thailand

P. Voottipruex
King Mongkut's Institute of Technology, North Bangkok, Thailand

T. Kanjananak
Royal Thai Air Force, Bangkok, Thailand

ABSTRACT: Construction of bridge approach highway embankment using reinforced light weight geomaterials over soft ground will alleviated problems of instability and long-term settlements. Backfills of retaining structure can also be constructed using lightweight materials resulting in lower earth pressure and, consequently, improved economics. The aim of this study is to investigate the behavior of lightweight geomaterials consisting of tire chip-sand mixture reinforced with geogrids for use as bridge approach in highway application. The weight mixing ratios of 30:70, 40:60, and 50:50% were used for the tire chip-sand mixtures. Different experiments including index tests, compaction tests, pullout tests, and large scale direct shear test were performed. Geogrids consisting of Saint-Gobain (Geogrid A) and Polyfelt (Geogrid B) were selected as reinforcing materials. The experimental results indicated that the mixing ratio of 30:70% is the most suitable fill material. The pullout resistance and interaction coefficients of Geogrid A were slightly higher than those of Geogrid B. In contrast, the direct shear resistance, the direct shear interaction coefficients, and the efficiency values of Geogrid B were higher than those of Geogrid A. The ultimate tensile strength of Geogrid A was slightly lower than that of Geogrid B. Finally, it was concluded that Geogrid B and the tire chip-sand mixture at the mixing ratio of 30:70% by weight were recommended for full scale test embankment. The test embankment was constructed in the campus of Asian Institute of Technology (AIT). The geogrid reinforced embankment system is extensively instrumented in the subsoil and within the embankment itself in order to monitor its behavior during construction and in the post-construction phases, and thereby evaluate its performance. Settlements of embankment were observed by surface and subsurface settlement plates. Excess pore water pressures were observed by open stand pipe piezometer. Lateral wall movement was monitored by digitilt inclinometer. The movements of the geogrid reinforcements were monitored by high strength extensometer wires and strain gages. Finally, the behavior of rubber tire-sand mixture and geogrid reinforced embankment were analyzed to obtain the performance for lightweight geomaterials for bridge approach utilizations.

1 INTRODUCTION

Generally, to improve the stability and performance of infrastructures on soft foundations two alternative methods are available, namely: 1) improve the strength and deformation characteristics of the foundation and 2) reduce the weight of the structure on the foundation. The latter method was first used in Oslo, Norway, where expanded polystyrene (EPS) was used in road embankments on soft ground (Freudelund and Aaboe, 1993) called "The dawn of the lightweight geomaterial"

45

(Yasuhara, 2002). Several materials and methods have been proposed to produce lightweight geo-materials and are classified into three categories as follows: i) use of lightweight material, ii) mix lightweight material with natural soil and cementing agent and iii) add air foam agent to reduce weight. The advantages of using a lightweight geomaterials are not only the reduction of vertical pressures on foundations but also the decrease of lateral earth pressure, and a decrease of traffic induced vibration.

Construction of highway embankments on soft ground faces problems of high settlements and stability. Lightweight materials can be used as backfills in retaining structures and in the construction of embankments resulting in lower earth pressure and greater stability on soft ground. In recent years, however, there has been a growing emphasis on using industrial by-products and waste materials in construction. Used rubber tire is one of waste material that can be used as lightweight backfills of wall embankments.

Because of the low specific gravity of scrap tire relative to that of the soil solids, tire chips alone or in mixtures with soil offer an excellent lightweight and strong fill material. The application of lightweight geomaterials on soft ground foundation has been summarized by Miki (1996) as follows:

- Reducing residual settlement of low embankment road built on soft ground
- Minimizing differential settlement between approach embankment and structure, to prevent lateral movement of piled structures
- Minimizing deformation when constructing near adjacent structure
- Minimizing residual settlement for high standard dikes and artificial islands
- Reducing the construction period substantially
- Achieving nearly maintenance-free infrastructure

Due to the advantage of lightweight geomaterials for geotechnical application on soft ground, the performance of full scale embankment test made of rubber tire-sand mixture reinforced with geogrid was constructed to study its behavior. The settlement of embankment was observed and analyzed with existing data. Excess pore water pressure during and after construction were also monitored to evaluate consolidation settlement. Lateral wall movement and geogrid movement were measured with the help of digitilt inclinometer and high strength extensometer wire respectively. Finally, the performance of lightweight embankment is evaluated for possible geotechnical application on soft ground area.

2 LABORATORY INVESTIGATION

The experimental work of this study consists of four phases. The first phase involves the determination of the physical properties (grain size distribution and specific gravity) for both of Ayutthaya sand and tire chips. In the second phase, the compaction tests on tire chip-sand mixture were performed to determine maximum dry unit weight and optimum moisture content of the different tire chip-sand mixtures (30:70, 40:60, and 50:50% by weight). The third phase concerns with the mechanical properties of the utilized geogrid (in-air tensile strength). In the last phase, large-scale direct shear and pullout tests were done to investigate the interaction between reinforcing and fill materials. In the following sections, the experimental program will be presented and the test results will be discussed.

This study was directed to investigate the interaction between tire chip-sand mixture and two different types of extensible grid reinforcement, namely: Saint-Gobain geogrids (DJG 120X120-1) denoted as Geogrid A, and Geogrid B referred to Polyfelt geogrids (GX 100/30). Geogrid A is made of the high tenacity polyester yarns knits into mesh coated with modified polymer mixture, high tensile strength, high modulus, and good creep resistance. It is generally used for soil reinforcement and stabilization such as steep slopes, retaining walls, bridge abutments etc. Geogrid B consists of high molecular, high strength polyester yarns that are knitted to a stable network and equipped with a polymeric coating protection. This product is suitable for both short-term and long-term soil reinforcement application. Geogrid B are designed for technical applications such as reinforcement in reinforced earth structures.

2.1 Testing program

The fill materials in this study consists of mixtures of tire chips and Ayutthaya sand at three different mixing ratios of 30:70, 40:60, and 50:50% by weight. The specific gravity tests of sand and tire chips were conducted according to ASTM D854-97 and ASTM C127-01, respectively. The grain size distributions of sand and tire chips were conducted according to ASTM D422-63. ASTM D689-91 test procedures were also adopted to obtain the optimum moisture content and maximum dry unit weight of the mixture fill materials. Two types of geogrid reinforcements and tire chip-sand mixtures were tested in both pullout and large-scale direct shear tests. For convenience, the mixing ratios of the tire chip-sand mixtures were based on the dry weight of each material in sample preparation. Each group of fill materials needs to be cured to its respective optimum moisture condition based on the results of standard Proctor compaction test with the modified mold.

The pullout tests program was mainly used for investigating the interaction between tire chip-sand mixture and geogrid reinforcements, and the relationship between pullout force and pullout displacement. In the entire tests, there were four normal stresses of 30, 60, 90, and 120 kPa applied on the fill materials. The purpose of applying these different values was to cover the range of possible reinforcement failures (i.e. slippage and breakage). The pullout machine used for evaluating the interaction between tire chip-sand mixture and geogrid reinforcements is shown schematically in Fig. 1. The pullout forces were usually generated by a 225 kN capacity electro-hydraulic controlled jack through the steel reaction frame. The normal pressures were applied by the inflated air bag installed between the flexible steel plate and the top cover of the pullout box. The load cell used in the pullout resistance measurement was connected to the 21X data logger to automatically record the resistances. The pullout displacements of a geogrids sample were monitored by using a Linear Variable Differential Transducer (LVDT). To determine the displacements along the longitudinal direction of geogrids sample during the pullout tests, four inextensible wires were mounted on the geogrids sample at predetermined positions. The pullout rate of 1 mm/min was adopted throughout the tests. The pullout forces and pullout displacements were measured and recorded by the data logger.

The large-scale direct shear apparatus was adapted from the pullout machine. Likewise, the measurement apparatus was set up similar to the in-soil pullout tests. The instrumented geogrids sample with the sizes of 500 mm × 700 mm was laid on the shear plane. The upper shear box was pulled at a constant rate of 1 mm/min throughout the test. The same test procedure was followed to determine the shear strength parameters of each fill material group except in the cases of the tests without any geogrid reinforcements placed on the shear plane.

2.2 Laboratory test results

2.2.1 Index properties of tire chip-sand backfills
The specific gravity of Ayutthaya sand is 2.65, while that of tire chips is 1.12. For Ayutthaya Sand, there was 1.64% passing through No. 200 sieve. The effective diameter (D_{10}) is 0.22 mm, D_{30} is 0.38 mm, D_{60} is 0.62 mm, the uniformity coefficient (C_u) is 2.82, and the gradation coefficient (C_c) is 1.06. According to the Unified Soil Classification System (USCS), the sand can be classified as poorly graded (SP). For tire chips, most of the particle size range between 12 and 50 mm with irregular shape due to the random cutting process. Compaction test results of the tire chip-sand mixtures are shown in Fig. 2. The maximum dry unit weight and the optimum moisture content of the tire chip-sand mixtures vary from 9.5 to 13.6 kN/m³ and from 5.7 to 8.8 kN/m³, respectively.

2.2.2 In-soil pullout test results
The in-soil pullout test results revealed that the pullout resistance normally increased while the displacement at the maximum pullout force tended to decrease as the normal stress increased.

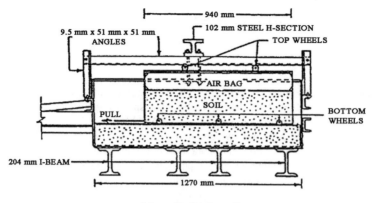

a) Longitudinal section

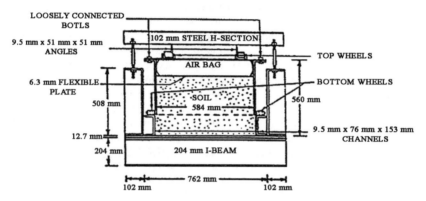

b) Cross section

Figure 1.   Schematic pullout test apparatus.

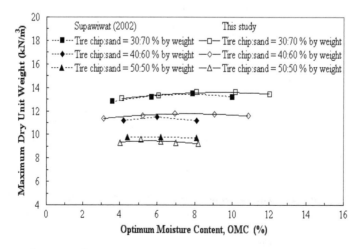

Figure 2.   Compaction test results.

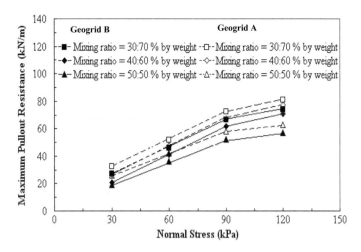

Figure 3.   Maximum pullout resistance vs normal stress.

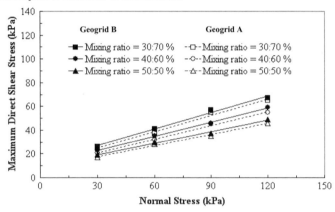

Figure 4.   Maximum direct shear stress versus normal stress.

Moreover, the pullout resistance increased with the increasing sand content in the mixture. The mixing ratio of 30:70% by weight yielded the highest pullout resistance for both geogrids as shown in Fig. 3. The frictional resistance affects the pullout resistance rather than bearing resistance. Hence, the sand content in the tire chip-sand mixtures directly affects the pullout resistance because the frictional angle of sand is higher than that of tire chips. Thus, the frictional resistance obtained from sand governs the pullout resistance rather than that obtained from tire chips. The comparison between the pullout resistance of two types of geogrids at the same mixing ratio and the same normal stress is illustrated in Fig. 4. The displacements at the maximum pullout force were measured along the length of geogrid reinforcements during the in-soil pullout tests. The results of both geogrid reinforcements indicated that the largest displacement occurred at the pullout face, which was connected to the in-soil pullout clamp. The displacement at the maximum pullout force along the entire geogrid reinforcements decreased with the increasing distance from the pullout face.

2.2.3   *Large-scale direct shear test results*

Under the same normal stresses and mixing ratios, the direct shear stresses of the tire chip-sand backfills were higher than those of both geogrid reinforcements because there were no any reinforcements blocking the contact area of the backfills at the shear plane. Therefore, the direct shear stresses were able to be mobilized fully at the shear plane. As shown in Fig. 4, at the same normal stresses and mixing ratios, the direct shear stresses of Geogrid B were higher than those of Geogrid

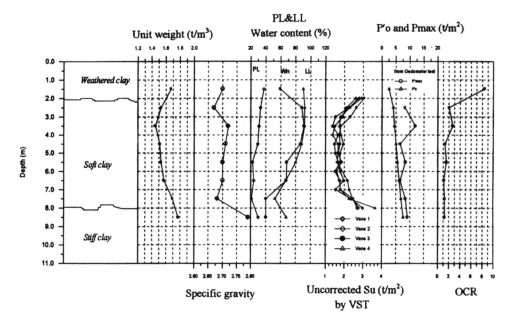

Figure 5.   Subsoil profile and relevant parameters.

A because the aperture sizes of Geogrid B were bigger than those of Geogrid A.. At the same normal stresses and mixing ratios, the adhesion and skin friction angles of Geogrid A and Geogrid B were found to be lower than those of the backfills. In comparison between geogrid reinforcements, the adhesion and skin friction angles of Geogrid B were found to be slightly higher than those of Geogrid A.

3   FULL SCALE EMBANKMENT TEST

The test embankment was constructed in the campus of Asian Institute of Technology (AIT). The general soil profile consists of weathered crust layer of heavily overconsolidated reddish brown clay over the top 2 m. This layer is underlain by soft grayish clay down to about 8.0 in depth. The medium stiff clay with silt seams and fine sand lenses was found at the depth of 8.0 to 10.5 m depth. Below this layer is the stiff clay layer. Figure 5 summarizes the subsoil profile and relevant parameters.

3.1   Subsoil investigation

Soil samples were obtained from the borehole at the construction site down to a depth of about 8 m to the bottom of the soft clay layer. Index tests, consolidation tests and unconfined compression tests were performed on the subsoil samples. The in-situ strength of the subsoil was measured by field vane shear test. Laboratory consolidation tests were performed on subsoil samples from to different depths to determine the coefficient of consolidation and compression index.

3.2   Instrumentation program

The geogrid reinforcement embankment/wall system was extensively instrumented both in the subsoil and within the embankment itself. Since the embankment was founded on a highly compressible and thick layer of soft clay which dictates the behavior of the embankment to a great extent, several field instruments were installed in the subsurface soils. The 3D illustration of full-scale field test embankment is shown in Fig. 6a. The instrumentation in the subsoil were installed prior to the

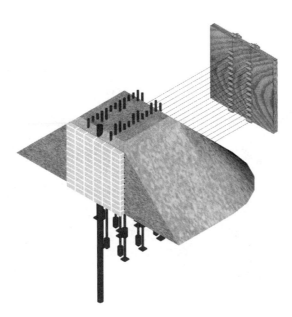

Figure 6a. Embankment drawing in 3D.

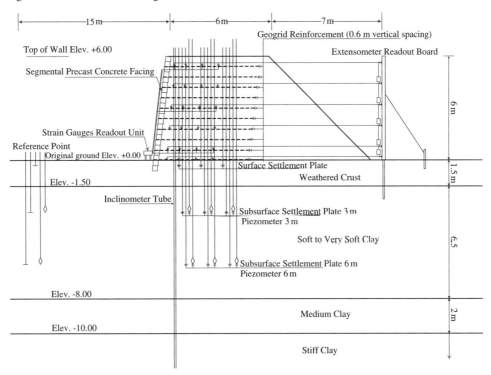

Figure 6b. Section view of embankment with instrumentation.

construction of the geogrid reinforcement wall and consisted of the surface settlement plates, subsurface settlement gauges, temporary bench marks, open standpipe, groundwater table observation wells, inclinometers, dummy open standpipe, dummy surface settlement plates and dummy subsurface settlement gauges (Fig. 6b). Six surface settlement plates were placed beneath the embankment

Figure 7. Completed full scale test embankment construction.

at 0.45 m depth below the general ground surface. Settlements were measured by precise leveling with reference to a benchmark.

The measurement of the subsurface settlements was similar to that of the surface settlements. Twelve subsurface gages, six of which were installed at 6 m depth, the rest at 3 m depth below the general ground surface at different locations.Two dummy gages were also installed at depths of 3 m and 6 m. The pore water pressure was monitored by the conventional open stand pipe piezometers. Six piezometer were installed in the soft clay subsoil at 3 m and 6 m depth from general level. Two of dummy open standpipes were installed at the area nearby temporary benchmarks.

The lateral movements of the subsoil and the embankment were measured by using a digitilt inclinometer. The inclinometer was installed vertically near the face to measure the lateral movements of the vertical face of the wall and the subsoil. The depths to which the inclinometers were installed were 12 m below original ground level.

Wire extensometers were used to measure the displacements of the geogrid reinforcement and the surrounding soil as well. The extensometer consisted of a 2 mm diameter high-strength stainless cable inside a flexible PVC tube. One end of the wire was fixed at the measured point and the other end was connected to a counterweight of about 0.8 kg through a pulley system at the readout board.

3.3 Construction of reinforced embankment/wall

The construction of the reinforced embankment/wall involved the precast concrete block facing unit with geogrid reinforcement. The rubber tire chips were mixed with sand in the ratio of 30:70 by weight. The vertical spacing of the geogrid reinforcement was 0.60 m. The backfill was compacted in layers of 0.15 m thickness to density of about 95% of standard proctor. The compactions were carried out with a roller compactor and with a hand compactor near instrumentation such as settlement plate, stand pipe and inclinometer. The degree of compaction and the moisture content were checked regularly at several points with a nuclear density gage. Wherever the degree of compaction was found to be inadequate, addition compaction was done until the desired standards were met. The sand backfill was used as the surface cover for the rubber tire chips-sand for reducing a self-heating reaction. The thickness of the cover was 0.6 m and a non-woven geotextile was used as the erosion protection on side slope. Hexagonal wire gabions were used on both side of the concrete facing at the fornt side slopes. Figure 7 illustrates the completed embankment construction (Kanjananak, 2006).

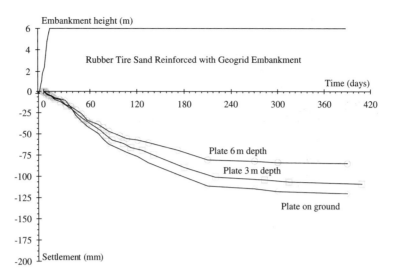

Figure 8.   Observed average settlements at different depths.

3.4   *Observed behavior of test embankment/wall on soft ground*

3.4.1   *Observed and predicted surface and subsurface settlements*

The observed surface and subsurface settlements of the test embankment are illustrated in Fig. 8. During the construction period, immediate elastic settlements were observed. The rate of settlement was low in all the surface and subsurface settlement plates during the construction period. After the construction, the rate of settlement increased. After 210 days from the end of construction, the maximum settlement was 122 mm as recorded in surface settlement plates near the facing. This is because the weight of the concrete facing is more than the lightweight embankment and the forward tilting of the embankment. Along the cross-section of the embankment, settlement decreased from front (122 mm) middle (112 mm) and back (104 mm). However, the settlements of the two sections of the embankment are almost the same. The average surface settlement on the ground after 210 days from the end of construction is about 111 mm. The settlements at 3 m and 6 m depths were lower than at ground surface, as expected.

The observed and predicted surface settlements of the test embankment are plotted together in Fig. 9. As expected the predictions from Asaoka (1978) closely follow the observed data while the predictions from one-dimensional method overpredicted.

Figure 10 demonstrates the comparison of the maximum settlements between conventional sand backfill reinforced embankment (Voottipruex, 2000) and lightweight backfill embankment in this study. The maximum settlement of lightweight embankment with unit weight of $13.6 \, kN/m^3$ is 130 mm compared to 400 mm for conventional embankment with unit weight of $18 \, kN/m^3$. The reduction of settlement amounted to 67.5%. The use of rubber tire chip-sand mixture as a lightweight geomaterials can alleviate the problem of settlement in soft ground area.

3.4.2   *Observed and predicted excess pore water pressure at 3 m depth*

The excess pore water pressure below the lightweight embankment was obtained from open stand pipe piezometer. Figure 11 shows the excess pore water pressures during and after construction at 3 m depth. The maximum pore water pressure of $57 \, kN/m^3$ occurred at 15 days after full height of embankment. The trend of excess pore water pressure dissipation is an indication of consolidation of soft foundation subsoil in the overconsolidation range when the load is below the maximum past pressure. After 50 days, the excess pore water pressure tends to dissipate very fast with time. The excess pore water pressure decreased to $18 \, kN/m^2$ and $25 \, kN/m^2$ at 3 m and 6 m depths, respectively. The excess pore water pressure become constant with time after 120 days from the end of construction.

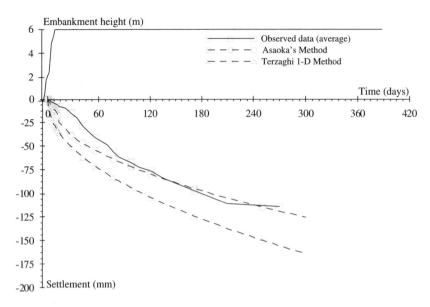

Figure 9.   Observed and predicted surface settlement at original ground level.

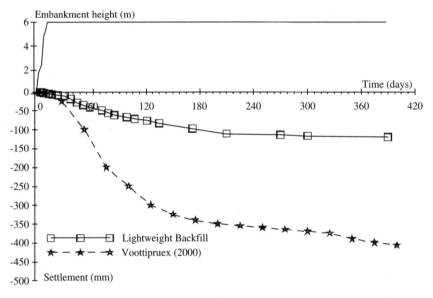

Figure 10.   Comparison of settlement between conventional and lightweight backfill.

The observed and predicted excess pore water pressures below the embankment are also plotted in Fig. 11. The 1-D method overpredicted excess pore water pressure while the predictions from the Skempton and Bjerrum (1957) method agree well with the observed data.

3.4.3   Observed lateral wall movement

The lateral wall movement was observed by digitilt inclinometer which was located near the embankment facing. The plots of lateral wall movement with depth from top of embankment to 12 m depth below original ground are shown in Fig. 12 The lateral wall movement was monitored once a week since end of construction and every month afterwards until 13 months. The lateral movement increased significantly until 4 months after construction and decreased to negligible

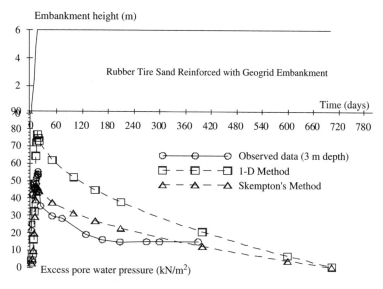

Figure 11.   Observed and predicted excess pore water pressure at 3 m depth.

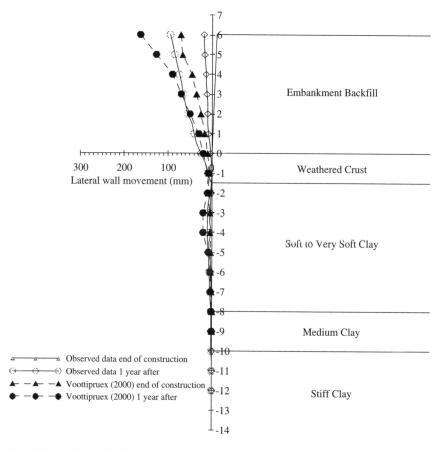

Figure 12.   Observed lateral wall movement.

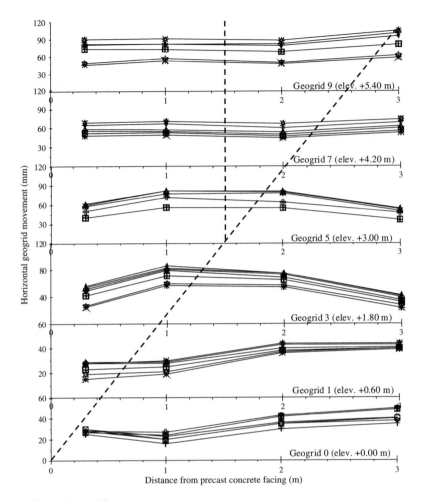

Figure 13. Observed geogrid movement.

amounts 13 months after embankment construction. The lateral movement occured in the short term after construction. The total wall movement is quite small from 95 mm to 100 mm (at top of embankment).

The maximum lateral movement in the soft clay subsoil occurred at about 3.00 to 5.00 m depth below the ground surface, corresponding to the weakest zone of the subsoil. The maximum lateral wall movement at the top was 100 mm which is less than the corresponding value of 350 mm as reported by Voottipruex (2000). Vootipruex (2000) used conventional backfill of silty sand reinforced with hexagonal wire. The lateral wall movement of this study is about 70% less than that of the conventional backfill material. This indicates that the use of lightweight backfill significantly reduces the lateral movement of the embankment.

3.4.4 Observed geogrid movement
High strength wire extensometers were used to measure the displacements of the geogrid reinforcement related to the surrounding soil. The measurement points were located at 0.3 m, 1.0 m, 2.0 m and 3.0 m from concrete facing to observe the geogrid movement in the rubber tire chips-sand backfill zone. The extensometers were installed at 10 layers of geogrid reinforcements with vertical spacing of 0.6 m. Geogrid movement was observed from the end of construction in the same period of time as the observations of the lateral wall movement. Figure 13 illustrates the geogrid

movement observed from the end of construction until 210 days after embankment construction. At points near concrete facing geogrid movement is less than at other points except near the base.

At the base of the embankment, the movements were higher near the facing than at other points. For the top geogrid, the maximum movement of 100 mm was measured. This value agreed well with the lateral wall movement observed from the inclinometer. Overall, the movement or deformation of the geogrid reinforcements corresponds to the bilinear type of maximum tension line.

4 CONCLUSIONS

4.1 Laboratory investigation

The percentage of sand mixed in tire chip-sand mixtures was the most significant factor controlling the unit weight of the mixtures. The pullout resistance of geogrid reinforcements depended on the sand content in the tire chip-sand mixtures, not the tire chip content. The pullout resistance increased with the increasing sand content in the mixture. The applied normal stresses were significant factors for pullout resistance which increased with the increasing normal stresses. The higher tensile strength of geogrids in longitudinal direction and the higher strength of the junctions could contribute to the pullout resistance of geogrids. The failure modes of geogrid reinforcements were confirmed to be slippage failure at the normal stresses of 30 and 60 kPa, and breakage failure at the high normal stresses of 90 and 120 kPa. The direct shear resistance of tire chip-sand mixtures and geogrid reinforcements depended on the sand content in the tire chip-sand mixtures which increased with the increasing sand content. It was confirmed that the aperture sizes of geogrids significantly affected the direct shear resistance of geogrids. The bigger the aperture size, the higher the direct shear resistance. The tire chip-sand mixture with the mixing ratio of 30:70% by weight yields the higher results in the pullout and direct shear resistance rather than the other mixtures. Therefore, the mixture with the mixing ratio of 30:70% can be recommended as lightweight tire chip-sand backfill material. Even though the tensile strength of Geogrid A was much higher than that of Geogrid B, the pullout resistance of Geogrid B in tire chip-sand backfills was only slightly lower than that of Geogrid A. Hence, Geogrid B was recommended as reinforcing material.

4.2 Full scale embankment test

The unit weight of rubber tire sand mixture 30:70% by weight is 13.6 kN/m^3 compare to conventional backfill sand of 18.0 kN/m^3, it is lighter by about 75%. This lightweight geomaterials can be used for embankment construction on soft ground area to reduce total settlement of structure. The total settlement magnitude of 122 mm at ground surface is 67.5% less when compared to the corresponding value for conventional backfill. This type of lightweight material is appropriate for highway bridge approach utilization. The excess pore water starts to build up after 15 days since the end of construction and start to dissipate after 50 days. The excess pore water pressure becomes constant after 120 days since the end of construction. The maximum lateral wall movement observed at 13 months after construction at top of wall is 100 mm which is 70% less compared to 350 mm of hexagonal gabion with sand backfill. Geogrid movements correspond well with the bilinear type of maximum tension line. The observed geogrid movement agrees well with lateral wall movement observed from inclinometer reading.

REFERENCES

Bergado, D. T. & Chai, J.C. (1994). Pullout Force/Displacement Relationship of Extensible Grid Reinforcements, *Geotextiles and Geomembranes*, Vol. 13, pp. 295–316.
Bergado, D. T. & Voottipruex, P. (2000). Interaction Coefficient between Silty Sand Backfill and Various Types of Reinforcements, *Proceedings of the 2nd Asian Geosynthetics Conference*, Vol. 1, pp. 119–151.
Chai, J. C. & Bergado, D. T. 1993. Performance of reinforced embankments on soft ground. *Soils and Foundation,* Vol. 33, No. 4, pp. 1–17.

Freudelund, T. E. & Aaboe, R. 1993. Expanpolystylene-A light way across soft ground. *Proceedings 14th International Conference of Soil Mechanics and Geotechnical Engineering,* New Delhi, India, pp. 1256–1261.

Hsieh, C. W., Huang, A. B., Chang, Y. C. & Chou, N. N. S. 1998. The construction of a test geogrid reinforced embankment. *Proceedings 13th Southeast Asian Geotechnical Conference,* Taipei, Taiwan, pp. 315–320.

Humphrey, D. N., Whetten, N., Weaver, J. & Recker, K. 2000. Tire shreds as lightweight fill for construction on weak marine clay. *Coastal Geotechnical Engineering in Practice,* Rotterdam, A. A. Balkema Publishers, pp. 611–616.

Kanjananak, T. 2006. Prediction and performance of reinforced rubber tire-sand test embankment, *M. Eng. Thesis,* Asian Institute of Technology, Bangkok, Thailand.

Kerr, J. R. Bennette, B., Perzia, O. N. & Perzia, P. A. 2001. Geosynthetic reinforced highway embankment over peat. *Proceedings Geosynthetics'2001,* Portland, Oregon, U.S.A., pp. 539–552.

Lockett, L. & Mattox, R. M. 1987. Difficult soil problems on Cochrane Bridge finessed with geosynthetics. *Proceedings Geosynthetic'87 Conferrence,* New Orleans, U.S.A., Vol. 1, pp. 309–319.

Mattox, R. M. & Fugua, D. A. 1995. Use of geosynthetics to construct the Vera Cruz access ramp of the bridge of the Americas, Panama. *Proceedings Geosynthetics'95,* Nashville, Tennessee, U.S.A., Vol. 1, pp. 67–77.

Miki, H. 1996. An overview of lightweight banking technology in Japan, *Proceedings of International Symposium on EPS Construction Method,* Tokyo, Japan, pp. 10–30.

Oikawa, H., Sasaki, S. & Fujii, N. 1996. A case history of the construction of a reinforced high embankment, *Proceedings of the International Symposium on Earth Reinforcement,* Fukuoka, Japan, pp. 261–266.

Prempramote, S. 2005. Interaction between geogrid reinforcement and tire chip-sand mixture, *M.Eng. Thesis No. GE-04-12,* Asian Institute of Technology, Thailand.

Rowe, R. K. & Li, A. L. 2005. Geosynthetic-reinforced embankments over soft foundations, *Geosynthetics International,* Special Issue on the Giroud Lectures, Vol. 12, No. 1, pp. 50–85.

Skempton, A. W. & Bjerrum, L. 1957. A contribution to the settlement analysis of foundations on clays, *Geotechnique,* Vol. 7, No. 4, pp. 168–178.

Supawiwat, N. 2002. Behavior of Shredded Rubber Tires with and without Sand, its Interaction with Hexagonal Wire Reinforcement and their Numerical Simulation, *M.Eng. Thesis No. GE-01-14,* Asian Institute of Technology, Thailand.

Tanchaisawat, T., Bergado, D. T., Prempramote, S., Ramana, G. V., Piyaboon, S. & Anujorn, P. 2006. Interaction between geogrid reinforcement and tire chips-sands mixture, *Proceedings 8th International Conference on Geosynthetics,* Yokohama, Japan, pp. 1463-1466.

Tweedie, J. J., Humphrey, D. N. and Sanford T. C. 1998. Tire shreds as light-weight retaining wall backfill: active conditions, *ASCE Journal of Geotechnical and Geoenvironmental Engineering,* Vol. 124, No. 11, pp. 1061–1070.

Voottipruex, P. 2000. Interaction of hexagonal wire reinforcement with silty sand backfill soil and behavior of full scale embankment reinforced with hexagonal wire, *D. Eng. Dissertation No. GE-99-1,* Asian Institute of Technology, Bangkok, Thailand.

Yasuhara, K. 2002. Recent Japanese experiences with lightweight geomaterials, *Proceedings International Workshop on Lightweight Geomaterial,* pp. 32–59.

Youwai, S. & Bergado, D. T. 2003. Strength and Deformation Characteristics of Shredded Rubber Tire-Sand Mixtures, *Canadian Geotechnical Journal,* Vol. 40, pp. 254–264.

Tire derived aggregate as lightweight fill for embankments and retaining walls

D.N. Humphrey

Department of Civil and Environmental Engineering, University of Maine, Orono, Maine, USA

ABSTRACT: Tire derived aggregate (TDA) is scrap tires that have been cut into 50 to 300-mm pieces. They have an in-place unit weight between 0.72 to 0.93 Mg/m^3. Due to their low unit weight, widespread availability, and relatively low cost, they are finding increasing use as lightweight fill for embankment construction on weak, compressible foundation soils. Moreover, they produce low earth pressure when used as backfill behind retaining walls. TDA has a shear strength that is comparable to or greater than the shear strength of typical embankment soils. Procedures to calculate the in-place unit weight and immediate compression of the TDA layer due to the weight of overlying soil are described. Two case histories are given. TDA was used to improve the stability of a 9.8-m high embankment constructed on weak marine clay. In addition, TDA was used as backfill to reduce lateral pressure on a cantilever retaining wall.

1 INTRODUCTION

Tire derived aggregate (TDA) is made from scrap tires that have been cut into pieces with a maximum size ranging from 50 to 300 mm. This material is lightweight, free draining, and compressible. Moreover they have a thermal resistivity that is about seven times higher than soil, they produce low earth pressure, and absorb vibrations. Because of their special properties TDA is increasingly being used as lightweight fill for embankments constructed on weak foundation soils (Edil and Bossher 1992, Bossher et al. 1993, 1997, Humphrey et al. 1998, Whetten et al. 1997, Humphrey et al. 2000, Dickson et al. 2001) and pile supported embankments (Tweedie et al. 2004), lightweight backfill for retaining walls and bridge abutments (Tweedie et al. 1997, 1998a, b, Humphrey et al. 1998), compressible inclusions behind integral abutment and rigid frame bridges (Humphrey et al. 1998; Reid & Soupier 1998), thermal insulation to limit frost penetration beneath roads (Humphrey & Eaton 1995, Lawrence et al. 1998), drainage layers for road (Lawrence et al. 1998) and landfill applications (Jesionek et al. 1998), and vibration damping layers beneath rail lines (Wolfe et al. 2004). TDA has also been mixed with soil to result in a high strength composite with applications in embankments and wall backfill (Tatlisoz et al. 1997, Abichou et al. 2001). In 1996, an estimated 10 million scrap tires were used for civil engineering applications (STMC 1997). By 2004, this figure had grown to 60 million tires, or 20% of the scrap tires that were generated in the United States in that year (Recycling Research Institute 2005). The reasons for this growth are simple – TDA has properties that civil engineers need and using TDA can significantly reduce construction costs.

The cost of TDA in the U.S. varies from about $5/tonne (approx. $4.50/$m^3$) to $44/tonne (approx. $39/$m^3$), with several projects priced at about $25/tonne ($22/m^3). This is a very economical price for lightweight fill and for other applications that can make use of the special properties of TDA. The reason for this low cost is that scrap tire processors are paid to collect scrap tires – thus, they are paid to take in the raw material, a unique situation in the construction industry. In some states the scrap tire processors are paid by tire retail outlets, while in other states they are paid from a state-run tire fund, but the effect is the same – scrap tire processors only have to make a small portion of their income by selling their product.

Civil engineering applications of TDA did experience a significant setback in 1995 and early 1996 when three thick TDA fills (each greater than 7.9 m thick) experienced a serious self-heating reaction, however, guidelines to limit self-heating are now available (Ad Hoc Civil Engineering Committee 1997, ASTM 1998). The effectiveness of these guidelines has been demonstrated by internal temperature measurements in several projects (Humphrey 2004).

This paper presents the properties of TDA that are needed to design lightweight fill applications. This includes gradation, unit weight, compressibility, time-dependent settlement, and shear strength. The guidelines to limit self-heating and water quality effects of TDA will be discussed. Case histories using TDA as lightweight fill for highway embankments and retaining wall backfill are presented. From the information presented below, it will be clear that TDA is a viable and economical material for use as lightweight fill for highway construction and other applications.

2 ENGINEERING PROPERTIES OF TDA

2.1 *Gradation*

Large pieces are desirable when the TDA zone is greater than 1 m thick, because they are less susceptible to self-heating (as will be discussed in more detail below). However, when the TDA contain a significant number of pieces larger than 300 mm, they tend to be difficult to spread in a uniform lift thickness. Thus, a typical specification requires that a minimum of 90% (by weight) of the TDA have a maximum dimension, measured in any direction, of less than 300 mm and 100% of the TDA have a maximum dimension less than 450 mm. Moreover, at least 75% (by weight) must pass a 200-mm sieve and at least one sidewall must be severed from the tread of each tire. To minimize the quantity of small pieces, which can contribute to self-heating, the specifications require that no more than 50% (by weight) pass the 75-mm sieve, 25% (by weight) pass the 38-mm sieve, and no more than 1% (by weight) pass the No. 4 (4.75-mm) sieve. In the U.S., pieces of this size are commonly referred to as Type B TDA. Samples for gradation analysis should be taken directly from the discharge conveyor of the processing machine. This ensures that the minus No. 4 fraction will be representative, which is not the case when samples are taken by shoveling pieces from a stockpile.

TDA meeting the size requirements given above generally have a uniform gradation. Typical results are shown in Figure 1. The combination of large size and uniform gradation generally dictates that a geotextile be used as a separator between the TDA and adjacent soil to prevent fines from migrating into the TDA over time.

2.2 *Compacted unit weight*

The compacted unit weight of TDA has been investigated in the laboratory for pieces up to 75-mm. These tests were generally done with 254-mm or 305-mm inside diameter compaction molds and impact compaction. Compacted dry unit weights ranged from 0.61 to 0.69 Mg/m^3 (ASTM 1998).

The effect of compaction energy and compaction water content was investigated. Increasing the compaction energy from 60% of standard Proctor to 100% of modified Proctor increased the compacted unit weight by only 0.02 Mg/m^3 showing that compaction energy has only a small effect on the resulting unit weight. Unit weights were about the same for air dried samples and samples at saturated surface dry conditions (about 4% water content) indicating that water content has a negligible effect on unit weight (Manion & Humphrey 1992, Humphrey & Manion 1992). A practical significance of this finding is that there is no need to control the moisture content of TDA during field placement.

Measuring the compacted unit weight of TDA with a 300-mm maximum size is impractical in the laboratory. However, results from a highway embankment constructed with large TDA suggest that the unit weight of 300-mm maximum size TDA is less than for 75-mm maximum size TDA. This will be discussed further in a later section.

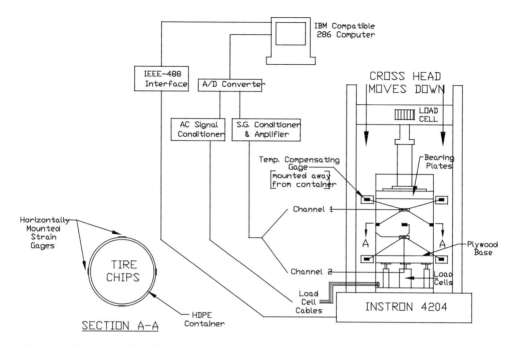

Figure 1.   Gradation of Type B TDA used as lightweight fill for Portland Jetport Interchange.

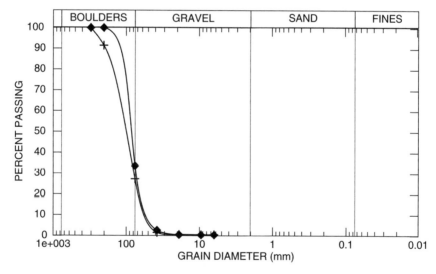

Figure 2.   Compressibility apparatus used by Nickels (1995).

2.3   *Compressibility*

The compressibility of TDA with a 75-mm maximum size has been measured in the laboratory. An apparatus used by Nickels (1995) had a 356-mm inside diameter and could accommodate a sample up to 356-mm thick. One challenge to measuring the compressibility of TDA is friction between the TDA and the inside wall of the test container. The apparatus used by Nickels (1995) had load cells to measure the load carried by the specimen both at the top and bottom of the sample as shown in Figure 2. Even though Nickels (1995) used grease to lubricate the inside of the container, up

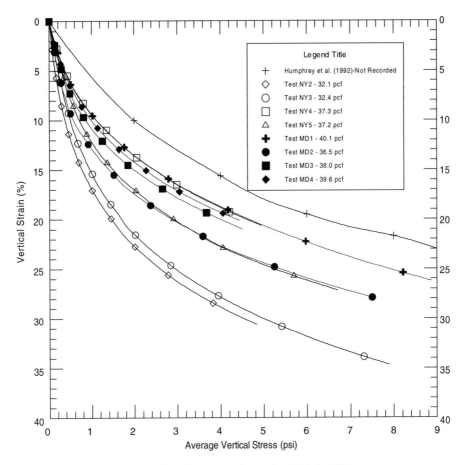

Figure 3.   Compressibility of TDA with a (75-mm maximum size (Nickels 1995).

to 20% of the load applied to the top of the sample was transferred to walls of the container by friction.

Compressibility of TDA with a 75-mm maximum size is shown in Figure 3. The initial unit weights ranged from 0.51 to 0.64 Mg/m³. There is a general trend of decreasing compressibility with increasing initial unit weight. The results for tests MD1 and MD4 are most applicable to using TDA as lightweight fill since they have an initial unit weight that is typical of field conditions. Compressibility for stresses up to 480 kPa are given by Manion & Humphrey (1992) and Humphrey & Manion (1992). Laboratory data on the compressibility of TDA with a maximum size greater than 75 mm are not available, however, field measurements indicates that TDA with a 300-mm maximum size are less compressible than smaller TDA. This will be discussed further in later sections.

2.4   Time dependant settlement

TDA exhibit a small amount of time dependant settlement. Time dependent settlement of thick TDA fills was measured by Tweedie et al. (1997). Three types of TDA were tested with maximum sizes ranging from 38 to 75 mm. The fill was 4.3 m thick and was surcharged with 36 kPa, which is equivalent to about 1.8 m of soil. Vertical strain versus elapsed time is shown in Figure 4. It is seen that time dependent settlement occurred for about 2 months after the surcharge was placed. During the first two months about 2% vertical strain occurred which is equivalent to more than

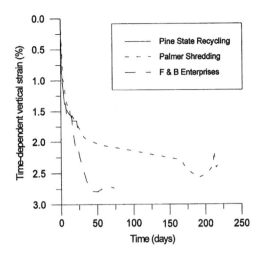

Figure 4. Time dependent settlement of TDA subjected to a surcharge of 36 kPa (Tweedie et al. 1997).

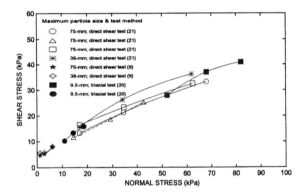

Figure 5. Failure envelopes for TDA with maximum sizes ranging from 9.5 to 75 mm.

75 mm of settlement for this 4.3-m thick fill. The measurements are in general agreement with time dependent laboratory compressibility tests conducted by Humphrey et al. (1992). When TDA is used as backfill behind a pile supported bridge abutment or other structure that will experience little settlement, it is important to allow sufficient time for most of the time dependent settlement of the TDA to occur prior to final grading and paving. Time dependant settlement is of less concern when the ends of the TDA fill can be tapered from the full thickness to zero over a reasonable distance.

2.5 Shear strength

The shear strength of TDA has been determined using direct shear and triaxial shear apparatus. The large TDA typically used for civil engineering applications requires that specimen sizes be several times greater than used for common soils. Because of the limited availability of large triaxial shear apparatus this method has generally been used for TDA 25-mm in size and smaller. Moreover, the triaxial shear apparatus is generally not suitable for TDA that have steel belts protruding from the cut edges of the pieces since the wires puncture the membrane used to surround the specimen.

The shear strength of TDA has been measured using triaxial shear by Bressette (1984), Ahmed (1993), and Benda (1995); and using direct shear by Humphrey et al. (1992, 1993), Humphrey & Sandford (1993), Cosgrove (1995), and Gebhardt et al. (1998). Failure envelopes determined from direct shear and triaxial tests for TDA with a maximum size ranging from 9.5 to 75 mm are shown in Figure 5. Data from Gebhard et al. (1998) on larger size (900 mm maximum size) TDA falls in

Table 1. Strength parameters for TDA.

| Supplier | Maximum shred size | | Test method | Applicable range of normal stress | | ϕ deg. | Cohesion intercept | |
	in.	mm		psf	kPa		psf	kPa
F&B	0.5	38	D.S.	360 to 1300	17 to 62	25	180	8.6
Palmer	3	75	D.S.	360 to 1300	17 to 62	19	240	11.5
Pine state	3	75	D.S.	360 to 1400	17 to 68	21	160	7.7
Pine state	3	75	D.S.	310 to 900	15 to 43	26	90	4.3
Dodger	35	900	D.S.	120 to 580	5.6 to 28	37	0	0
Unknown	0.37	9.5	Triaxial	1100 to 1700	52 to 82	24	120	6.0
Unknown	0.37	9.5	Triaxial	230 to 400	11 to 19	36	50	2.4

Note: D.S. = Direct shear.

the same range. Available data suggest that shear strength is not affected by TDA size. Moreover, results from triaxial and direct shear tests are similar. Overall, the failure envelopes appear to be concave down. Thus, best fit linear failure envelopes are applicable only over a limited range of stresses. Friction angles and cohesion intercepts for linear failure envelopes for the data shown in Figure 5 are given in Table 1. TDA require sufficient deformation to mobilize their strength (Humphrey and Sandford, 1993). Thus, a conservative approach should be taken when choosing strength parameters for TDA embankments founded on sensitive clay foundations.

3 DESIGN CONSIDERATIONS

3.1 Final in-place unit weight

The final in-place unit weight of the TDA must be estimated during design. This must consider compression of the TDA under its own self-weight and the weight of overlying soil and pavement. The calculation procedure is straight forward and is outlined below:

Step 1. From laboratory compaction tests or typical values, determine the initial uncompressed, compacted dry unit weight of TDA (γ_{di}) (for Type A TDA with a 75-mm maximum size, use 0.64 Mg/m^3).

Step 2. Estimate the in-place water content of TDA (w) and use this to determine the initial uncompressed, compacted total (moist) unit weight of TDA: $\gamma_{ti} = \gamma_{di} (1 + w)$. Unless better information is available, use w = 3 or 4 %.

Step 3. Determine the vertical stress in center of TDA layer ($\sigma_{v\text{-center}}$). To do this, make a first guess of the compressed unit weight of TDA (γ_{tc}) (0.80 Mg/m^3 is suggested for the first guess).

$$\sigma_{v\text{-center}} = t_{soil}(\gamma_{t\text{-soil}}) + (t_{TDA}/2)(\gamma_{tc})$$

where: t_{soil} = thickness of overlying soil layer
$\gamma_{t\text{-soil}}$ = total (moist) unit weight of overlying soil
t_{TDA} = compressed thickness of TDA layer
(Note: In the equation, the thickness of the TDA layer is divided by 2 since we are computing the stress in the center of the layer).

Step 4. Determine the percent compression (ε_v) using $\sigma_{v\text{-center}}$ and the measured laboratory compressibility of the

64

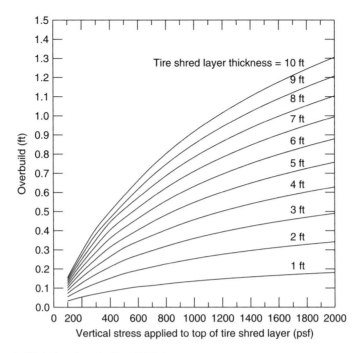

Figure 6.   Overbuild design chart for Type B TDA.

TDA; for TDA with a 75-mm maximum size, use the results for
test MD1 or MD4 in Figure 3. (Note: 1 kPa = 0.145 psi.)

Step 5.   Determine the compressed moist unit weight of the TDA:
$\gamma_{tc} = \gamma_{ti}/(1 - \varepsilon v)$. If necessary return to step 3 with a better
estimate of the compressed moist unit weight.

This procedure was used to predict the compressed unit weight of a 4.3-m thick TDA fill covered
by 1.8 m of soil built in Topsham, Maine. TDA with a 75-mm maximum size were used in the
upper third of the fill while TDA with a 150-mm maximum size were used in the lower part of the
fill. The predicted compressed moist unit weight was $0.91\,Mg/m^3$. The actual in-place unit weight
determined from the final volume of the TDA zone and the weight of TDA delivered to the project
was also $0.91\,Mg/m^3$. This validates the reliability of the laboratory compressibility tests and the
procedure to estimate the compressed moist unit weight for TDA with maximum sizes between 75
and 150 mm. However, when the procedure was applied to TDA with a 300-mm maximum size,
the predicted unit weight was greater than determined in the field. For a highway embankment
built in Portland, Maine with TDA with a 300-mm maximum size, the predicted compressed moist
unit weight was $0.93\,Mg/m^3$ compared to an actual unit weight of $0.79\,Mg/m^3$. The reasons for the
difference appear to be a lower initial uncompressed unit weight and lower compressibility for the
larger TDA. It is recommended that for 300-mm maximum size TDA, the unit weight calculated
using the procedure outlined above should be reduced by 15%.

3.2   Calculation of overbuild

TDA experiences immediate compression under an applied load, such as the weight of an overlying
soil cover. The top elevation of the TDA layer(s) should be overbuilt to compensate for this compression. The overbuild is determined using the procedure given below with the aid of a design chart
(Figure 6). The design chart was developed using a combination of laboratory compressibility tests

and compression measured from field projects. Figure 6 is applicable to Type-B TDA (300-mm maximum size) that have been placed and compacted in 300-mm thick layers. To use this procedure with smaller Type A TDA (75-mm maximum size), increase the calculated overbuild by 30%.

The overbuild for a single TDA layer is determined directly from Figure 6. First, calculate the vertical stress that will be applied to the top of the TDA layer as the sum of the unit weights times the thicknesses of the overlying layers. Second, enter Figure 6 with the calculated vertical stress and the final compressed thickness of the TDA layer to find the overbuild. Consider the following example:

$$0.229 \text{ m pavement at } 2.56 \text{ Mg/m}^3$$

$$0.610 \text{ m aggregate base at } 2.00 \text{ Mg/m}^3$$

$$0.610 \text{ m low permeability soil cover at } 1.92 \text{ Mg/m}^3$$

$$3.05\text{-m } (10 \text{ ft}) \text{ thick TDA layer}$$

The vertical stress applied to the top of the TDA layer would be:

$$(0.229 \text{ m} \times 2.56 \text{ Mg/m}^3 \times 9.81 \text{ m/s}^2) + (0.610 \text{ m} \times 2.00 \text{ Mg/m}^3 \times 9.81 \text{ m/s}^2) +$$

$$(0.610 \text{ m} \times 1.92 \text{ Mg/m}^3 \times 9.81 \text{ m/s}^2) = 29.1 \text{ kPa} \times 20.885 \text{ psf/kPa} = 610 \text{ psf}$$

Enter Figure 6 with 610 psf and using the line for a TDA layer thickness of 10 ft results in an overbuild of 0.68 ft ÷ 3.28 ft/m = 0.21 m. Round to the nearest 0.1 m, thus use an overbuild of 0.2 m.

The overbuild for the bottom TDA layer of a two layer cross section is also determined directly from Figure 6. The procedure is the same as described above for a single TDA layer. Consider the following example.

$$0.229 \text{ m pavement at } 2.56 \text{ Mg/m}^3$$

$$0.610 \text{ m aggregate base at } 2.00 \text{ Mg/m}^3$$

$$0.610 \text{ m low permeability soil cover at } 1.92 \text{ Mg/m}^3$$

$$3.05\text{-m } (10 \text{ ft}) \text{ thick TDA layer at } 0.80 \text{ Mg/m}^3$$

$$0.915 \text{ m soil separation layer at } 1.92 \text{ Mg/m}^3$$

$$3.05\text{-m } (10 \text{ ft}) \text{ thick lower TDA layer}$$

The vertical stress applied to the top of the TDA layer would be:

$$(0.229 \text{ m} \times 2.56 \text{ Mg/m}^3 \times 9.81 \text{ m/s}^2) + 0.610 \text{ m} \times 2.00 \text{ Mg/m}^3 \times 9.81 \text{ m/s}^2) +$$

$$(0.610 \text{ m} \times 1.92 \text{ Mg/m}^3 \times 9.81 \text{ m/s}^2) + (3.05 \text{ m} \times 0.80 \text{ Mg/m}^3 \times 9.81 \text{ m/s}^2) +$$

$$(0.915 \text{ m} \times 1.92 \text{ Mg/m}^3 \times 9.81 \text{ m/s}^2) = 70.4 \text{ kPa} \times 20.885 \text{ psf/kPa} = 1470 \text{ psf}$$

Enter Figure 6 with 1470 psf (70.4 kPa) and using the line for a TDA layer thickness of 10 ft (3.05 m) results in an overbuild of 1.13 ft (0.34 m). Round to the nearest 0.1 m, thus, use an overbuild of 0.3 m for the lower TDA layer.

The overbuild of the top elevation for the upper TDA layer a two layer cross section must include both the compression of the upper TDA layer when the pavement, base, and soil cover is placed, and the compression of lower TDA layer that will still occur under the weight of these layers. In other words, the lower TDA layer has not yet compressed to its final thickness. This will only occur once the embankment reaches final grade. So the question is, "How much compression of the lower TDA layer will occur due to placing the pavement, base and soil cover?" Consider the same two-layer example used above.

$$0.229 \text{ m pavement at } 2.56 \text{ Mg/m}^3$$

$$0.610 \text{ m aggregate base at } 2.00 \text{ Mg/m}^3$$

$$0.610 \text{ m low permeability soil cover at } 1.92 \text{ Mg/m}^3$$

$$3.05\text{-m } (10 \text{ ft}) \text{ thick TDA layer at } 0.80 \text{ Mg/m}^3$$

$$0.915 \text{ m soil separation layer at } 1.92 \text{ Mg/m}^3$$

$$3.05\text{-m } (10 \text{ ft}) \text{ thick lower TDA layer}$$

Step 1. The final vertical stress applied to the top of the upper TDA layer would be: $(0.229 \text{ m} \times 2.56 \text{ Mg/m}^3 \, x \, 9.81 \text{ m/s}^2) + (0.610 \text{ m} \times 2.00 \text{ Mg/m}^3 \times 9.81 \text{ m/s}^2) + (0.610 \text{ m} \times 1.92 \text{ Mg/m}^3 \times 9.81 \text{ m/s}^2) = 29.1 \text{ kPa} \times 20.885 \text{ psf/kPa} = 610 \text{ psf}$. Enter Figure 6 with 610 psf (29.1 kPa) and using the line for a TDA layer thickness of 10 ft (3.05 m) results in a compression of 0.68 ft (0.21 m).

Step 2. Once the upper TDA layer (but not the top soil cover) is in place, the vertical stress applied to the top of the lower TDA layer would be: $(3.05 \text{ m} \times 0.80 \text{ Mg/m}^3 \times 9.81 \text{ m/s}^2) + (0.915 \text{ m} \times 1.92 \text{ Mg/m}^3 \times 9.81 \text{ m/s}^2) = 41.1 \text{ kPa} \times 20.885 \text{ psf/kPa} = 860 \text{ psf}$. To determine the compression of the lower TDA layer that has occurred up to this point, enter Figure 5 with 860 psf (41.1 kPa) and using the line for a TDA layer thickness of 10 ft (3.05 m) results in a compression of 0.84 ft (0.26 m).

Step 3. Once the embankment reaches its final grade, the vertical stress applied to the top of the lower TDA layer would be 70.4 kPa = 1470 psf, as calculated previously. Enter Figure 6 with 1470 psf (70.4 kPa) and using the line for a TDA layer thickness of 10 ft (3.28 ft) results in an overbuild of 1.13 ft (0.34 m). (Note: rounding to 0.3 m would give the overbuild of the lower TDA layer.)

Step 4. Subtract the result from Step 2 from Step 3 to obtain the compression of the lower TDA layer that will occur when the pavement, base, and soil cover is placed. 0.34 m – 0.26 m = 0.08 m.

Step 5. Sum the results from Steps 1 and 4 to obtain the amount the top elevation of the upper TDA layer should be overbuilt. 0.21 m + 0.08 m = 0.29 m (0.95 ft). Round to the nearest 0.1 m. Thus, the elevation of the top of the upper TDA layer should be overbuilt by 0.3 m.

Final result: Overbuild the top elevation of the lower TDA layer by 0.3 m and the upper TDA layer by 0.3 m.

3.3   *Thickness of overlying soil cover*

Sufficient soil cover must be placed over the compressible TDA layer to preserve the durability of the overlying pavement. Cover thickness is defined as the combined thickness of base course and soil measured from the bottom of the asphaltic concrete pavement to the top of the TDA layer. This was investigated by Nickels (1995) and Humphrey & Nickels (1997). Computer modeling showed that the tensile strain at the base of asphaltic concrete pavement with 760 mm of soil cover over the TDA layer was the same as a control section underlain by conventional aggregate and soil. They also predicted that the tensile would be similar for as little as 457 mm of soil cover. Soil cover thickness between 0.5 to 1.2 m are recommended depending on the traffic loading. For applications with low truck traffic, such as parking lots and rural roads, the lower end of this range may be acceptable, however, the surface deflections under heavy vehicles immediately after placing the soil cover will be noticeable, but will decrease with additional trafficking.

3.4 *Guidelines to limit heating*

Three thick TDA fills (greater than 7.9 m thick) have undergone a self-heating reaction. Two of these projects were located in Washington State and one was in Colorado. These projects were constructed in 1995 and each experienced a serious self-heating reaction within 6 months after completion (Humphrey, 1996). The lessons learned from these projects were condensed into design guidelines developed by the Ad Hoc Civil Engineering Committee (1997), a partnership of government and industry dealing with reuse of scrap tires for civil engineering purposes. The overall philosophy behind development of the guidelines was to minimize the presence of factors that could contribute to self-heating. The guidelines were subsequently published by ASTM (1998) and distributed by the Federal Highway Administration. For TDA layers ranging in thickness from 1 to 3 m, the guidelines give the following recommendations:

- TDA shall be free of contaminants such as oil, grease, gasoline, diesel fuel, etc., that could create a fire hazard
- In no case shall the TDA contain the remains of tires that have been subjected to a fire
- TDA shall have a maximum of 50% (by weight) passing 75-mm sieve, 25% (by weight) passing 38-mm sieve and a maximum of 1% (by weight) passing no. 4 (4.75-mm) sieve
- TDA shall be free from fragments of wood, wood chips, and other fibrous organic matter
- TDA shall have less than 1% (by weight) of metal fragments that are not at least partially encased in rubber
- Metal fragments that are partially encased in rubber shall protrude no more than 25 mm from the cut edge of the TDA on 75% of the pieces and no more than 50 mm on 90% of the pieces
- Infiltration of water into the TDA fill shall be minimized
- Infiltration of air into the TDA fill shall be minimized
- No direct contact between TDA and soil containing organic matter, such as topsoil
- TDA should be separated from the surround soil with a geotextile
- Use of drainage features located at the bottom of the fill that could provide free access to air should be avoided

The guidelines further recommend that a TDA layer be no greater than 3 m thick. The guidelines also give less stringent requirements for TDA layers less than 1 m thick (Ad Hoc Civil Engineering Committee 1997, ASTM 1998).

3.5 *Water quality effects*

Several studies have shown that TDA can be used in most applications with negligible effects on ground water quality (Bosscher et al. 1992, Humphrey & Katz 2000, 2001). Humphrey et al. (1997) studied the effect of TDA placed above the water table in a test project in North Yarmouth, Maine. The final results of this study were published by Humphrey and Katz (2000). In this study, two 3 m x 3 m geomembrane-lined collection basins were used to collect water after it has passed through a 0.61-m thick layer of TDA. The TDA had a 75-mm maximum size and a significant amount of exposed steel belts. A third basin was overlain only by soil and served as a control. Water samples were taken quarterly for a 5.5 year period. Samples were analyzed for metals with primary and secondary drinking water standards. Metals with a primary standard are a known or suspected health risk. Metals with a secondary standard are of aesthetic concern, which means that they may impart some taste, odor, and/or color to water but they do not pose a health risk. For metals with a primary drinking water standard, the levels are about the same for basins overlain by TDA compared to the control basin overlain by soil. As an example, the results for chromium (Cr) are shown in Figure 7. Similar results were found for metals with secondary (aesthetic) standards except for manganese (Mn) and, to a lesser extent, iron (Fe). Manganese consistently had higher levels in the basins overlain by TDA compared to the control basin (Figure 8). The source of the manganese is thought to be the exposed steel belts, which contain 2 to 3% manganese by weight. On some sampling dates, iron was higher in the basins overlain by TDA as shown in Figure 9.

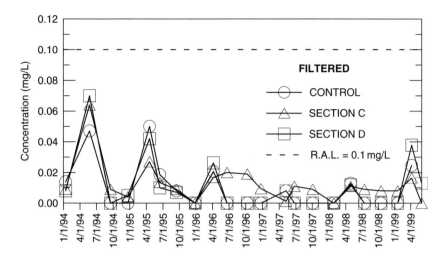

Figure 7. Chromium levels for filtered samples at North Yarmouth field trial.

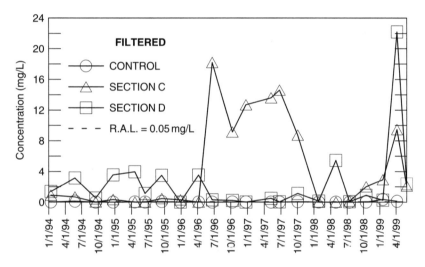

Figure 8. Manganese levels for filtered samples at North Yarmouth field trial.

Volatile and semi-volatile organics were tested on three dates. On the first two dates all substances were below the test method detection limits. On the third date trace levels of organics were detected, but the only significant concentrations were found in the control basin. Similar results for metals and organics were found for the Witter Farm Road test project (Humphrey, 1999a).

TDA placed below the water table were the subject of another study at the University of Maine (Downs et al. 1996, Humphrey 1999b, Humphrey & Katz 2001). In this study, 1.4 metric tons of TDA was buried below the water table in a 0.5-m wide trench in three different soil types (marine silty clay, glacial till, and peat). Samples were taken over a three year period from the TDA filled trench and from wells located 3.0 m up-gradient (control wells), 0.6 m down-gradient, and 3.0 m down-gradient. For samples taken from the TDA filled trench, the levels of metals with a primary standard were below their respective regulatory limits and were similar to background levels taken from control wells. However, the levels of manganese and iron were above their secondary (aesthetic) standards. A few organic compounds were present at detectable levels in the TDA filled trench. However, after flowing through only 0.6 m of soil to the first down gradient

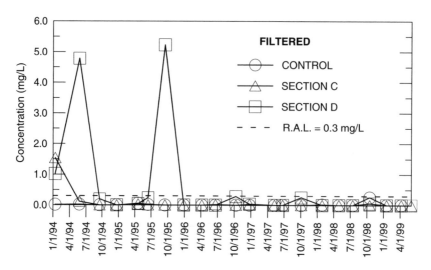

Figure 9.   Iron levels for filtered samples at North Yarmouth field trial.

wells, the levels in the groundwater had decreased to below the detection limit except for three compounds (benzene, cis-1,2-dichloroethene, and toluene) which were present at levels below their respective drinking water standards. 1,1-dichloroethane was present in some down-gradient wells at the marine clay and till sites but the highest concentration was 6.9 parts per billion. Concentrations of 1,1-dichloroethane were below detection limit in all down-gradient wells on the most recent sampling date. A drinking water standard has not been established for this compound.

These studies show that TDA has a negligible impact on groundwater quality and can be used for most civil engineering applications provided the pH of the groundwater is near neutral. Further field studies would be needed to establish the effect of TDA on water quality for acidic or basic conditions.

4   CASE HISTORY – DIXON LANDING INTERCHANGE

In Milpitas, California, the site for a new on-ramp connecting Dixon Landing Road to the south bound lane of Interstate Highway I-880 was underlain by of soft San Francisco Bay Mud. California Department of Transportation (Caltrans) geotechnical engineers were concerned about the magnitude of settlements that would be caused by the 9.8 m of embankment fill, so they specified lightweight fill for the project. With technical and logistical support from the California Integrated Waste Management Board, IT Corporation, and the University of Maine, Caltrans designed an embankment with a core of TDA. TDA was selected because its unit weight of 0.8 Mg/m^3 was less than the 1.3 Mg/m^3 of lightweight aggregate typically used for these types of projects. Moreover, TDA was less expensive. The 213-m long embankment would used the equivalent of 600,000 passenger tires.

To meet the guidelines to limit self heating discussed above, the Dixon Landing design called for the core of the embankment to be constructed of two layers of TDA separated by 1 m of low permeability soil. Each TDA layer was up to 3 m thick and the side slopes are covered by 1.8 m of soil with a minimum of 30% fines. A typical cross section is shown in Figure 10. As a final design feature, the TDA layers were wrapped in non-woven geotextile.

The TDA for this project had no pieces longer than 457 mm, 90% by weight smaller than 300 mm in length, and tight restrictions limiting the amount of small pieces and exposed steel belt. In the U.S., this is commonly referred to as "Type B" TDA. These relatively large size pieces are needed

70

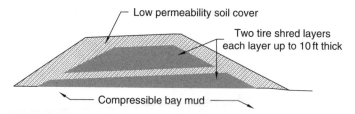

Figure 10. Typical cross section of Dixon Landing project.

Figure 11. Close up of TDA used for Dixon Landing interchange.

to minimize the exposed steel and freshly cut surfaces on the pieces, thereby reducing the potential for self-heating. A close up of the TDA used for this project is shown in Figure 11.

The TDA was delivered by tractor-trailers with up to 28 m^3 per load. The large volume per load was possible because of the lightweight of the TDA. The TDA was easily spread into 300 mm thick lifts with a bulldozer and compacted with six passes of a vibratory smooth drum roller weighing 9.1 tonnes. Compaction of TDA is unaffected by water content, no water content control was necessary. Typical construction photographs are shown in Figures 12 and 13.

The TDA zone was instrumented with thermisters to monitor the internal temperature of the TDA fill. The TDA was about 27°C at the time of placement, which is typical of the early summer temperatures in the region. Within 8 months of placement the internal temperatures had dropped to about 20°C. Ongoing monitoring shows that the temperatures are slowly dropping with time. Thus, the measures that were taken to limit internal heating were completely effective. Temperature measurements for similar projects in Maine, New York, Pennsylvania, and southern California confirm the effectiveness of the ASTM D6270 guidelines to limit heating (Humphrey 2004).

The contractor provided bid prices for two types of lightweight fill. The placement cost of the TDA, including the surrounding geotextile, was $4.89/m^3. The TDA was supplied by three tire processors, and had an average purchase and delivery price of $30.93/m^3. The market conditions for TDA in California are such that it was economically justifiable for one processor located in the Los Angeles area to make a 1,100 km round-trip haul to supply TDA to the project. Combining the purchase and placement costs results in an in-place cost of the TDA of $35.82/m^3. This price is much higher than common borrow for this project ($9.78/m^3), however, it is approximately half

Figure 12. Unloading TDA from tractor trailer.

Figure 13. Spreading TDA with bulldozer.

the in-place bid price of $65/m³ for lightweight aggregate imported from Oregon. In the end, the savings to Caltrans by using TDA rather than lightweight aggregate was $230,000.

5 CASE HISTORY – WALL 119, RIVERSIDE, CALIFORNIA

TDA was used as lightweight backfill for a cantilever retaining wall constructed in the summer of 2003. The wall is part of a major Caltrans project to widen State Route 91 in Riverside, California. This section of highway passes through a heavily built up section of downtown Riverside. The highway is elevated about 6 m above the surrounding grade to allow the roadway to pass over bridges spanning the many city streets. Between bridges the highway is constructed on an earth embankment that, prior to reconstruction, had 2:1 (H:V) sideslopes. Increased traffic volume necessitated adding additional travel lanes in each direction. Taking the additional right of way needed for embankments with 2:1 sideslopes was out of the question because of the large number of existing businesses that

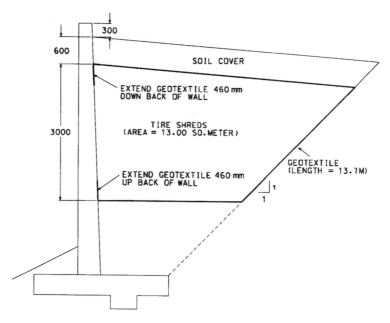

Figure 14. Typical section.

would have to be removed. Thus, the solution was to construct retaining walls near the existing right of way line, creating the additional embankment width needed for the additional lanes.

The wall was instrumented at four stations to measure performance. Three stations have TDA backfill and one has conventional backfill. Each instrumented station has three total pressure cells to measure horizontal stress produced by the backfill, four strain gages to measure deformation of vertical reinforcing bars, and a tiltmeter to measure tilt of the wall during backfilling. The California Integrated Waste Management Board sponsored use of TDA on the project.

5.1 Project design

TDA was used as backfill for a 80-m length of wall. The wall height varied from about 3.4 to 5.4 m over the section with TDA backfill. The wall was a conventional reinforced concrete cantilever retaining wall conforming to the requirements for a Caltrans Type 1 Retaining Wall (Caltrans, 2000). In preparation for construction, the existing slope as excavated to about 1:1 (H:V). This created a sufficient width to allow construction of the wall footing. The backfill consisted of 0.6 m of compacted soil cover underlain a trapezoidal shaped zone of TDA with a thickness of about 3 m. The TDA was encapsulated in geotextile. The TDA was underlain by compacted soil to the top of the footing. A typical cross section is shown in Figure 14.

The wall was instrumented at four stations to monitor performance. Stations A, B, and C have TDA backfill. A fourth station, Station D, has conventional soil backfill to serve as a comparison. The wall heights, measured from the top of the footing to the top of the backfill are 3.90 m for Station A, 4.54 m for Station B, 4.80 m for Station C, and 3.21 m for Station D. Each instrumented station had three vibrating wire pressure cells (Roctest Model TPC), two vibrating wire strain gages (Roctest Model SM-2W) mounted on the tension reinforcement 0.45 m above the footing, two vibrating strain gages mounted on the compression reinforcement 0.45 m above the footing, one tilt meter (Soil Instrument, Ltd, Electrolevel Model TLT2) placed near the top of the wall, and three temperature sensors (Roctest Model TH-T) placed in the TDA backfill. A typical instrumented section is shown in Figure 15. Each pressure cell was calibrated by the research team. A

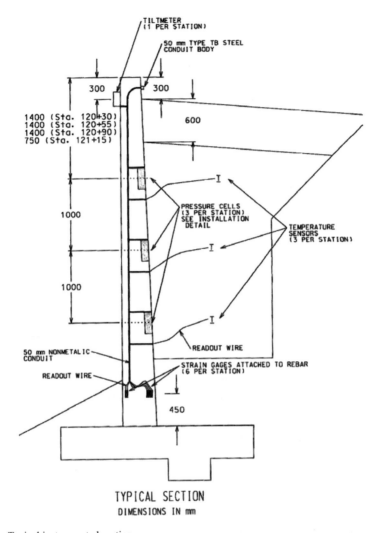

TILTMETER
(1 PER STATION)

50 mm TYPE TB STEEL
CONDUIT BODY

300

300

600

1400 (Sta. 120+30)
1400 (Sta. 120+55)
1400 (Sta. 120+90)
750 (Sta. 121+15)

1000

PRESSURE CELLS
(3 PER STATION)
SEE INSTALLATION
DETAIL

TEMPERATURE
SENSORS
(3 PER STATION)

1000

50 mm NONMETALIC
CONDUIT

READOUT WIRE

READOUT WIRE

STRAIN GAGES ATTACHED TO REBAR
(6 PER STATION)

450

TYPICAL SECTION
DIMENSIONS IN mm

Figure 15. Typical instrumented section.

temperature correction factor was developed for each pressure cell after installation using data from several years of seasonal temperature change. Calibration factors provided by the manufacturers were used for the other instrument types. Initially, the Roctest instruments were read manually with a Roctest Model MB-6TL vibrating wire readout unit. The tiltmeters were read with a Soil Instrument, Ltd, HELM readout unit. About 4 months after construction was completed, the instrumentation was attached to a Campbell Scientific datalogger and reading were taken and stored hourly.

Wall backfill was placed during August and September, 2003. The initial layer of soil backfill was placed using conventional placement methods. Then, geotextile was placed on this base and extended up the 1:1 sideslope of the excavation. The TDA was spread with a large tracked front-end loader as shown in Figure 16. The TDA was placed in 0.3-m thick lifts and compacted with six passes of a vibratory roller with an operating weight of about 9 tonnes. Once the TDA was brought to grade, geotextile was placed followed by the compacted soil cover. Throughout the construction process, heavy construction equipment operated immediately behind the wall. After completion of the TDA section, a large rubber-tired loader operated immediately behind the wall to place soil

74

Figure 16.   Large tracked loader operating immediately behind the wall.

Figure 17.   Large wheeled loader operating immediately behind the wall (Photo courtesy of J. Wright, Kennec Engineering).

fill over the TDA and to haul soil to other parts of the project as show in Figure 17. This heavy equipment affected the earth pressures as will be discussed later.

5.2   *Pressure cell and strain gage results*

The pressure distributions measured by the pressure cells at Station C (TDA backfill) and Station D (soil backfill) are shown in Figures 18 and 19, respectively. The active earth pressures predicted using Rankine's method for soil with a unit weight of 2.00 Mg/m^3 and a friction angle of $\phi = 30°$ and at rest pressure calculated using the equation 1-sin ϕ is also shown. In Section C, the high stresses at shallow depths are thought to be due to over compaction by heavy equipment operating

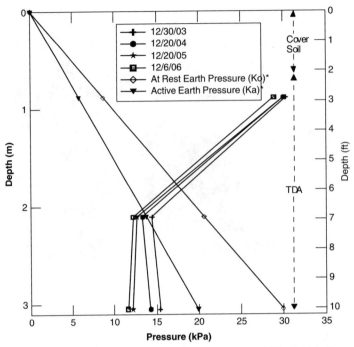

* Theoretical Values based on assumed soil properties (125pcf, 30deg)

Figure 18. Backfill pressure distribution for Station C with TDA backfill.

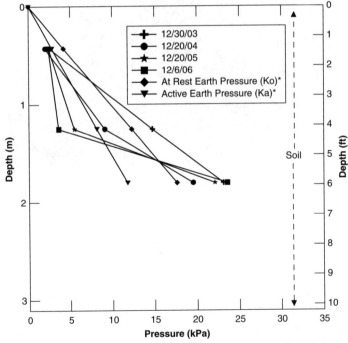

* Theoretical Values based on assumed soil properties (125 pcf, 30deg)

Figure 19. Backfill pressure distribution for Station D with soil backfill.

76

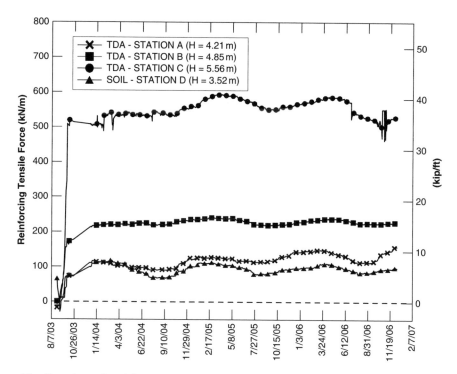

Figure 20. Force in tensile reinforcement steel per unit length of wall.

immediately behind the wall. At the mid-thickness of the TDA layer, the stresses are slightly less than calculated at-rest and active earth pressures, while they are significantly less near the bottom of the TDA layer. In Section D, the stresses range from slightly below to slightly above at-rest and active earth pressures.

The force in the tensile reinforcement was calculated using the measured strains and the modulus of the reinforcing steel. The results are expressed as the force per unit length of the wall in Figure 20. The force increased during backfill placement and then has remained approximately constant thereafter. Given the wide range of wall heights at the instrumented sections, it was necessary to develop a method of comparison. This was done as follows. The tensile reinforcing force is created by the overturning moment generated by the backfill. Calculation of earth pressures using Rankine's method show that the resultant horizontal force generated by the backfill is a function of the square of the fill height and that the overturning moment is a function of the cube of the fill height. Accordingly, the reinforcing force was plotted as a function of the cube of the fill height above the gage locations as shown in Figure 21 for data from April 3, 2004. There is a nearly linear relationship between the tensile reinforcing force and the cube of the fill height for the three stations backfilled with TDA. The result for Station D with soil backfill falls significantly to the left of the trend line for TDA indicating that use of TDA backfill reduced the tensile force in the reinforcement.

The moment in the wall stem taken about the neutral axis was calculated using data from the strain gages on both the tension and compression reinforcement. This moment would be expected to increase as the cube of the fill height above the gages based on the same reasoning as discussed above. The results are plotted in Figure 22. The results for the three TDA stations vary linearly with the cube of the fill height above the gages, while the result for Station D falls significantly to the left. This shows that use of TDA backfill reduced moment wall stem.

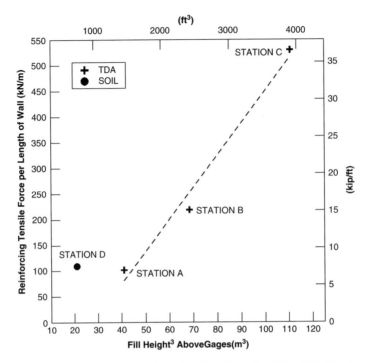

Figure 21. Force in tensile reinforcement steel as a function of the cube of the fill height above the gages.

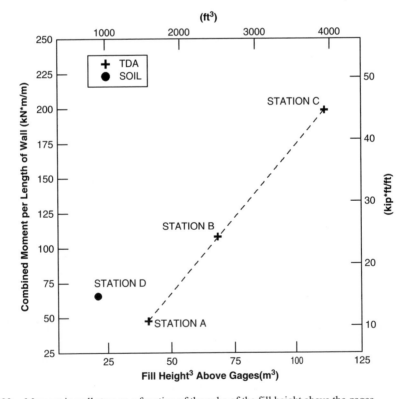

Figure 22. Moment in wall stem as a function of the cube of the fill height above the gages.

The improved performance of the TDA stations were observed in spite of the high earth pressures at the top of the wall due to over compaction. At second instrumented wall also located in Riverside, California, the contractor was prohibited from using heavy equipment immediately behind the wall. Backfill behind this wall is nearing completion. At present, the TDA is covered by 0.6 m of compacted soil, the same as the project discussed above. High earth pressures at shallow depths have not been measured. When these results are confirmed by data from other ongoing instrumented projects, it will be possible to developed revised structural designs for canteliver retaining walls that have reduced tension reinforcing steel requirements.

6 CONCLUSIONS

The low unit weight, widespread availability, and low cost of TDA has led to their being used as lightweight fill for embankments constructed on weak foundation soils and retaining wall backfill. The engineering properties of TDA is known including gradation, unit weight, compressibility and shear strength. When the special properties of TDA is needed for a project they are often the lowest cost alternative. Thus, civil engineers are choosing TDA because they offer both the properties needed to solve special problems and lower costs to satisfy the demands of their clients for the most economical project possible. In the next few years, continued increases in the number of scrap tires used for civil engineering applications is possible because of their growing record of successful performance.

ACKNOWLEDGEMENTS

This paper presents the results from several projects. It took many dedicated people to make these projects happen. Special thanks go to Stacey Patenaude from the California Integrated Waste Management Board and Joaquin Wright from Kennec Engineering.

Several University of Maine students put extraordinary effort into these projects. Will Manion and Michelle Cribbs are thanked for their laboratory work to determine the engineering properties of TDA. Aaron Smart and Lisa Downs are thanked for performing most of the water quality tests. Jeremy Labbe is thanked for his careful analysis of the data from the California Route 91 wall project. The Maine Turnpike Authority, Maine Department of Transportation, New England Transportation, and California Integrated Waste Management Board are thanked for funding these projects.

REFERENCES

Abichou, T., Edil, T.B., Benson, C.H. & Berilgen, M. 2001. Mass Behavior of Soil-Tire Chip Backfills. *Proceedings of the Conference on Beneficial Use of Recycled Materials in Transportation Applications.* Arlington, Virginia: 689–698.

Ad Hoc Civil Engineering Committee 1997. *Design Guidelines to Minimize Internal Heating of Tire Shred Fills.* Washington, DC: Scrap Tire Management Council.

Ahmed, I. 1993. Laboratory Study on Properties of Rubber Soils. *School of Civil Engineering Report No. FHWA/IN/JHRP-93/4.* West Lafayette, Indiana: Purdue University.

ASTM1998. Standard Practice for Use of Scrap Tires in Civil Engineering Applications. *ASTM D6270-98.* W. Conshohocken, Pennsylvania: Am. Soc. Testing & Mat.

Benda, C.C. 1995. Engineering Properties of Scrap Tires Used in Geotechnical Applications. *Materials and Research Division Report 95-1.* Montpelier: Vermont Agency of Transportation.

Bosscher, P.J., Edil, T.B. & Eldin, N.N. 1993. Construction and Performance of a Shredded Waste Tire Test Embankment. *Transportation Research Record 1345*, 44–52. Washington, DC: Transportation Research Board.

Bossher, P.J., Edil, T.B. & Senro, K. 1997. Design of highway embankments using tire chips. *Journal of Geotechnical and Geoenvironmental Engineering.* 123(4): 295–305.

Bressette, T. 1984. Used Tire Material as an Alternative Permeable Aggregate. *Office of Transportation Laboratory Report No. FHWA/CA/TL-84/07*. Sacramento: California Department of Transportation.

Caltrans 2000. *Standard Plans July 1999, Erratum No. 99-1*. Sacramento: California Department of Transportation.

Cosgrove, T.A. 1995. Interface Strength Between Tire Chips and Geomembrane for Use as a Drainage Layer in a Landfill Cover. *Proceedings of Geosynthetics '95*, 3: 1157–1168. St. Paul, Minnesota: Industrial Fabrics Association.

Cosgrove, T.A. & Humphrey, D.N. 1999. Field Performance of Two Tire Shred Fills in Topsham, Maine. Report for the Maine Department of Transportation, by Dept. of Civil and Environmental Engineering, Orono: University of Maine.

Dickson, T.H., Dwyer, D.F. & Humphrey, D.N. 2001. Prototype Tire Shred Embankment Construction by the New York State Department of Transportation. *Transportation Research Record No. 1755*: 160–167. Washington, DC: Transportation Research Board.

Downs, L.A., Humphrey, D.N., Katz, L.E. & Rock, C.A. 1996. Water Quality Effects of Using Tire TDA Below the Groundwater Table. Augusta: Maine Department of Transportation.

Edil, T.B. & Bossher, P.J. 1992. Development of Engineering Criteria for Shredded Waste Tires in Highway Applications. *Research Report No. GT-92-9*, Madison, Wisconsin: University of Wisconsin.

Gebhardt, M.A., Kjartanson, B.H. & Lohnes, R.A. 1998. Shear Strength of Large Size Shredded Scrap Tires as Applied to the Design and Construction of Ravine Crossings. *Proceedings of the 4th International Symposium on Environmental Geotechnology and Global Sustainable Development, Danvers, Massachusetts.*

Humphrey, D.N. 1996. *Investigation of Exothermic Reaction in Tire Shred Fill Located on SR 100 in Ilwaco, Washington*. Washington, DC: FHWA.

Humphrey, D.N. 1999a. *Water Quality Results for Witter Farm Road Tire Chip Field Trial*. Dept. of Civil and Env. Eng. Orono, Maine: University of Maine.

Humphrey, D.N. 1999b. *Water Quality Effects of Using Tire Chips Below the Groundwater Table – Final Report*. Augusta: Maine Department of Transportation.

Humphrey, D.N. 2004. Effectiveness of Design Guidelines for use of Tire Derived Aggregate as Lightweight Embankment Fill. *Recycled Materials In Geotechnics*: 61–74. Arlington, Virginia: ASCE.

Humphrey, D.N. & Eaton, R.A. 1995. Field Performance of Tire Chips as Subgrade Insulation for Rural Roads. *Proceedings of the Sixth International Conference on Low-Volume Roads*. 2: 77–86. Washington, DC: Transportation Research Board.

Humphrey, D.N., Katz, L.E. & Blumenthal, M. 1997. Water Quality Effects of Tire Chip Fills Placed Above the Groundwater Table. In Mark A. Wasemiller and Keith B. Hoddinott, eds. *Testing Soil Mixed with Waste or Recycled Materials, ASTM STP 1275*: 299–313. W. Conshohocken, Pennsylvania: American Society for Testing and Materials.

Humphrey, D.N. & Katz, L.E. 2000. Five-Year Field Study of the Effect of Tire TDA Placed Above the Water Table on Groundwater Quality. *Transportation Research Record No. 1714*: 18–24. Washington, DC: Transportation Research Board.

Humphrey, D.N. & Katz, L.E. 2001. Field Study of the Water Quality Effects of Tire TDA Placed Below the Water Table. *Proceedings of the International Conference on Beneficial Use of Recycled Materials in Transportation Applications, November 13, 2001*. Arlington, Virginia.

Humphrey, D.N. & Manion, W.P. 1992. Properties of Tire Chips for Lightweight Fill. *Grouting, Soil Improvement, and Geosynthetics*, 2: 1344–1355. Arlington, Virginia: ASCE.

Humphrey, D.N. & Nickels, W.L., Jr. 1997. Effect of Tire Chips as Lightweight Fill on Pavement Performance. *Proceedings of the Fourteenth International Conference on Soil Mechanics and Foundation Engineering*, 3: 1617–1620. Rotterdam: Balkema.

Humphrey, D.N. & Sandford, T.C. 1993. Tire Chips as Lightweight Subgrade Fill and Retaining Wall Backfill. *Proceedings of the Symposium on Recovery and Effective Reuse of Discarded Materials and By-Products for Construction of Highway Facilities*: 5–87 to 5–99. Washington, DC: Federal Highway Administration.

Humphrey, D.N., Sandford, T.C., Cribbs, M.M., Gharegrat, H. & Manion, W.P. 1992. Tire Chips as Lightweight Backfill for Retaining Walls – Phase I. A Study for the New England Transportation Consortium, Dep. of Civil Eng., Orono: University of Maine.

Humphrey, D.N., Sandford, T.C., Cribbs, M.M., Gharegrat, H., & Manion, W.P. 1993. Shear Strength and Compressibility of Tire Chips for Use as Retaining Wall Backfill. *Transportation Research Record No. 1422*: 29–35. Washington, DC: Transportation Research Board.

Humphrey, D.N., Whetten, N., Weaver, J., Recker, K. & Cosgrove, T.A. 1998. Tire TDA as Lightweight Fill for Embankments and Retaining Walls. *Proceedings of the Conference on Recycled Materials in Geotechnical Application*: 51–65. Arlington, Virginia, ASCE.

Humphrey, D.N., Whetten, N., Weaver, J. & Recker, K. 2000. Tire TDA as Lightweight Fill for Construction on Weak Marine Clay. *Proceedings of the International Symposium on Coastal Geotechnical Engineering in Practice*: 611–616. Balkema: Rotterdam.

Jesionek, K.S., Humphrey, D.N. & Dunn, R.J. 1998. Overview of Shredded Tire Applications in Landfills. *Proceedings of the Tire Industry Conference, March 4–6, 1998*. Clemson, South Carolina: Clemson University.

Lawrence, B.K., Chen, L.H. & Humphrey, D.N. 1998. Use of Tire Chip/Soil Mixtures to Limit Frost Heave and Pavement Damage of Paved Roads. A Study for the New England Transportation Consortium, by Dept. of Civil and Environmental Engineering. Orono, Maine: University of Maine.

Manion, W.P. & Humphrey, D.N. 1992. Use of Tire Chips as Lightweight and Conventional Embankment Fill, Phase I – Laboratory. *Technical Services Division Technical Paper 91-1*. Augusta: Maine Department of Transportation.

Nickels, W.L., Jr. 1995. The Effect of Tire Chips as Subgrade Fill on Paved Roads *M.S. Thesis*. Department of Civil Engineering, Orono, Maine: University of Maine.

Recycling Research Institute 2005. *Scrap Tire and Rubber Users Director 14th Edition*. Leesburg, Virginia: Recycling Research Institute.

Reid, R.A. & Soupir, S.P. 1998. Mitigation of Void Development Under Bridge Approach Slabs Using Rubber Tire Chips. *Proceedings of the Conference on Recycled Materials in Geotechnical Applications*, 37–50. Arlington, Virginia: ASCE.

STMC 1997. *Scrap Tire Use/Disposal Study – 1996 Update*. Washington, DC: Scrap Tire Management Council.

Tatlisoz, N., Benson, C.H. & Edil, T.B. 1997. Effect of fines on mechanical properties of soil-tire chip mixtures. In Mark A. Wasemiller and Keith B. Hoddinott, eds. *Testing Soil Mixed with Waste or Recycled Materials, ASTM STP 1275*: 93–108. W. Conshohocken, Pennsylvania: American Society for Testing and Materials.

Tweedie, J.J., Humphrey, D.N. & Sandford, T.C. 1997. Tire Chips as Lightweight Backfill for Retaining Walls – Phase II. A Study for the New England Transportation Consortium, Dept. of Civil and Env.Eng. Orono, Maine: University of Maine.

Tweedie, J.J., Humphrey, D.N. & Sandford, T.C. 1998a. "Full Scale Field Trials of Tire Shreds as Lightweight Retaining Wall Backfill, At-Rest Conditions. *Transportation Research Record No. 1619*: 64–71. Washington, DC: Transportation Research Board.

Tweedie, J.J., Humphrey, D.N. & Sandford, T.C. 1998b. Tire Shreds as Retaining Wall Backfill, Active Conditions. *Journal of Geotechnical and Geoenvironmental Engineering*. 124(11): 1061–1070.

Tweedie, J.J., Humphrey, D.N., Wight, M.H., Shaw, D.E. & Stetson, J.H. 2004. Old Town – Mud Point Inlet Bridge, Rehabilitation of a Timber Structure with Tire Shreds. *Geotechnical Engineering for Transportation Projects: Proceedings of Geo-Trans 2004*: 740–749. Arlington, Virginia: ASCE.

Whetten, N., Weaver, J. Humphrey, D. & Sandford, T. 1997. Rubber Meets the Road in Maine. *Civil Engineering*. 67(9): 60–63.

Wolfe, S.L., Humphrey, D.N. & Wetzel, E.A. 2004. Development of Tire Shred Underlayment to Reduce Groundborne Vibration from LRT Track. *Geotechnical Engineering for Transportation Projects Proceedings of Geo-Trans 2004*: 750–759. Arlington, Virginia. ASCE.

Scrap Tire Derived Geomaterials – Opportunities and Challenges – Hazarika & Yasuhara (eds)
© 2008 Taylor & Francis Group, London, ISBN 978-0-415-46070-5

Engineering characteristics of tire treads for soil reinforcement

Yeo Won Yoon
Inha University, Incheon, Korea

ABSTRACT: Tire treads as material for soil reinforcement was reviewed. Treads obtained by elimination of sidewalls can be used to make plain tread-mat or to make 3-dimensional cell-type tire. In order to utilize the waste tires for geotechnical engineering purpose, laboratory test, plate load test in a chamber and pull-out test under embankment were carried out. Geogrid can be replaced partly by tread-mat which was combined by treads and geocell by cell-type tire. From laboratory and field pull-out tests, tire treads proved to be useful material for soil reinforcement without special environmental problems.

1   INTRODUCTION

Waste tire disposal has become a major environmental issue in many countries. Each year more than 250 million waste tires are stockpiled in the United States(Rma, 2004) and Canada generates over 28 millions of passenger car tires per year(Garga & O'shaughnessy, 2000). Korea has generated approximately 20 millions of waste tires per year since 1998, and some of the tires are utilized for rubber tiles and blocks or for cement materials. However, the cost of making rubber powder from tires is very high. Therefore, several beneficial uses of waste tires have been proposed in the last decade, and some of them have already been applied in construction. Waste tires are desirable as construction material because of their excellent mechanical properties and durability. Fine-ground tire powders can be used as partial replacement for asphalt in asphaltic concrete; tire chips can be used for lightweight fill(Humphrey & Manion, 1992; Foose et al., 1996; Humphrey et al., 1998; Reid et al., 1998); treads can be used as a form of grid(Yoon et al., 2004); tires with one sidewall removed(Garga & O'Shaughnessy, 2000; Nguyen, 1996; Ecoflex, 2006) can be used. In some cases, environmental assessments of toxic compounds in the tire-embedded earth fill were performed previously(O'Shaughnessy & Garga, 2000; Humphrey and Katz, 2000; Humphrey and Katz, 2002; Moon, 2003). Humphrey and Katz(2002) monitored the effluents and the results indicated that toxic compounds had no significant adverse effects on ground water quality over a period of 5 years.

In this paper, engineering characteristics of tire treads were reviewed. Treads can be used to make 3-dimensional cell-type tire(Yoon, 2006) which can be used in the same way as a commercial geocell or to make plain tread-mat for soil reinforcement(Yoon & Cha, 2002). A large number of plate load tests for both tire mat and cell-type tire were performed to study reinforcement effects on the bearing capacity increase and settlement reduction in a test chamber filled with sands. And pull-out tests for tread-mat and cell-type tire under embankment were performed. In addition tensile tests for both tread connection and cell connection and large direct shear tests between soil and tread surface were carried out.

2   REINFORCEMENT WITH WASTE TIRES

Tires are fabricated with vulcanized rubber which contains textile, high-strength steel and reinforcing bead. As shown in Figure 1, main components of a tire are tread, sidewalls and bead.

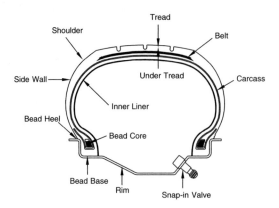

Figure 1. Typical section of tire.

Table 1. Typical composition of tire material.

Compound	Weight (%)
Rubber polymer (SBR)	62.1
Carbon black	31.0
Extender oil	1.9
Zinc oxide	1.9
Stearic acid	1.2
Sulphur	1.1
Accelerator	0.7

Tire rubber (Williams et al., 1990).

As shown in Table 1, tire is typically composed of synthetic rubber, rubber compound and others. Tires are not susceptible to corrosion. And deterioration is negligible under pH ranges of 4 to 5 of usual groundwater condition(O'Shaughnessy & Garga, 2000). Though tire is a combustible and susceptible to UV radiation material, the use of tires under buried condition will not have any problem. Use of whole tires is difficult to use for soil reinforcement because the compaction of soil inside the tire is not easy. In order to improve the utilization or disposal, one sidewall is removed or whole tires are shredded into tire chips. However use of tires with removal of one sidewall as done by Garga & O'Shaughnessy(2000) causes some soil deformation and making tire chips for light weight fill and other purposes cost much.

Improved method for soil reinforcement was devised and combination process of waste tires for soil reinforcement are shown in Figure 2(Yoon; 2004, 2006). To make cell-type tire both sidewalls were removed from a tire and a shallow, large diameter, cylinder tire was folded to make small two cells forming an Arabic number 8 type(called Tirecell). Many cell-type tire units can be combined to complete a cell-type tire. In case of tread-mat was woven treads cut across tire eliminated both sidewalls. Both sidewalls can be made sidewall-mat with connected to each other, and they were overlapped in four locations of the sidewalls. The mat made of sidewalls can be formed by connecting each sidewalls in the four locations mentioned. The long treads are crossed at inside an empty circle within the sidewall in order to make tire mats composed of plurality of sidewalls and treads as shown in Figure 2.

The mechanism of soil reinforcement of cell-type tire and tread-mat is same to geocell and geogrid, respectively. Therefore cell-type tire and tread-mat can be utilized in roadway,

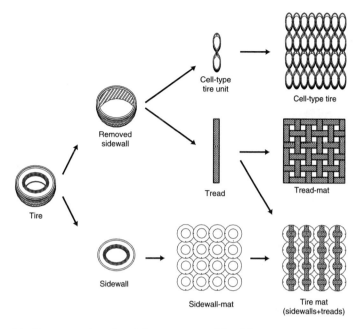

Figure 2.   Combination process to make reinforcement materials of three types.

embankment, reinforced wall, and foundation as the same way to geocell and geogrid for soil reinforcement, respectively.

3   LABORATORY TESTS

For utilization of waste tires in soil reinforcement, it is important to assess internal friction angle of soil, interface friction angle between soil and tread and tensile strength of tire connection joints. Also, reinforcement effect on bearing capacity is necessary to know, because the embedded tire is confined by surrounding soil and resists to shear by friction.

Large direct shear tests for soil-soil and soil-tread were performed. Tensile tests using universal testing machine(max. cap. 490 kN) for various conditions of connection joint between tire treads were performed either. Also plate load tests with variation of relative densities of sands were performed to know mechanical properties as reinforcement material.

3.1   *Physical properties of sand and weathered soil*

Two type of soil was used for the tests. One is silica sand and the other is weathered soil with some silts. The index properties and grain size distribution of the sand and the weathered soil used are shown in Table 2 and Figure 3, respectively. Soil compaction tests were performed by using both the standard proctor method(KS F 2312 A) and the modified proctor method(KS F 2312 D). The results are given in Table 3 and Figure 4, respectively.

3.2   *Large direct shear test*

Large direct shear tests for both outside and inside of tread were performed to know frictional characteristics between soil and surface of treads as soil reinforcement material. The soil specimen was placed in shear box (300 mm × 300 mm × 180 mm) with 3 layers. In the tests, relative densities

Table 2. Index properties of sample.

Sample	G_s	D_{10} (mm)	D_{30} (mm)	D_{60} (mm)	C_u	C_z	$\gamma_{d\,max}$ (kN/m^3)	$\gamma_{d\,min}$ (kN/m^3)	USCS
Sand	2.64	0.19	0.34	0.53	2.79	1.15	15.86	13.27	SP
Weathered soil	2.71	0.005	0.08	0.25	50	5.12	–	–	SM

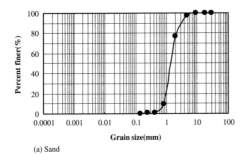

(a) Sand

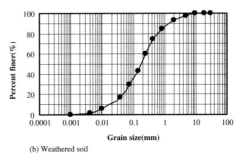

(b) Weathered soil

Figure 3. Grain size distribution of sand and weathered soil.

Table 3. Laboratory results for compaction test of weathered soil.

	Standard proctor compaction		Modified proctor compaction	
W_n (%)	$\gamma_{d\,max}$ (kN/m^3)	OMC (%)	$\gamma_{d\,max}$ (kN/m^3)	OMC (%)
11.7–12.3	17.25	13.01	18.42	10.66

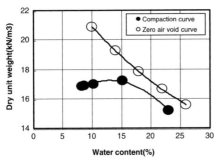

(a) Standard proctor compaction

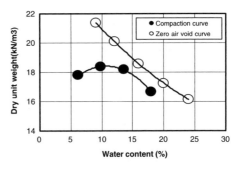

(b) Modified proctor compaction

Figure 4. Compaction curve of the sample.

of sand were 55% and 70%, respectively. And for the tests of weathered soil, the degree of relative compaction was 90% of max. dry density of modified proctor compaction method.

Table 4 shows the list of large direct shear tests. The tests were performed with 1.0 mm/min speed under the vertical pressure of 39.2–313.8 kN/m^2. In the test for soil-tread, as shown in the Figure 5, dummy was placed below the tread to fix the tread and to adjust shear surface.

Figure 6 and Table 5 show the results of large direct shear tests. For the relative densities of 55% and 70%, the friction angles of sand were 39.8°, 42.1°, respectively and shear angles of friction

Table 4. The list of large direct shear tests.

Soil	Preparation of sample		Type of sample	Vertical pressure (kN/m^2)	Strain rate (mm/min)
Sand	relative density	55%, 70%	sand-sand(ϕ) sand-tread (δ) (outside, inside)	39.2, 78.5,	
Weathered soil	relative compaction	90%	soil-soil(ϕ) soil-tread(δ) (outside, inside)	156.9, 313.8	1.0

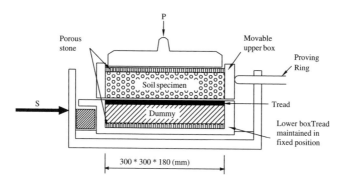

Figure 5. Schematic diagram of direct shear test for soil-tread.

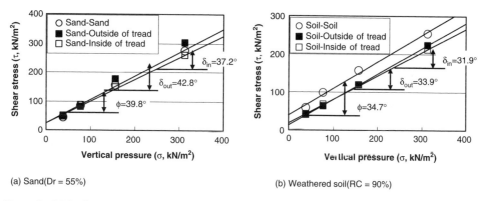

(a) Sand(Dr = 55%)

(b) Weathered soil(RC = 90%)

Figure 6. Mohr diagrams for sand and weathered soil.

between the sand and both outside and inside surfaces of tread for the relative density of 55% were 42.8°, 37.2°, respectively. And interface friction angles for both sides at the relative density of 70% were 44.5°, 39.4°, respectively. In other words the ratios of friction angle (δ/ϕ) for sand-outside surface of tread were more than 1.06. But the ratios of friction angle (δ/ϕ) for sand-inside surface of tread were 0.93, 0.94 for respective relative densities. The different friction can be explained from the roughness of the surfaces. Outside surface is softer than that of inside because of the material composition. Furthermore outside tread has groove that contributes additional shear resistance except for the penetration of soil grain into the soft surface. The inside of tread has smoother and harder surface than the outside.

Table 5. Test results for various conditions of two soils.

Soil	Preparation of sample		Type of sample	Strength parameter		
				Friction angle	Cohesion (c, kN/m^2)	Friction angle ratio (δ/ϕ)
Sand	Relative density	55%	Sand-sand(ϕ)	39.8°	13.67	–
			Sand-outside surface of tread (δ_{out})	42.8°	14.75	1.08
			Sand-inside surface of tread (δ_{in})	37.2°	22.99	0.93
		70%	Sand-sand(ϕ)	42.1°	16.71	–
			Sand-outside surface of tread (δ_{out})	44.5°	20.04	1.06
			Sand-inside surface of tread (δ_{in})	39.4°	18.96	0.94
Weathered soil	Relative compaction	90%	Soil-soil(ϕ)	34.7°	39.70	–
			Soil-outside surface of tread (δ_{out})	33.9°	13.20	0.98
			Soil-inside surface of tread (δ_{in})	31.9°	18.38	0.92

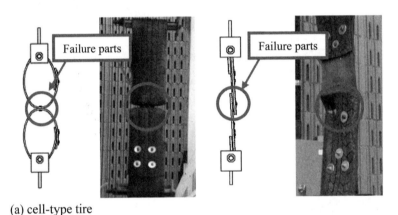

(a) cell-type tire

Figure 7. Tensile test for connection joint of small cell-type tire units and treads.

For the weathered soil which contains smaller-size fraction than the sand, the shear angle of friction was lower than the sand. For the degree of relative compaction of 90%, shear angles of friction were 34.7° for soil-soil, 33.9° for soil-outside surface of tread and 31.9° for soil-inside surface of tread. It means that the ratios of friction angle (δ/ϕ) for soil-outside and soil-inside surface of tread were 0.98, 0.92, respectively. The weathered soil contains silt and clay and the fines makes lower friction angle than the sand. Therefore the ratios of friction angle (δ/ϕ) in weathered soil for outside and inside surface of tread are similar to each other. Considering the δ/ϕ ratio for concrete is approximately 2/3 or so, these ratios are generally very high though. According to Koerner(2005), the ratios of friction angle (δ/ϕ) for geogrid of various type and dense sand of well graded angular sand(SW) were observed 0.72–1.07 in the large shear box(450 × 450 mm).

3.3 Tensile test of connection joint of tires

3.3.1 Test program

In order to utilize the waste tires as soil reinforcement material, tires were treated to make the cell-type tire and tread-mat tire. Because the tensile strengths of cell-type tire and tread-mat are mainly governed by the strength of the connection joint rather than the strength of the tread itself, tensile tests for the connection joint are essential. For the tensile strength tests, UTM equipment was used. Because the deformation of cell-type tire is large, as shown in Figure 7, a small cell-type tire suitable to the machine was made.

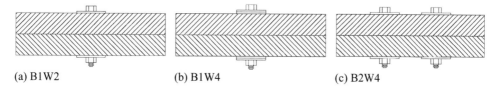

(a) B1W2   (b) B1W4   (c) B2W4

Figure 8.   Connection method with bolts.

Table 6.   The list of tensile tests for cell-type tire units and treads.

| | | Connection method | | |
Form of tire	Connection material	Number of bolts	Number of washers	Inside-outside diameter of washer (mm)
Cell-type tire unit	High strength bolt, Stainless bolt Plastic bolt	1–2	0–4	8–17–10–40
	PP rope	1–2 wraps		
Tread	High strength bolt Plastic bolt	1–2	0–4	8–30, 8–40

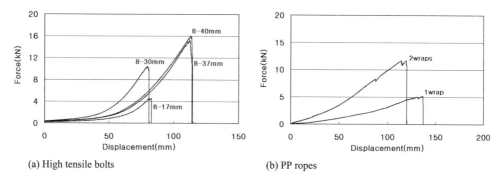

(a) High tensile bolts   (b) PP ropes

Figure 9.   Tensile test results for connection joint of cell-type tire units.

As shown in Figure 8 and Table 6, by changing the connection materials, the number of bolts and washers, inside-outside diameter of washer, the effect of the connection material and method for tensile strength was studied. In Figure 8, B1W2 denotes 1 bolt and 2 washers, B1W2 imply 1 bolt and 4 washers, B2W4 means 2 bolts and 4 washers for the connection joint. And in case of polypropylene rope(PP rope) was tested with wrap times.

3.3.2   *Test results*
3.3.2.1   Test results for the connection joint of cell-type tire units

Figure 9 shows tensile behavior for the connection material of cell-type tire units and the results are given in Table 7. The shape of the cell-type tire units changes from oval cylinder type to plate type as the tensile force increases. Therefore, although tension force is small, the large initial displacement develops. After some nonlinear deformation, the displacement curve becomes linear due to the stiffness of the wires inside the tread. When the cell-type tire units developed ultimate tensile force, the wires in the tread around the bolts started cutting off. As a result, bolts and washers were out of the positions and cell-type tire units were broken. In case of PP rope is similar

Table 7. Tensile strength of cell-type tire units connection.

The quality of bolt material	Connection method			
	The number of bolts	The number of washers	The inside-outside diameter of a washer (mm)	Tensile strength (kN)
High strength bolt	1	2	8–17	4.57
	1	2	8–30	10.47
	1	2	8–37	14.80
	1	2	8–40	15.99
	1	4	8–40	15.51
	2	4	8–40	31.64
	1	2	10–40	17.92
Stainless bolt	1	2	10–40	13.37
Plastic bolt	1	–	–	0.62
	2	–	–	2.01
PP rope	1 wrap			5.19
	2 wraps			11.75

to the behavior of cell-type tire unit connected bolts, too. According to the results of the tensile test for the high strength bolts, stainless bolts, plastic bolts and PP rope, although tensile behavior is similar to each other, different tensile strengths appeared. The high strength bolt showed the largest tensile strength value among connection materials. It is clear that tensile strength for connection joint of the plastic bolt is weaker than that of the wires in the tread. Therefore, the plastic bolts were broken before failure of the wires in the tread. The tensile strengths for PP ropes were 5.87 kN and 11.75 kN when wrapped once and twice, respectively.

As the outside diameter of washer increased from 17 mm to 40 mm, tensile strength increased proportionally from 4.57 kN to 15.99 kN. However, comparing B1W2 and B1W4 which used 8–40 mm washer, the number of washers has no effect on the strength of connection joint. If the hole is larger than outside diameter of washer after failure of wire inside the tread, in spite of increasing the number of washer, bolt and washer don't resist pull-out force. However, from the comparison of B1W2 and B2W4, tensile strength increased approximately 2 times as the number of bolts increased from 1 to 2. It is due to the load distribution to the wider area of washer by increasing outside diameter of washer or the number of bolts.

3.3.2.2 Test results for the connection joint of treads
Figure 10 and Table 8 show test results for the connection joint of the treads. Figure 10 shows tensile behavior for the connection material of tread. Because the shape of the tread is plain, the displacement curve varied linearly from the beginning of the tests. However at around the ultimate tensile force, the graph shows saw-type failure and that came from the gradual failure of the wires inside the tread. From the results, two types for failure pattern of the connection joint of treads were observed. One type is that as the load increases the wires in the tread and the rubber start tearing toward the direction of the tension and bolts and washers pulled out toward the end of the tread afterwards. The other type is that bolts and washers were pulled out from the hole of the tread which was torn toward the direction of the tension. The tensile strength of the tread with a high strength bolt was higher than that of plastic bolt as cell-type tire unit.

As the outside diameter of washer and the number of bolts increased like cell-type tire unit, tensile strength increased, too. This is because the outside diameter of washer and the number of bolts increase, bolts and washers don't pulled out from the tearing hole of the tread until the ultimate tensile strength of the tread is developed. It was observed that the strength of the tread

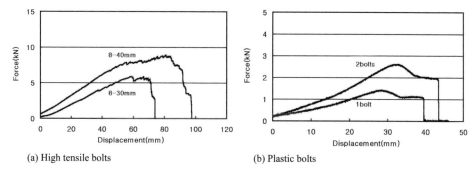

(a) High tensile bolts   (b) Plastic bolts

Figure 10.   Tensile test for treads.

Table 8.   Tensile strength of the treads connection.

| | Connection method | | | |
Type of Bolts	Number of bolts	Number of washers	Inside-outside diameter of washer (mm)	Tensile strength (kN)
High strength bolt	1	2	8–30	5.90
	1	2	8–40	8.94
	2	4	8–40	15.76
Plastic bolt	1	–	–	1.45
	2	–	–	2.63

connection joint was 0.5–0.56 times smaller than that of cell-type tire, because the direction of tension is similar to the array of the wires in the tread. According to Koerner(2005), average single rib strength of various types of geogrid were 0.61–8.90 kN, for average junction strength indicated 0.18–2.68 kN.

3.4   Plate load tests

As shown Figure 11, in this section plate load tests for both tire mat made by treads and sidewalls and cell-type tire were presented. The chamber was filled with sands of various densities and reinforcement was done by tire mat and cell-type tire for various densities.

Three sizes of tire mat and cell-type tire were used to study boundary influence by mat width. Tires used here were for small-size light truck and treads and sidewalls and cell-type tire units were combined with plastic bolts.

The mat may be used either as an underlying material over unstable ground surfaces, or as reinforcing elements between soil layers for embankment or for soil reinforcement of foundations.

3.4.1   Experimental program

The plate load tests were carried out in a test chamber of 2.0 m width, 2.0 m length, and 1.5 m height. The chamber was placed under a loading frame and was filled with sand up to the depth of 150 cm. A 350 mm load plate was used and oil jacks, dial gages, and a measuring system were involved for the test (Figure 12). The chamber has rigid boundaries, and the size was determined through the finite element analysis in order to check boundary influence. No boundary influence was confirmed from the pilot plate load tests preceding the main tests.

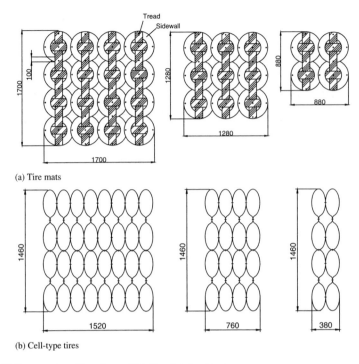

(a) Tire mats

(b) Cell-type tires

Figure 11.   Different ratios of length to width for plate load tests.

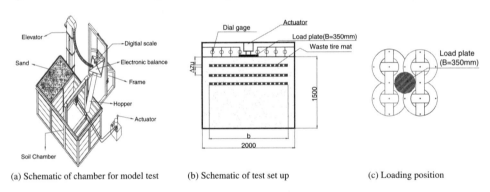

(a) Schematic of chamber for model test   (b) Schematic of test set up   (c) Loading position

Figure 12.   Schematic for plate load test.

For comparison of test results, the bearing capacity ratio (BCR) and settlement reduction factor (SRF) were used as previously described by Guido & Christou(1988) and others.

$$BCR = q_r / q_o \qquad (1)$$

$$SRF = (s/B)_r / (s/B)_o \qquad (2)$$

where q_r and q_o are the ultimate bearing capacities for the reinforced and unreinforced sands, respectively and s is settlement at the corresponding ultimate bearing capacity.

Test programs for tire mat and cell-type tire are given in Table 9. Five parameters in the bearing capacity and settlement of tire-reinforced sand were investigated. The parameters include relative density, Dr, of the sand, number of layers of tire-reinforcement, N, interval depth between reinforced layers, Δz, expressed in non-dimensional form of $\Delta z/B$ where B is the diameter of the plate, depth

Table 9. Scheme of plate load tests.

Test series	Dr (%)	No. of tests	Remarks
Unreinforced	40	1	
	55	1	
	70	1	
Relative density (Dr)	40	2	– tested for tire mat, cell-type tire with various Dr
	55	2	– N = 1, u/B = 0.4, b/B = 4.86 for tire mat
	70	2	– N = 1, u/B = 0.2, b/B = 4.17 for cell-type tire
Number of reinforced layers (N)	40	3	– tested for only tire mat
	55	3	– N = 1, 2, 3
	70	3	– u/B = 0.4, Δz/B = 0.4, b/B = 4.86
Interval of reinforced layer (Δz/B)	55	8	– tested for only cell-type tire – N = 2, 3 and Δz/B = 0.2, 0.3, 0.4, 0.5 – u/B = 0.2 and b/B = 4.17
Embedded depth (u/B)	55	8	– N = 1 – u/B = 0.2, 0.4, 0.8 and b/B = 4.86 for tire mat – u/B = 0.2, 0.3, 0.4, 0.5, 1.0 and b/B = 4.17 for cell-type tire
Mat size (b/B)	55	6	– N = 1 – b/B = 2.50, 3.66, 4.86 and u/B = 0.4 for tire mat – b/B = 1.09, 2.17, 4.17 and u/B = 0.2 for cell-type tire

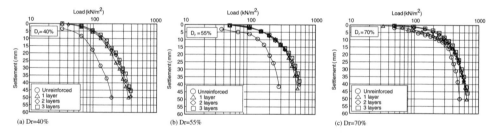

Figure 13.   Load-settlement curves for tire mat reinforced sands (u/B = 0.4, Δz/B = 0.4, b/B = 4.86).

below the bearing plate to the top of the first layer of tire-reinforcement, u, in non-dimensional form of u/D, and the width, b, of a square area of tire mat reinforcement in non-dimensional form of b/B, the width ratio.

Sand compaction was tamping technique applying undercompaction concept (Ladd, 1978). This method has the advantage to obtain almost same density throughout the layers consistently. The relative densities of 40, 55 and 70% of sand were prepared to investigate the effect of soil improvement by tire mat. For more uniform density, the wooden frame divided into 9 cells was used. The height was divided into 5 levels. The weight of sand in each section was measured and poured carefully into every square cell of a wooden frame. The frame was removed, and tamping was conducted with a tamper with 25 cm by 25 cm size of a square plate.

3.4.2   Results and discussion for tire mat
3.4.2.1   Bearing capacity with relative density
Figures 13 shows typical load-settlement curves for unreinforced and tire mat reinforced sands of relative density 40, 55 and 70%, respectively. It can be observed that there were no pronounced ultimate points in the load-settlement curves of reinforced sands and that the curve patterns of

Table 10.   Ultimate bearing capacities of tire mat reinforced sands (kN/m²).

Dr (%)	Unreinforced	1 layer	2 layers	3 layers
40	132.4	282.0	297.1	308.9
55	230.5	472.5	509.9	519.8
70	426.6	511.9	529.6	563.9

Table 11.   BCR and SRF for different densities and number of layers of tire mat.

	1 layer		2 layers		3 layers	
Dr (%)	BCR	SRF	BCR	SRF	BCR	SRF
40	2.13	0.31	2.24	0.30	2.33	0.26
55	2.05	0.40	2.21	0.34	2.26	0.29
70	1.20	0.66	1.24	0.63	1.32	0.57

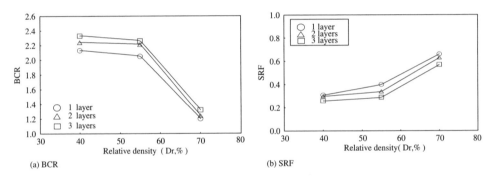

Figure 14.   Variation of BCR and SRF with relative density.

reinforced sand were quite similar to un-reinforced sands except for higher bearing capacities for reinforced soils.

Plate load test results are expressed in terms of BCR and SRF for different densities. The results are listed in Tables 10 and 11, and shown in Figures 14, respectively. The bearing capacity increases steadily from 132.4 kPa to 426.6 kPa for the unreinforced sand, and the bearing capacity increases from 282.0 kPa to 511.9 kPa for the single layer tire mat reinforced sand. It is evident that the percent increase in bearing capacity is substantially greater for the unreinforced sand than for the tire mat reinforced sand.

In order to show more clearly the increase of bearing capacity and the decrease of settlement by inclusion of tire mat, BCR and SRF are used in the following Figures 14. From Figure 14 (a) it is clear that BCR is higher at lower density than at higher density irrespective of the number of reinforced layers. There is a slight difference between 40% and 55% relative densities. For single layer reinforcement the bearing capacity increases about 113% (BCR = 2.13) at relative density 40%, whereas there is only 20% (BCR = 1.2) increase at relative density of 70%. Figure 14 (b) shows the variation of SRF with different densities. For single layer reinforcement, settlement reduces 69% (SRF = 0.31) at relative density 40%, whereas there is 34% (SRF = 0.66) reduction at 70% relative density. Reinforcement of a layer reduces the settlement remarkably at loose soil than at dense soil.

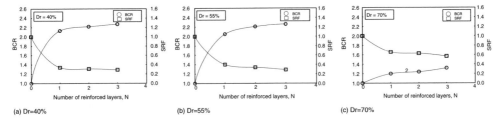

(a) Dr=40%   (b) Dr=55%   (c) Dr=70%

Figure 15.   Variation of BCR and SRF with the number of reinforced layers.

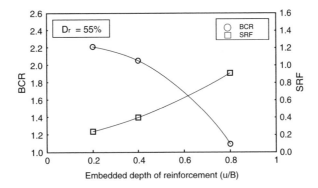

Figure 16.   Variation of BCR and SRF with embedded depth.

Table 12.   BCR and SRF with different embedded depths of tire mat reinforced layers.

Dr (%)	u/B = 0.2		u/B = 0.4		u/B = 0.8	
	BCR	SRF	BCR	SRF	BCR	SRF
55	2.21	0.34	2.05	0.29	1.06	0.91

3.4.2.2   Number of reinforced layers

Figures 15 shows the variation of the BCR and SRF of relative densities 40, 55 and 70% respectively, as the number of tire mat layers varies. The results are listed in Table 11. From Figures 15, it can be observed that the increases in the bearing capacity and decreases in the settlement due to additional layers of reinforcement begin to converge at around the 2nd layer and almost constant at the 3rd layer of reinforcement. The depth of the 3rd layer corresponds to z/B = 1.23.

3.4.2.3   Embedded depth of the first reinforced layer

Figure 16 and Table 12 show the variation of the BCR and SRF with the embedded depth to the top of the first reinforced layer, u/B, respectively. As the depth of the reinforced layer of tire mat decreases BCR increases and the SRF decreases. It indicates that the bearing capacity increases while the settlement decreases.

3.4.2.4   Width size of tire mat

Different sizes of mats were made as shown in Figure 11 in order to know the maximum width ratio where width ratio is defined by the ratio of mat width(b) to width of load plate(B). Figure 17 and Table 13 show the variation of BCR and SRF with tire mat width b/B, respectively. As b/B increases, the BCR increases and the SRF decreases, reaching a maximum value of BCR and a minimum value of SRF at b/B value of approximately 5.0.

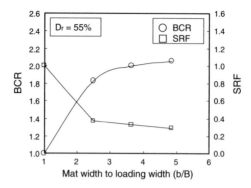

Figure 17. Variation of BCR and SRF for different tire mat widths.

Table 13. BCR and SRF for different tire mat widths.

	2 × 2 (b/B = 2.05)		3 × 3 (b/B = 3.66)		4 × 4 (b/B = 4.86)	
Dr(%)	BCR	SRF	BCR	SRF	BCR	SRF
55	1.83	0.37	2.00	0.33	2.05	0.29

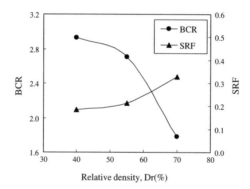

Figure 18. Variation of BCR and SRF with relative density.

3.4.3 Results and discussion for cell-type tire

The reinforcement of both tire mat and cell-type tire shows similar trend of the increase of bearing capacities and the decrease of settlements due to the parameters mentioned at previous section.

3.4.3.1 Bearing capacity with relative density

The results of cell-type tire with different relative densities are shown in Figure 18 and Table 14, respectively. The BCR decreases from 2.93 at Dr = 40% to 1.78 at Dr = 70%, although the bearing capacity increases with the relative density of a single layer of cell-type tire reinforced sand. It indicates that the reinforcement of well-compacted ground in order to increase of bearing capacity is ineffective.

For single layer reinforcement, settlement reduces by as much as 81% (SRF = 0.19) at relative density 40%, whereas it reduces by as much as 67%(SRF = 0.33) at 70% relative density. Reinforcement of one layer reduces the settlement remarkably at loose sand than at dense sand. Settlement reduction effect is higher at lower densities. At higher density, the reinforcement layer

Table 14. BCR and SRF for different densities.

Dr (%)	q_r (kN/m^2)	BCR	SRF
40	402.09	2.93	0.19
55	637.46	2.71	0.21
70	784.56	1.78	0.33

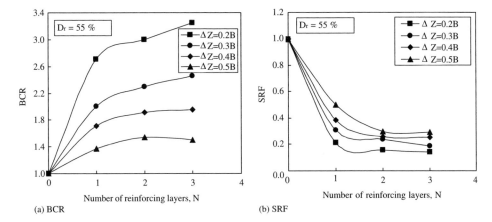

(a) BCR   (b) SRF

Figure 19. BCR and SRF with the number of reinforced layer and interval of reinforcement.

Table 15. BCR and SRF with the number of reinforced layer and interval of reinforcement.

N	$\Delta z/B$	q_r (kN/m^2)	BCR	SRF
1	0.2	637.46	2.71	0.21
	0.3	470.74	2.00	0.31
	0.4	402.09	1.71	0.39
	0.5	323.63	1.38	0.50
2	0.2	706.10	3.00	0.16
	0.3	539.39	2.29	0.24
	0.4	451.12	1.92	0.26
	0.5	362.86	1.54	0.30
3	0.2	764.95	3.25	0.14
	0.3	578.61	2.46	0.19
	0.4	460.93	1.96	0.25
	0.5	353.05	1.50	0.29

has less effect on settlement reduction because soil stiffness itself is enough to support the imposed load. The inclusion of waste tires seems to increase the stiffness of the sand remarkably. Therefore, as in the case of commercial geosynthetics such as geogrid, geotextile and geocell, benefits can also be obtained from the inclusion of tire mat and cell-type tire for lower densities.

3.4.3.2 Number of reinforced layers and interval of reinforcement
Figure 19 shows the variation of the BCR and SRF with the number of layers and interval of reinforcement at relative density of 55%. The results are listed in Table 15, and $\Delta z/B$ is equal to u/B at N = 1. The bearing capacity increased substantially and the settlement decreased substantially

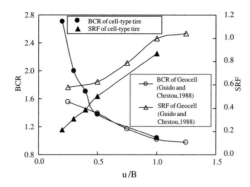

Figure 20.   Variation of BCR and SRF with embedded depth.

Table 16.   BCR and SRF with different embedded depths of reinforced layers.

u/B	q_r (kN/m^2)	BCR	SRF
0.2	637.46	2.71	0.21
0.3	470.69	2.00	0.31
0.4	402.09	1.71	0.39
0.5	323.63	1.38	0.50
1.0	245.15	1.04	0.87

because of the simple addition of a layer of cell-type tire. Figure 19 also shows that the increases in the bearing capacity and decreases in the settlement due to additional layers of cell-type tire reinforcement and the BCRs begin to converge at a certain layer depending on the interval of reinforcement. When more than 2 layers of cell-type tire were used, the upper and the lower layers were lain perpendicularly each other. At 0.2B interval, the BCR increases almost up to the 4th layer of reinforcement, whereas it converges at the 2nd layer of reinforcement at the 0.5B interval. This result does not much differ from the effective depth of z/B = 1 observed by Guido et al.(1985), Guido & Christou(1988) and Omar et al.(1993). This convergence meant that the reinforcement was effective within the depth of the stress influence zone and that the bearing capacity can be expected to increase within the depth. However the increase rate of BCR and the decrease rate of SRF by of the addition of the reinforcement layer decreased steadily until they converged to a certain value. It shows that the efficiency decreases with the increase of the number of reinforced layers. Therefore, to obtain effective reinforcement by cell-type tire and tire mat, one layer of reinforcement is recommended.

3.4.3.3   Embedded depth of the top layer of reinforcement

Figure 20 shows the variation of the BCR and SRF with the embedded depth of the cell-type tire reinforcement, u/B and the test results are listed in Table 16. As the depth of the reinforced layer decreases, BCR increases and SRF decreases. These phenomena indicate that the bearing capacity increases while the settlement decreases. Guido et al.(1985), Guido & Christou(1988) and Omar et al.(1993) gave similar results. At u/B of about 1.0, the reinforced layers had no influence on the improvement of bearing capacity. In addition, the settlements associated with this u/B value were essentially those of unreinforced sand. This result implies that a high concentration of cell-type tire reinforcement within this depth of sand sufficiently can reinforce the sand to produce higher bearing capacities and lower settlements. However, when u/B is large enough, the cell-type tire reinforcement does not interfere with the formation of the failure planes, and, in turn, shear failure of the sand occurs above the top of the uppermost layer of the cell-type tire reinforcement.

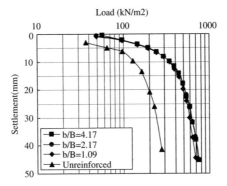

Figure 21.   Variation of load-bearing capacities with cell widths.

Table 17.   BCR and SRF for different widths.

b/B	q_r (kN/m^2)	BCR	SRF
1.09	598.2	2.54	0.22
2.17	637.5	2.71	0.21
4.17	637.5	2.71	0.21

Figure 20 also shows the variation of BCR and SRF for both geocell(h = 230 mm) and cell-type tire(h = 135 mm). Test results from geocell by Guido and Christou(1988) were used for comparison. They performed plate load tests on the geocell-reinforced sands in a chamber of 48in. × 48in. × 36in. The sand was poorly graded with Cu = 1.90 and Cc = 1.23 whereas the sand in this research is poorly graded with Cu = 2.81 and Cc = 1.0. The test condition of geocell was N = 2, Δz/B = 0.25, b/B = 2, Dr = 55% and cell-type tire was N = 1, b/B = 4.17, Dr = 55%. Main differences between test conditions were the number of reinforced layers, N, and the ratio of cell-type tire width to plate width, b/B. However the widths of cell-type tire could be considered the same because both widths were beyond the boundary effect range. The variation of bearing capacities with b/B is discussed at the following section. The comparison of the test results showed the reinforcing effects by cell-type tire and was meaningful because both BCR and SRF are normalized values.

As shown in Figure 20, the BCR of cell-type tire are higher than those of geocell, especially, at lower normalized embedded depth(u/B) in spite of 1 layer reinforcement. The higher BCR of cell type tire at shallow depth is probably due to the good condition of the tire treads. The settlement reductions are higher(i.e. low SRF values) at shallow depth of the embedment and the degree of settlement reduction is higher with cell-type tire reinforcement than with geocell. Therefore from the view point of bearing capacities, cell-type tire might be useful as recycling material for soil reinforcement if there is no environment problem by inclusion of tires into soil. The BCR and SRF became similar with increasing embedded depth, thus, decreasing reinforcement effect.

According to Seo(2003) the condition of waste tire, i.e. the thickness of the tread, affects the bearing capacities. The stiffness of waste tire change with the degree of wearing down of the tire treads. The variation of stiffness with the degree of wearing can be studied further in the future.

3.4.3.4   Width of cell-type tire

A series of plate load tests for different sizes of cell-type tire were conducted to determine the maximum width ratio that would have no influence the size of the cell-type tire, where width ratio is defined by the ratio of the cell-type tire width(b) to load plate width(B). Figure 21 and Table 17 show the variation of BCR and SRF with cell-type tire width ratio, b/B. As b/B increased, the BCR increased until it reached the maximum value and the SRF decreased until it reached the minimum value, both at b/B value of more than about 2.0. This result was different from those

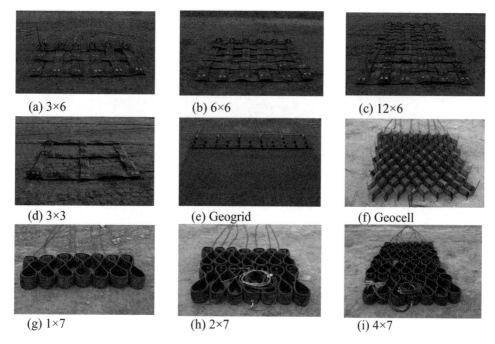

(a) 3×6	(b) 6×6	(c) 12×6
(d) 3×3	(e) Geogrid	(f) Geocell
(g) 1×7	(h) 2×7	(i) 4×7

Figure 22.   Plain tread-mat, geogrid and cell-type tire for pull-out tests.

at b/B = 2.5 of Guido & Christou(1988) in case of geocell but similar to b/B ≈ 6 of Fragaszy and Lawton(1984) in case of reinforcing strips. High BCR at smaller b/B ratio may come from the stiffness of the cell-type tire. The higher the stiffness of reinforcement material, the smaller the size of the reinforcement is necessary. The result indicated that width ratio of 4.17 in cell-type tire, which is the natural result of connection of small truck tires, does not affect the size of the cell-type tire because the width ratio was larger than the minimum value. This result indicates that a mat width 5.0 times larger than the width of the footing can be ineffective. Also the direction of the cell-type tire shows almost no difference. The reason is probably due to the cover thickness of soil.

4   FIELD PULL-OUT TESTS

4.1   *Experimental program*

4.1.1   *Plain tread-mat and cell-type tire*
In order to know pull-out characteristics with reinforcement material field tests were carried out for various configurations of tread-mat and cell-type tire as shown Figure 22. For the same field condition commercial geogrid and geocell was also tested. The notation 3 × 6 means 3 lines are perpendicular tread to pull-out direction and 6 lines are parallel tread to the pull-out direction. For consider effect by reinforcement length, 3 × 6, 6 × 6, 12 × 6 in case of tread-mat and 1 × 7, 2 × 7, 4 × 7 in case of cell-type tire were prepared for pull-out tests. And 3 × 3, 4 × 4 of tread-mat was tested to evaluated effect by net area between tire and soil. The size of geogrid and geocell were same to 6 × 6 and 2 × 7, respectively.

4.1.2   *Construction of test embankment for pull-out tests*
The index and strength properties of weathered soil in field tests, sieve analysis results are given at the previous section. Maximum dry density was 18.42 kN/m^3 at an optimum moisture content of 10.7% by modified Proctor method (KD F 2321 D). The soil contains about 30% fines with mostly

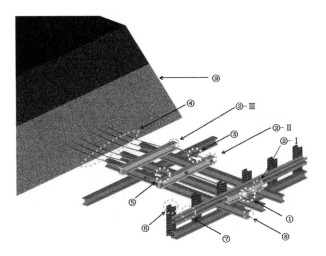

Figure 23. Schematic of pull-out tests.

Table 18. Principal component of pull-out tests.

No. in Figure 23	Remarks
①	Oil Jack (Cap.:1570 kN, Stroke: 200 mm)
②	Cross beam (H beam − 300 × 300 × 12 × 10)
③	Load cell (Cap.: 196 kN, Standard error: 0.098 kN)
④	Strand (Cap.: 78 kN/ea.)
⑤	LVDT, (Sens.:0.01 mm, Stroke: 200 mm)
⑥	Reaction pile (H beam − 300 × 300 × 12 × 10)
⑦	Channel
⑧	Supplementary beam(H − 300 × 300 × 12 × 10)
⑨	Test embankment

sandy soil. Direct shear test results were performed on the soil samples compacted at optimum moisture content of modified proctor compaction method are given in Table 3. For pull-out tests embankment was designed to use conventional compaction techniques. 0.5 m height of embankment was completed before setting tire-made reinforcement material and 1.5 m of embankment surcharge was placed after careful array of tread-mat, cell-type tire geogrid and geocell. A relative compaction of 90% or more on the basis of a laboratory test commenced with the leveling of 300 mm thick soil using 98 kN vibratory roller.

4.2 *Pull-out test*

Schematic and principal components for pull-out tests are shown in Figure 23 and Table 18, respectively. The pull-out force is transmitted form oil jack(①) to cross beam(②-II,III), strands(④) and reinforcement material embedded in test embankment. The reaction piles(⑥) of 10 m length penetrated to 8.5 m depth by driving are resist to pull-out force. For distribution of pull-out forces of reaction piles channel(⑦) was used. To ensure an equal transfer of pull-out force to each frontal tire element, load cell(③) and LVDT(⑤) system were installed on two side between cross beam(②-II,III). Supplementary beam were used to control the height of the equipment to the level of reinforcement material. And rigid rods (10 mm dia., 300 mm length) were put on to reduce the friction between cross beam and supplementary beam.

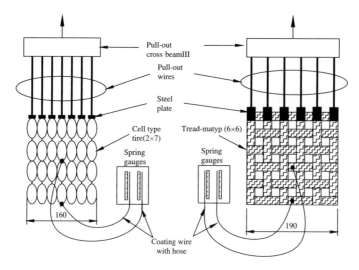

Figure 24.   The tire pull-out assembly, rear displacement gauges.

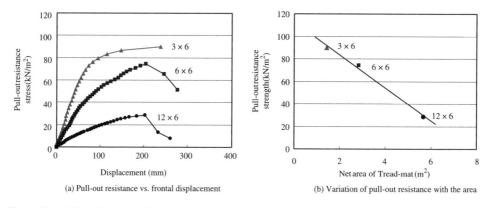

(a) Pull-out resistance vs. frontal displacement

(b) Variation of pull-out resistance with the area

Figure 25.   Pull-out behavior of tread-mat.

The pull-out force was measured every 4.9 kN by load cell with a capacity 196 kN and a standard error of 0.098 kN. The test was performed until 15% displacement of length of reinforcement materials except for the case of clear peak value before 15%. To measure rear displacement and elongation of the tire assembly a stiff steel wires of 3 mm diameter covered by polyurethane hose were attached to some points of the tire while the free end was connected to dial gage as shown Figure 24.

4.3   *Test results and discussion*

4.3.1   *Pull-out resistance with the length of reinforcement*

Figure 25 (a) shows the pull-out resistance with frontal displacement and the pull-out resistance is represented pull-out force to the net contact area of tread-mat. Frontal displacement mean displacement of cross beam. In Figure 25 (a) 3 × 6 tread-mat does not show peak value but reached at almost peak value. In that case 15% displacement was considered as failure. At 6 × 6 tread-mat end of tread-mat displaced at peak resistance whereas 12 × 6 tread-mat did not shows the movement of at end of the mat due to failure in front part of parallel treads to pull-out force. Figure 25 (b)

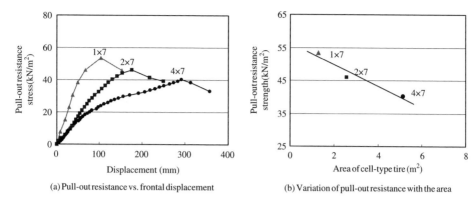

(a) Pull-out resistance vs. frontal displacement   (b) Variation of pull-out resistance with the area

Figure 26.   Pull-out behavior of cell-type tire.

Table 19.   Characteristics of geogrid.

Materials	Weight (g/m²)	Tensile strength (kN/m)		Tensile elongation (%)	
		MD	CD	MD	CD
PET/PVC	470	101.0	29.4	14	N.A

Table 20.   Characteristics of geocell.

Materials	Weight (g/m²)	Dimension(mm)			
		Width	Length	Height	Thickness
HDPE	1430	140	160	13.5	1.25 ± 0.64

showss the pull-out resistance for net contact area of tread- mat. The pull-out forces are transferred from the front part of the reinforcement to the rear part of the mat.

Figures 26 shows the pull-out resistance for cell-type tire. In case of cell-type tire every tests occurred peak resistance. From Figure 26 (b) it can also be observed that the resistance decreased with the increase of area of the cell-type tire. Area of the cell-type tire is not net contact area of tire but is plane area by width time length.

4.3.2 *Pull-out resistance of geogrid and geocell*

The characteristics of geogrid made by Samyang are given in Table 19. MD means machine direction and CD represents cross direction. Table 20 shows the characteristics of geocell made by Presto. Figure 27 shows the pull-out test results of geogrid and geocell to compare with the tread-mat 6 × 6 and cell-type tire 2 × 7. In Figure 27 pull-out loads are represented for the unit width of the reinforcement. The curves show clear peaks for every reinforcement materials. From the results tread-mat 6 × 6 shows approximately 2 times higher than pull-out load of geogrid and in case of cell-type tire 2 × 7 is approximately 1.3 times of geocell. After the tests it can be seen that geogrid and geocell was failed material at front parts connected with pull-out strands whereas tread-mat 6 × 6 and cell-type tire 2 × 7 was not failed treads.

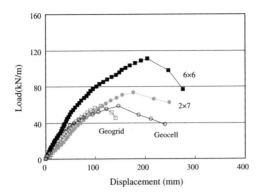

Figure 27. Comparison of behavior for commercial reinforcement material.

5 SUMMARY

This paper discussed a series of tests to utilize waste tire as reinforcement materials and compare to geosynthetics such as geogrid and geocell. From the test results the following can be summarized.

(1) The utilization of waste tires as soil-reinforcing material proved to be excellent. Therefore, the treatment method as form of tire mat, tread-mat and cell-type tire could be useful not only as reinforcing material but also as a recycling of waste material.
(2) The ratios of friction angle(δ/ϕ) for sand-outside surface of tread were 1.06–1.08 and 0.93–0.94 for the inside surface of tread. For the case of weathered soil the values were 0.98 for outside and 0.92 for inside surfaces of tread.
(3) The strength of connection joints of cell-type tire unit increased with the strength of bolts, the number of bolts and sizes of washers. The strength of cell-type tire connection joint was higher than that of tread at same connection condition. In order to simplified the connection of tires further study should be considered.
(4) The effect of bearing capacity increase and settlement decrease in the sand reinforced by waste tire showed more excellent than the commercial geosynthetics. The result may comes form the stiffness of tread.
(5) Increasing of the BCR or decreasing of the SRF was largely related with the increase of relative density, the number of reinforcing layers, the reinforcement width ratio, or the decrease of reinforcement interval, embedded depth of reinforcement layer.
(6) The pull-out load to unit width of tread-mat and cell-type tire was higher than that of geogrid and geocell, respectively.

ACKNOWLEDGEMENTS

The author wish to express his sincere thanks to Mr. Kyoung-soon Choi, Young-il Cha and other former graduate students and Mr. Keun-soo Kim, Sung-soo Cho, Seung-won Lee and other graduate students for their contributions.

REFERENCES

Ecoflex, 2006. Scour protection mat & head and wing walls. Shoalhaven. NSW. http://www.ecoflex.com.au/erosiongallery.
Foose, G.J., Benson, C.H. & Boscher P.J. 1996. Sand reinforced with shredded waste tire. *Journal of Geotechnical Engineering. A*SCE. Vol. 122, 760–767.

Fragaszy, R.J. & Lawton, E. 1984. Bearing capacity of reinforced sand subgrades. *Journal of Geotechnical Engineering*. ASCE. Vol. 110. No. 10, 1500–1587.

Garga, V.K. & O'Shaughnessy, V. 2000. Tire-reinforced earth fill. Part1: Construction of a test fill, performance and retaining wall design. *Canadian Geotechnical Journal*. Vol. 37, 75–96.

Gudio, V.A. & Christou, S.N. 1988. Bearing capacity and settlement characteristics of Geoweb-reinforced earth slabs. *Proc. ASCE 1988 Spring Meeting*. Nashville. TN, 21–36.

Gudio, V.A., Bisiadecki, G.L. & Sullivan, M.J. 1985. Bearing capacity of a geotextile-reinforced foundation. *Proc. Of the 11th International Conference on Soil Mechanics and Foundation Engineering*. Vol. 3. San Francisco. CA. 1777–1780.

Humphrey, D.N. & Katz, L.E. 2000. Water quality effects of tire shreds placed above the water table – five-year field study. Transportation Research Record No. 1714. Transportation Research Board. Washington. D.C. 18–24.

Humphrey, D.N. & Katz, L.E. 2002. Water quality effects if using tire shreds below the ground water table. *Final Repor., Dept. of Civil and Environmental Engineering. University of Main*. Orono. ME.

Humphrey, D.N. & Manion, W.P. 1992. Properties of tire chips for light weight fill. Grouting, Soil Improvement and Geosynthetics. *Geotechnical Special Publication*. No. 30. Vol. 2. ASCE. New York. N.Y. 1345–1355.

Humphrey, D.N., Whetten, N,. Weaver, J., Recker, K. & Cosgrove, T.A. 1998. Tire shreds as lightweight fill for embankments and retaining walls. Recycled Materials in Geotechnical Applications. Geotechnical Special Publications. No. 79, 51–65.

Koerner, R.M. 2005. Designing with geosynthetics. 5th ed. Pearson Education Inc.

Ladd, R.S. 1978. Preparing test specimens using undercompaction. *Geotechnical Testing Journal*. ASTM. Vol. 1. No. 1, 16–23.

Moon, C.M. 2003. Environmental effect of waste tires as earth reinforcing material. *Master thesis*. Inha university(in Korean).

Nguyen T.H. 1996. Utilization of used tyres in civil engineering – The Pneusol 'Tyresoil'. *Proceedings of the 2nd International Congress on Environmental Geotechnics*. Rotterdam. Netherlands. 809–814.

Omar, M.T., Das, B.M., Yen, S.C., Puri, V.K. & Cook, E.E. 1993. Ultimate bearing capacity of rectangular foundations on geogrid-reinforced sand. *Geotechnical Testing Journal*. ASTM. Vol. 16. No. 2, 246–252.

O'Shaughnessy V. & Garga V.K. 2000. Tire-reinforced earth fill. Part3: Environmental assessment. *Canadian Geotechnical Journal*. Vol. 37, 117–131.

Reid, R.A., Soupir, S.P. & Schaefer, V.R. 1998. Mitigation of void development under bridge approach slabs using rubber tire chips. Recycled Materials in Geotechnical Applications. Geotechnical Special Publications. No. 79, 37–50.

Rma, 2004. U.S scrap tire markets 2003 edition. Rubber Manufacturer Association. http://www.rma.org/scrap_tires.

Seo, D.S. 2003. Bearing capacity of Tirecell and geocell in Sand. Master thesis. Inha university (in Korean).

Williams, P.T., Besler, S. & Taylor, D.T. 1990. The pyrolysis of scrap automotive tyres: the influence of temperature and heating rate on product composition. Fuel 69, 1474–1482.

Yoon, Y.W. & Cha, Y.I. 2002. Improvement of bearing capacity of sands by using waste tires. *6th International Symposium on Environmental Technology and Global Sustainable Development*. 615–620.

Yoon Y.W., Cheon S.H. & Kang D.S. 2004. Bearing capacity and settlement of tire-reinforced sands. *Geotextiles and Geomembranes* 22, 439–453.

Yoon Y.W., Heo S.B. & Kim K.S. 2007. Geotechnical performance of waste tires for soil reinforcement from chamber tests. *Geotextiles and Geomembranes* (acceptted).

Yoon, 2006. The manufacturing method of tire mat and construction method of retaining wall using waste tires (in Japanese). Patent pending.

Theme lectures

Scrap Tire Derived Geomaterials – Opportunities and Challenges – Hazarika & Yasuhara (eds)
© 2008 Taylor & Francis Group, London, ISBN 978-0-415-46070-5

Evaluation of environmental impact of tire chips by bioassay

N. Tatarazako
National Institute for Environmental Studies, Tsukuba, Japan

Miwa Katoh
Graduate school, Ibaraki University, Hitachi, Japan

Kiwao Kadokami
University of Kitakyushu, Fukuoka, Japan

ABSTRACT: It is important to reuse materials, but it is necessary, on the other hand, to consider effect to the environment of recycling materials. When we are going to reuse a tire chips as the recycle material, some kind of chemical substances were discharged in environment and were concerned about affecting ecosystem. Therefore we measured the chemical substances eluted from tire chips by GC/MS and evaluated the eluate of the tire chips by bioassay at the same time. Daphnia acute toxicity test and Microtox were used for the bioassay. As a result, we made clear that some chemical substances to elute from tire chips might affect biology.

1 INTRODUCTION

When tire chips are recycled, their effects on the environment should be considered. For example, chemicals eluted to groundwater may affect the organisms in an ecosystem. We eluted samples under conditions that tire chips might be subjected to if they were used in the environment and predicted the effects of the chips on the ecosystem by analysis and bioassay of their chemical components. There are currently no techniques or regulations in Japan for evaluating the effects of recycled materials on the environment. We used simple biological assays based on testing methods specified in standards from the USA, Canada, the EU, and Japan.

2 METHODS

2.1 *Selection of bioassay methods*

We selected two bioassays because of their simplicity of use and relatively high sensitivity.

The Microtox test (Strategic Diagnostics Inc.) evaluates inhibition of light emission by luminous bacteria (Chang etc, 1981),(US EPA,1990),(Barton,1996). The *Daphnia* test assesses acute immobilization of water fleas (OECD, 1996) and is used in OECD countries, the USA, Canada, and Japan (Table 1) (Muna etc, 1995).

The Microtox test uses the luminescent bacterium *Vibrio fischeri* NRRL B-11177 to measure the toxicity of environmental samples. Bacterial bioluminescence is tied directly to cell respiration, and any inhibition of cellular activity by toxicity results in a decreased rate of respiration and a corresponding decrease in luminescence. A toxicity value, EC_{20} (effective concentration that reduces bioluminescence by 20%), is calculated from the reduction of light emitted after exposure to the test solution for 15 min. This test method needs short time for an examination, but, not a limit tst, and it is used for evaluation of the water quality in the world.

In the *Daphnia* test, neonates (less than 24 h old) are exposed to the test chemical and those that cannot move at 48 h later, as judged by the naked eye, are counted. The proportion that cannot move is used to calculate the EC_{20} from the dilution of the test solution that had a 20% effect. Probit method was used for statistics processing (Finney, 1978). TU_{20} (toxic units giving 20% impact) is then calculated from EC_{20} ($TU_{20} = 100/EC_{20}$). A large TU_{20} indicates a high toxicity. The details of examination technique obeyed an OECD test guideline (TG202). This examination is very sensitive and responds for various chemical substances. In addition, the water flea occupies the extremely important position in ecosystem, and the whole ecosystem may be broken when bad effect gets up to this creature. A lot of creatures of 120 are used for one examination, and effect concentration is calculated after it was handled the result statistically. It is admitted that this examination is extremely objective, and plasticity is high in the world. The detail conditions of each bioassay are shown in Table 1.

2.2 *Elution method*

The tire chip eluate was basically made with reference to the 46th Environmental Agency notification in Japan. We examined the eluate quality, the period for elution, and the solid to liquid ratio, and we then selected the optimum conditions tentatively for each bioassay. The details are shown in Table 2. The solid to liquid ratio was selected as 1:5 not 1:10 shown in notification 46 of Ministry of the Environment. A strong sample is better for the initial concentration to do step-by-step dilution for examination liquid in the case of the bioassay.

2.3 *Component analysis of test solutions*

By gas chromatography – mass spectrometry (GC–MS) we analyzed about 760 chemicals. The GC–MS measurement conditions are shown in Table 3.

Table 1. Conditions of each test.

Bioassay	Microtox	*Daphnia* test
Method	Microtox® MAN	OECD TG202
Test biology	*Photobacterium phosphoreum (Vibrio fischeri)*	*Daphnia magna* neonates ($\leq$24 h)
Exposure	Static	Static
Exposure time	15 min	48 h
Number	–	20/dose
Concentration	Control + 4 doses	Control + 5 doses
Observation time	5, 15 min	24, 48 h
Endpoint	Inhibition	Immobilization
Value	EC_{20} (%)	EC_{20} (%)
Feeding	No	No

Table 2. Preparation of tire chip eluent.

	Solution 1	Solution 2
Size of sample	1–4 mm	
Water	filtered tap water	
Elution	Batch	Shake
Time	1 month	6 h
Solid to liquid ratio	1:5	
Revolutions (rpm)	–	200
Amplitude (mm)	–	40–50
Centrifuged	No	3000 rpm, 20 min
Percolation	Filtration	Suction filtration
Pore size (μm)	2.7	0.45

3 RESULTS

3.1 *Bioassay*

The results of Microtox test and the Daphnia test were showed in figures 1 and 2. The Microtox test gave approximately equal TU_{20} values in both test solutions (Fig. 1), even though they were prepared by different methods (Table. 2). The *Daphnia* test gave a density-dependent result in test

Table 3. GC–MS measurement conditions.

GC/MS: Shimadzu GCMS-QP 2010

Column: Agilent J&W DB-5 ms (5% phenyl, 95% methyl silicone) fused silica
capillary column, 30 m × 0.25 mm i.d., 0.25-μm film

Temperatures
Column: Temperature-programmed: 2 min at 40°C, 8°C/min to 310°C, 5 min at 310°C
Injector: 250°C
Transfer line: 300°C
Ion source: 200°C

Injection method: Splitless, 1 min for purge-off time

Carrier gas: He

Linear velocity: 40 cm/s, constant flow mode

Ionization method: EI

Tuning method: Target tuning for US EPA method 625

Scan range: 33–600 amu

Scan rate: 0.3 s/scan

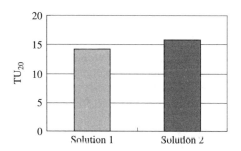

Figure 1. TU_{20} of Microtox test of both test solutions.

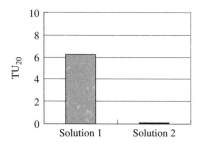

Figure 2. TU_{20} of *Daphnia* test of both test solutions.

Table 4. Chemicals likely to affect the Microtox test and their measured values of EC_{20} (μg/L).

Name	Sol. 1	Sol. 2	EC_{20}
n-$C_{13}H_{28}$	56	65	Unknown
Acetophenone	11	14	>1000
2-Methoxyphenol	4.7	5.9	>1000
2,6-Dimethylphenol	0.27	0.31	>1000
alpha-Terpineol	0.87	1.1	>1000
Aniline	160	130	130
Benzothiazole	1800	1900	390
Manganese	110	60	>1000

solution 1, but no result in test solution 2 (Fig. 2). It was shown that effect was not meaningful statistically with N.D. (not detected) for control.

A TU_{20} value of 15 means that a 20% biological effect was shown at a $15\times$ dilution of the test solution. It is likely that tire chip leachate would be diluted more than 15 times in rivers and seas, so the Microtox test would show little effect.

Because the TU_{20} value of test solution 1 was 6 in the *Daphnia* test, the solution's effect was very weak, so tire chip leachate would be unlikely to affect *Daphnia* in the wild.

Our results suggest that luminous bacteria and *Daphnia* are not likely to be influenced in the wild. However, tire chip shape, quantity, elution method, dissolution time, temperature, and pH could affect the quality or quantity of leachate, and thus affect the results. Therefore, test solutions should be made under scenarios that simulate the environment where tire chips are used, and a detailed investigation will be necessary.

Further examination will be necessary in the case of long-lasting contaminants that are difficult to resolve, or where high-accumulating chemicals are present in eluate.

3.2 *Chemical analysis of elution components*

We used chemical analyses to identify the chemicals that might affect ecosystems. Among 760 chemicals tested, we detected 77 in either or both test solutions. The types of chemicals and their concentrations were approximately equal in both solutions.

Benzothiazole and aniline were detected at high concentrations in both, and phenols and other organic chemicals were detected at low concentrations. Fluorine, boron, zinc, and manganese were detected also.

We assayed these components by Microtox and *Daphnia* tests, but recognize that the tests consider only single chemicals, not combined effects.

The two solutions gave similar results in the Microtox test (Fig. 1). Therefore, any chemical substance that gave a positive result in the Microtox test was likely to be present in both solutions at approximately at the same level. The candidates are shown in Table 4.

Only test solution 1 affected the *Daphnia* (Fig. 2). Therefore, any chemicals that affected *Daphnia* were present only in solution 1. The candidates are shown in Table 5.

Because benzothiazole and aniline were present at relatively high concentrations and we felt that they could have biological effects, we tested each by Microtox test. The sum of their toxicities accounted for around 40% of the total toxicity of test solutions 1 and 2 (Note 1).

We did a similar *Daphnia* test with bisethylhexylphthalate, phenol, and zinc (Table 5). Bisethylhexylphthalate had no appreciable effect. The combined toxicities of phenol, zinc, and aniline accounted for more than 50% of the toxicity of test solution 1.

In each test solution, we identified the components responsible for about half of the toxicity. The other half might be explained by the effects of very small amounts of as-yet unidentified materials and combinations of materials. It is important to clarify the environmental effects of many kinds

Table 5. Chemicals likely to affect *Daphnia* and their measured values of EC_{20} (μg/L).

Name	Sol. 1	Sol. 2	EC_{20}
n-C_9H_{20}	1.7	–	Unknown
n-$C_{15}H_{32}$	3.6	–	Unknown
n-$C_{17}H_{36}$	0.96	–	Unknown
Phenol	20	–	11
Bis(2-ethylhexyl)phthalate	47	2.5	340
Octanol	0.54	–	>5000
N,N-Dimethylaniline	1.4	–	1600
2-Anisidine	0.88	–	>5000
Zinc	180	–	160
Benzothiazole	1800	1900	>5000
Aniline	160	130	280

– Not detected.

of materials present in trace amounts, and we must also consider the environmental impacts of chemicals that are proximate causes. The components that had the highest EC_{20} values in the Microtox test were benzothiazole and aniline, and those that affected *Daphnia* were phenol, zinc, and aniline. However, the total biological effect of both test solutions was small, and effects are not likely to be seen in the environment. For example, bisethylhexylphthalate and benzothiazole had little effect on *Daphnia* (Note 2).

Because benzothiazole, phenol, and phthalic acid are easily volatilized, they do not accumulate in the environment. Although the concentration of zinc in test solution 1 was 180 μg/L, similar to the *Daphnia* acute toxicity value (EC_{20}), zinc eluted from tire chips in the environment is unlikely to have a serious influence on environmental biology, as it would be quickly diluted.

4 SUMMARY

We extracted tire chips with water and tested the eluates by the Microtox and *Daphnia* tests. If conditions of elution were different, it was become clear different chemicals were eluted, and the results of bioassay changing with the chemicals. In our elution conditions, the biological effects were small, and most toxic substances are not likely to pose a serious problem in the environment. Therefore, the environmental load imposed by tire chips is likely to be low, although other components and components with a combined effect remain to be examined. The biological effects of eluate leached from tire chips under conditions likely to occur in the environmental scenario, and including any chronic toxic effects, should be examined in the future.

Note 1: The EC_{20} value in the Microtox test of test solution 1 was 7%, and the quantity of aniline present in 7% of test solution 1 was 11.5 μg/L ($= 164 \times 0.07$). On the other hand, the EC_{20} value of aniline was 130 μg/L. Therefore, 130 μg/L of aniline would be needed to account for the whole effect of test solution 1 in the Microtox test. However, the contribution of aniline was more likely 8.8% ($= 11.5/130 \times 100$), because only 11.5 μg/L was present. The EC_{20} value of benzothiazole was 390 μg/L, and its contribution was 32.3% ($= 1800 \times 0.07/390 \times 100$). The sum of both contributions was 41.1% (38.1% in test solution 2).

However, this estimate of toxicity is not necessarily correct, because we assumed a linear dose–response relationship and also assumed that the toxicities were additive. Nevertheless, it gives us some ideas for future study directions.

Note 2: The EC_{20} value in the *Daphnia* test of test solution 1 was 16%, and the quantity of zinc present in 16% of the test solution 1 was 28.8 μg/L ($= 180 \times 0.16$). On the other hand, the EC_{20} value of zinc was 160 μg/L. Therefore, 160 μg/L of zinc would be needed to account for the whole effect of test solution 1. However, the contribution of zinc was more likely 18.0% ($= 28.8/160 \times 100$), because only 28.8 μg/L was present. The EC_{20} value of phenol was 11 μg/L, and its contribution was 29.1% (=20 × 0.16/11 × 100). Likewise, the EC_{20} value of aniline was 280 μg/L, and its contribution was 9.1% (=160 × 0.16/280 × 100). The sum of these three contributions was 56.2%. Therefore, most biological effect of test solution 1 can be explained by these three materials. The effect of bisethylhexylphthalate was small (EC_{20} = 340 μg/L; contribution = 2.2%). The contributions of octanol, anisidine, and benzothiazole could not be calculated because they had no effect on *Daphnia* (EC_{20} > 5000 μg/L).

REFERENCES

Chang, J.C., Taylor, P.B. & Leach, F.R. 1981. Use of the Microtox assay system for environmental samples. Bulletin of Environmental Contamination and Toxicology 26(2): 150–156.

US EPA. 1990. Revised Technical Support Document for Water Quality-Based Toxics Control, EPA 440/4-85-032. Washington DC: US EPA.

Barton, A.P. & Delhaize, A. 1996. The Application of Luminescent Bacteria for Monitoring the Toxicity of a Biocide Inhibitor.

OECD. 1996. Guideline for testing of chemicals, Section 2–4, OECD.

Muna, L., Guido, P., Colin, J., Wim, D.C. & Karl, S. 1995. Toxicity evaluations of wastewaters in Austria with conventional and cost-effective bioassays. Ecotoxicology and Environmental Safety 32(2): 139–146.

Guerra, R. 2001. Ecotoxicological and chemical evaluation of phenolic compounds in industrial effluents. Chemosphere 44(8): 1737–1747.

Finney, D.J. 1978. Stastical Methods in Biological Assay, 3rd edn., Chales Griffin & Company, Bucks, 508.

Structural stability and flexibility during earthquakes using tyres (SAFETY) – A novel application for seismic disaster mitigation

H. Hazarika
Akita Prefectural University, Akita, Japan

ABSTRACT: In this research, an innovative cost-effective disaster mitigation technique is developed, wherein various potentials of tire derived geomaterial, such as tire chips, have been exploited as a tool to reduce the dynamic load and the permanent displacements of structures during earthquakes. The technique renders flexibility and stability to structures, and is named SAFETY (Stability And Flexibility of Structures during Earthquake using TYres). A series of small-scale model shaking table tests and large-scale 1G underwater shaking table tests were conducted to confirm the validity of the technique using sinusoidal loading, actual earthquake loadings (Level 1 and Level 2 earthquakes) as well as synthetic earthquake loading (Level 2 earthquake). Test results have indicated that the use of the technique leads not only to reduction of the seismic load, but also curtailment of the seismically induced permanent displacement of structures.

1 INTRODUCTION

Statistics (JATMA/JTRA, 2003; RMA, 2004) reveal that in most of the countries scrap tires are mostly used for thermal recycling, which produces greenhouse gases such as CO_2 jeopardizing the effective implementation of the Kyoto Protocol (http://unfccc.int, 2005). Increasing attention has, therefore, been paid in recent years to reduce the thermal recycling of scrap tires and instead use them as materials. Use of scrap tires as materials reduces the environmental impact (CO_2 emission in material recycling is 1/4th that of the thermal recycling) from scrap tires, and thus contributes towards a sustainable global environment.

On the other hand, with the introduction of the performance based design (SEAOC, 1995) in various design codes around the globe, cost, environment and structural performance are becoming the three most decisive factors in the design and retrofitting of infrastructural facilities against earthquake loading. Concerns have been growing on the seismic stability of the existing and newly built port and harbor facilities in Japan after her bitter experiences of the 1995 Hyogoken-Nanbu earthquake, Kobe. The earthquake caused severe damage to more than 90% of the waterfront structures (Inagaki et al., 1996; Ishihara, 1997; JGS/JSCE, 1996; Kamon et al., 1996; Towhata et al., 1996). Figure 1 shows one such example of devastating damage of a quay wall located in the Kobe port, Japan.

Typical waterfront structure such as gravity type quay wall has rubble backfill immediately behind the wall. One of the reasons for using rubble backfill is to reduce the earth pressure due to friction. However, such granular material is vulnerable to deformation under seismic load, and hence can cause large permanent deformations to the structures. Fears loom large over three large-scale devastating earthquakes (Tokai Earthquake, Tonankai-Nankai Earthquake, Strong metropolitan Earthquake) that are predicted to strike the Tokai area, the Nankai area and the Kanto area of Japan any time in the near future. The central disaster management council, government of Japan (http://www.bousai.go.jp, 2006), have been making concerted efforts designed to mitigate the disasters and minimize the economic implications from these earthquakes. With tough Japanese

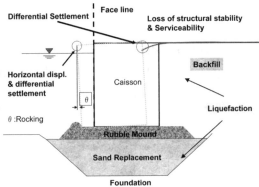

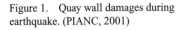

Figure 1. Quay wall damages during earthquake. (PIANC, 2001)

Figure 2. Typical damage pattern.

government policy of cutting expenditure on infrastructural projects, demands on engineers are mounting for developing novel and cost-effective disaster mitigation technique that can protect and reduce the damages of geotechnical structures during devastating earthquakes.

Figure 2 shows the damages that the waterfront structures typically suffer, resulting in the loss of serviceability of port immediately after an earthquake. If we can substitute the rubble backfill (Figure 1) with some other lightweight granular materials with other beneficial characteristics, then the earth pressure during earthquake can be reduced to a greater extent along with the curtailment of earthquake induced permanent deformation of the structures. Emerging tire derived geomaterial, known as tire chips, can be an answer to this. Tire chips is a recycled material that is lightweight, elastic, compressible, highly permeable, earthquake resistant, thermally insulating and durable. Due to such myriad of beneficial properties, this material has been coined a *smart geomaterial* by Hazarika et al. (2006). Various applications of tire derived geomaterials can be found in Abichou et al. (2004), Bosscher et al. (1997), Edil (2006), Humphrey (1998), Humphrey & Tweedie (2002), Tweedie et al. (1998) and Yasuhara et al. (2006).

In this research, an innovative cost-effective disaster mitigation technique is developed using tire derived geomaterial, which can be utilized as a seismic performance enhancer of geotechnical structures. The technique is called SAFETY (**S**tability **A**nd **F**lexibility of structures during **E**arthquake using **TY**res). The objective of this research is to examine whether SAFETY technique can reduce the earthquake related damages to structures. To that end, a small-scale and a large-scale 1G underwater shaking table test was conducted to validate the performances of structures subjected to various types of seismic loadings.

2   DESCRIPTION OF SAFETY TECHNIQUE

The underlying principle of SAFETY involves placing a cushion layer made out of tire chips as a vibration absorber immediately behind the structures (Figure 3). One function of the tire chips cushion is to reduce the load against the structure, due to energy absorption capacity of the cushion material. Another function is to curtail permanent displacement of the structure due to inherited flexibilities derived from using such elastic and compressible material. The principle of SAFETY lies in imparting flexibilities to the structure due to cushioning action of the tire chips layer, which in turn ultimately provides stability to the structure by absorbing the energy and restraining excessive structural displacement during earthquakes.

3   TEST MODEL DESCRIPTION

In order to validate the SAFETY technique described in Figure 3, a small scale model shaking table test (SS Test) for submerged backfill condition and a large scale model test (LS test) in underwater

116

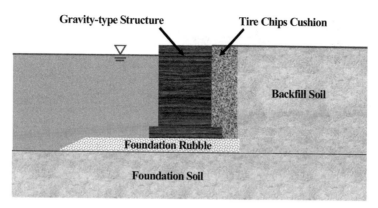

Figure 3. SAFETY technique.

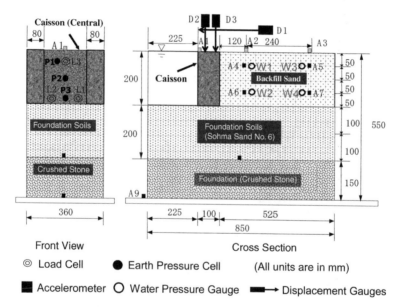

Front View   Cross Section

◎ Load Cell   ● Earth Pressure Cell   (All units are in mm)

■ Accelerometer   O Water Pressure Gauge   ■➤ Displacement Gauges

Figure 4. Cross section of the model in SS test.

condition were performed. The model shaking table assembly of Soil Dynamics Division, Port and Airport Research Institute (PARI) was used in the small scale testing program. On the other hand, the three dimensional underwater shaking table assembly of Structural Dynamics Division of PARI was used in the large scale underwater testing program.

3.1 Small scale 1G model shaking table test (SS test)

Figure 4 shows the cross section and front view of the model caisson and the locations of various devices (load cells, pore water pressure cells, accelerometers and displacement gauges) used in the testing program. The soil box was made of a steel container 0.85 m long, 0.36 m wide and 0.55 m deep. The model caisson is 20 cm high and 10 cm wide as shown in the figure. All the monitoring devices were installed at the central caisson (20 cm long) to eliminate the effect of sidewall friction on the measurements.

The caisson was modeled in a prototype to model ratio of 35. The backfill and the foundation were prepared using Sohma sand (No. 6). The dense saturated foundation sand was prepared in

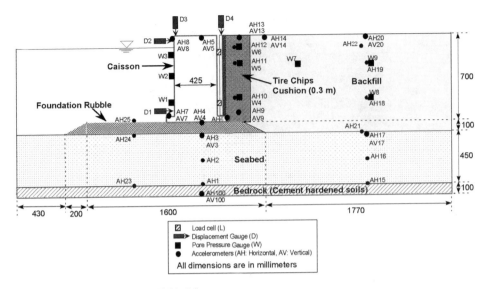

Figure 5.   Cross section of the model in LS test.

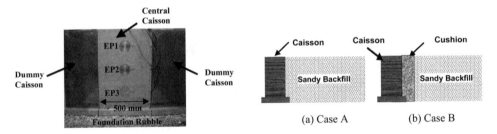

Figure 6.   Model caisson with pressure gauges.   Figure 7.   Test cases.

four layers using vibratory compaction technique. After preparing each layer, the whole assembly was shaken with 300 Gal of vibration of frequency 7 Hz for 4 sec. The process was repeated for 4 to 8 times depending on the requirements. Backfill (under submerged condition) was prepared in five layers using free falling technique. Compaction was achieved by shaking the soil box and the wall assembly 3 to 4 times (as needed) with 200 Gal of vibration of frequency 7 Hz for 4 sec. The relative densities of the foundation soil were 95% to 100%, while that of the backfill soils were between 60% to 70%.

3.2   Large scale underwater 1G shaking table test (LS test)

The three dimensional underwater shaking table assemblies, which was used in the LS test, consists of a shaking table of diameter 5.65 m, which is installed on a 15 m long by 15 m wide and 2.0 m deep water pool. The details of the shaking table can be found in Iai and Sugano (2000). The soil box was made of a steel container 4.0 m long, 1.25 m wide and 1.5 m deep.

A gravity type caisson quay wall (prototype to model ratio of 10) was used in the testing. The backfill and the seabed layer were prepared using Sohma sand (No. 5). Figure 5 shows the cross section of the soil box, the model caisson and the locations of the various measuring devices. The model caisson (425 mm in breadth) was made of steel plates filled with dry sand and sinker to bring its center of gravity to a stable position. As shown in Figure 6, the caisson consists of three parts; the central part (width 500 mm) and two dummy parts (width 350 mm each). All the monitoring devices were installed at the central caisson to eliminate the effect of sidewall friction on the measurements.

The dense foundation sand representing the seabed layer was prepared in two layers. After preparing each layer, the whole assembly was shaken with 300 Gal of vibration starting with a frequency of 5 Hz and increasing up to 50 Hz. Backfill was also prepared in stages using free falling technique, and then compacting using a manually operated vibrator. After constructing the foundation and the backfill, and setting up of the devices, the pool was filled with water gradually elevating the water depth to 1.3 m to saturate the backfill. This submerged condition was maintained for two days so that the backfill attains a complete saturation stage.

Cushion layer was prepared by filling the tire chips inside a geotextile bag. Geotextile was used to wrap the tire chips so that they do not mix with the surrounding soils. The average dry density of the tire chips achieved after filling and tamping was 0.675 t/m^3. Relative densities that were achieved after backfill preparation were about 50% to 60%, implying that the backfill is partly liquefiable. On the other hand, the foundation soils were compacted with mechanical vibrator to achieve a relative density of about 80%, implying a non-liquefiable foundation.

4 TESTING PROCEDURES

Two test cases were examined as shown in Figure 7. Case A involves a caisson with conventional sandy backfill behind it. Case B involves a caisson protected by the SAFETY technique. The size of tire chips used as a cushion material varies depending on the type of test (small scale or large scale). In the large scale test (LS test) tire chips of average grain size 20 mm was used. However, tire chips of average grain size 2 mm was used in the small scale test (SS test), in order to avoid instability and unreliability of the measured data (Miura et al., 2003) by the small size earth pressure cells installed in the model caisson.

The thickness of the cushion in the case of LS test was 30 cm (0.4H, H is the height of the structure) and in the case of SS test was 8 cm. In actual practice, the design thickness will depend upon a lot of other factors such as height and rigidity of the structure, compressibility and stiffness of the cushion material. The test results with different cushion thickness using the small-scale model shaking table test has been reported in Hazarika et al. (2006).

In the case of SS test, a sinusoidal acceleration was imparted to the soil-structure system for a period of 2 sec (frequency = 20 Hz). They were applied in several stages starting with 100 Gal and increasing up to 700 Gal at an increment of 100. In the case of LS tests, earthquake loadings of different magnitudes were imparted to the soil-structure system. The input motions selected were: (1) Port Island wave: the N-S component of the strong motion acceleration record at the Port Island, Kobe, Japan during the 1995 Hyogo-ken Nanbu earthquake (M 7.2), (2) Hachinohe wave: the N-S component of the earthquake motion recorded at the Hachinohe port, Japan during the 1968 Tokachi-Oki earthquake (M 7.9), and (3) Ohta Ward wave: a scenario synthetic earthquake motion assuming an earthquake that is presumed to occur in the southern Kanto region with its epicenter at Ohta ward, Tokyo, Japan (M 7.9). The wave records of these input motions are shown in Figure 8. Different series of tests was performed by changing the maximum acceleration of these records and they are summarized in Table 1. For a given input acceleration, the response accelerations, the dynamic loads on the caisson, displacement of the caisson, and the pore water pressures at various locations of the backfill were measured.

The similitudes of various parameters in 1G gravitational field for the soil-structure-fluid system, that were calculated using the relationship proposed by Iai (1989) for the respective prototype (P) to model (M) ratio, are shown in Table 2. It is worthwhile mentioning here that, the material particles size and compressibility of the material are assumed to remain unchanged, for the model and the prototype. Durations of the shaking in the large scale model testing were based on the time axes of the accelerograms shown in Figures 8(a)–(c), which were reduced by a factor of 5.62 according to the similitude shown in Table 2.

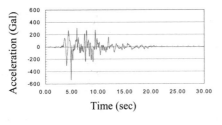

(a) Port Island (PI) wave

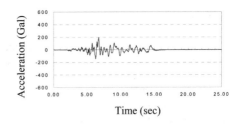

(b) Hachinohe (HN) wave

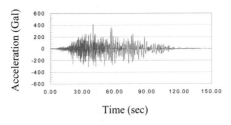

(c) Ohta Ward (OW) wave

Figure 8. Applied earthquake records.

Table 1. Loading sequences in the LS test.

Series	Earthquake code names	Acceleration ratio (Max Acc.)
No. 1	HN 0.5	0.5 (100 Gal)
No. 2	HN 1.0	1.0 (200 Gal)
No. 3	PI 0.5	0.5 (339 Gal)
No. 4	PI 1.0	1.0 (678 Gal)
No. 5	OW 1.0	1.0 (486 Gal)
No. 6	PI 1.5	1.5 (1018 Gal)

Table 2. Similitudes for 1G field.

Items	P/M ratio	Scale factor (SS test)	Scale factor (LS test)
Length	λ	35	10
Time	$\lambda^{0.75}$	14.09	5.62
Density	1	1	1
Stress	λ	35	10
Pressure	λ	35	10
Displacement	$\lambda^{1.5}$	207.06	31.62
Acceleration	1	1	1

5 TEST RESULTS

5.1 *Small scale test (SS test)*

Figure 9 shows the time histories of the dynamic load acting on the caisson for the two test cases at the acceleration amplitude of 600 Gal. It can be observed that as compared to conventional caisson, SAFETY technique yields a significant reduction of the dynamic load acting on the caisson. The maximum amplitude of the latter is much lower than that of the former (almost 1/4th). This implies that, the earthquake resistant capability of the caisson improves with the use of the tire chips cushion. Figure 10 shows the maximum dynamic load as a function of acceleration. With increasing acceleration amplitude, while conventional caisson (without any protective cushion) experiences higher dynamic thrust, the caisson protected with SAFETY technique shows only a mild increase. For the conventional caisson, after 400 Gal of acceleration, the dynamic load increases sharply. The dynamic load, however, can be reduced to more than half by protecting the soil-structure system with tire chips cushion, even beyond 400 Gal of acceleration. This implies that even at the higher acceleration levels, the presence of the cushion could render stability to a soil-structure system.

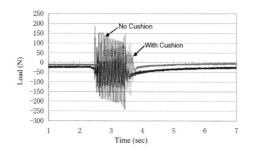

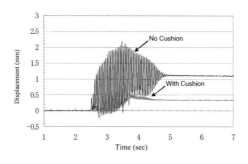

Figure 9. Time histories of the seismic load.

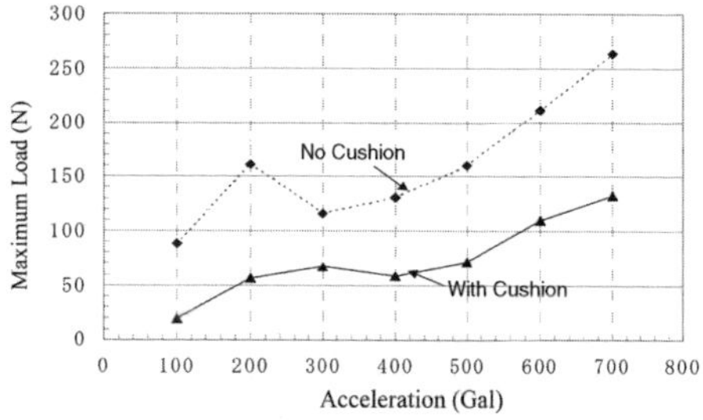

Figure 10. Seismic load Vs. Acceleration.

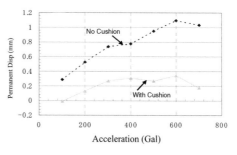

Figure 11. Time histories of the displacement.

Figure 12. Residual displacements Vs. Acceleration.

Figure 11 compares the time histories of the horizontal displacements of the caisson at the acceleration amplitude of 600 Gal. In this figure, positive displacement indicates a seaward displacement of the caisson. It can be observed that the unprotected caisson experiences about 2.2 mm of maximum displacement. However, the maximum displacement experienced by the caisson with cushion (SAFETY Technique) was only 0.91 mm. This figure also indicates that, the permanent seaward displacement of the caisson also is quite high for the unprotected caisson than the cushion protected caisson. Figure 12 shows the residual seaward displacements of the caisson at each acceleration level. It can be observed that the residual displacement for the caisson protected by SAFETY technique is less than 1/3rd than that of the unprotected caisson.

5.2 Large scale test (LS test)

As shown in Figures 8(a)~8(c), tests were performed for different earthquake loading conditions. However, the results presented in this paper are confined only to series No. 4 in Table 1. For details about the effect of earthquake types and repeated loading history, the readers are referred to Hazarika et al. (2007).

Figures 13–15 show the time histories of the measured horizontal seismic thrust acting on the top, the middle and the bottom of the caisson for both the test cases (Cases A and B) when subjected to the PI type wave. It can be observed that, as compared to conventional backfill, use of SAFETY technique yields a significant reduction of the seismic earth pressure acting on the caisson at each depth. While the caisson without any protective cushion experiences high fluctuation of the earth pressure with a predominant peak, the earth pressure on the cushion-protected caisson stabilizes soon. The peak amplitude of the latter (Case B) also attenuated considerably as compared to that of the former (Case A). The reduction in the earth pressure at a common peak ground acceleration was 75, 25, and 66% at the top, middle and bottom of the caisson respectively. This implies that the seismic performance of the caisson improves with the use of the SAFETY technique.

One interesting observation here is that, for the sandy backfill condition, the pressure does not come to a stabilized state immediately even at the end of the earthquake load application (3.0 sec).

121

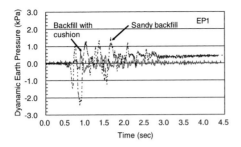

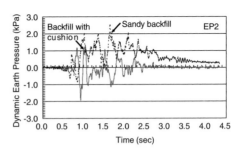

Figure 13. Seismic increment of pressure (top).

Figure 14. Seismic increment of pressure (middle).

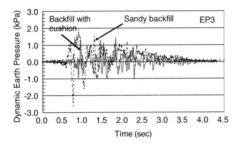

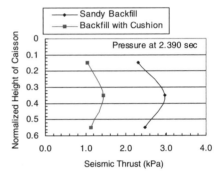

Figure 15. Seismic increment of pressure
(bottom).

Figure 16. Distribution of total seismic pressure.

On the contrary, the backfill improved by SAFETY comes to the stabilized state immediately upon stopping the load. This may be due to the residual earth pressure, which takes time to come to the original state even after ceasing of the seismic load. The residual pressure generated is the result of the excess pore water pressure, which takes time to stabilize in Case A.

At a particular time ($t = 2.39$ sec) of the time history, the total seismic thrusts acting on the caisson quay wall are plotted as in Figure 16. The total seismic thrust was obtained by adding the total static force (minus the static water pressure) to the seismic incremental thrust obtained from Figures 13–15. It can be observed that the total seismic force on the wall could be reduced to almost half in this particular case. Reduction of the seismic thrust implies a lower design load resulting in a smaller caisson width, which in turn will lead to a low material costs. Thus, the SAFETY technique can lead to a cost-effective design not only in terms the backfill material, but also in terms of the structural material as well.

In actual practice, it is of great importance whether the permanent structural deformation will lead to halting of the port operation in the event of a destructive earthquake. In order to evaluate, whether the SAFETY technique can minimize the maximum horizontal displacement as well as the residual horizontal displacement experienced by the caisson, the time histories of the horizontal displacements (D1 and D2 in Figure 5) during the PI type earthquake loading for the two test cases are compared in Figures 17 and 18. Comparisons reveal that the maximum displacement experienced by the caisson with tire chips cushion is toward the backfill in contrast to the caisson without cushion, in which case it is seaward. Compressibility of the tire chips renders flexibility to the soil-structure system, which allows the caisson to bounce back under its inertia force, and this tendency ultimately (at the end of the loading cycles) aids in preventing the excessive seaward deformation of the caisson. However, the caisson without any protective cushion experiences very high seaward displacements right from the beginning due to its inertia. Consequently, the caisson can not move back to the opposite side and ultimately undergoes a huge seaward permanent displacement. It is to be noted that the lesser displacement magnitudes (1.5 mm at the top) experienced by the cushion

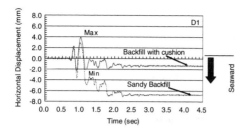

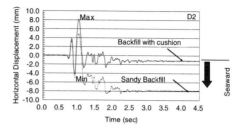

Figure 17.   Time history of displacement (bottom).   Figure 18.   Time history of displacement (top).

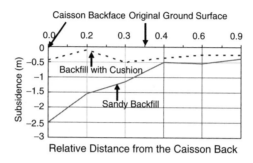

Figure 19.   Comparison of backfill subsidence.

protected structure are not solely due to the compressibility of tire chips, but also the result of strong earthquake resistant characteristics of tire chips as compared to the backfill soil.

Figure 19 compares the average subsidence of the backfill (prototype equivalent calculated using Table 2) from the start of the first earthquake loading in Table 1 (HN 0.5) to the end of the last earthquake loading (PI 1.5), which were measured by installing 10 targets on the surface of the backfill (along the line bordering the central caisson). It can be seen that as compared to the conventional sandy backfill, the SAFETY technique could substantially reduce (less than 50 cm) the backfill subsidence. This indicates that the developed technique can contribute towards the safe operation of port facilities even after experiencing many devastating earthquakes, since the differential settlements were significantly less and uniform.

6   CONCLUSIONS

A series of 1G model shaking table tests were conducted to examine the performance of a newly developed technique, called SAFETY, intended for earthquake disaster mitigation. Test results have indicated that the use of the technique leads not only to reduction of the seismic load, but also the seismically induced permanent displacement of the structure. The SAFETY technique also could prevent the bumpiness of the backfill after an earthquake, thus maintaining the performance of infrastructural facilities after strong earthquakes. Reduction of the load against structure implies lowering of the design seismic load, which in turn yields a slim structure with reduced material cost. Such applications of scrap tire derived material, thus, not only reduce considerably the execution and construction cost of a project, but also contributes towards a sustainable environment by recycling of the scrap tires as materials. SAFETY technique can also be applied for upgrading (retrofitting) of the existing structures, which do not satisfy the current seismic design criteria and, thus, run the risk of damages during devastating future earthquakes. SAFETY, thus, is expected to have a great potential in the cost-effective seismic design and retrofitting of structures.

123

ACKNOWLEDGEMENTS

The author gratefully acknowledges the financial support for this research provided by the Japan Society of Promotion of Science (JSPS) and Ministry of Education, Culture, Sports, Science and Technology (MEXT), Japan under the Grant-in-Aid for scientific research (Grant No. 18206052, Principal Investigator: Hemanta Hazarika). The author also would like to acknowledge the supports extended by Port and Airport Research Institute (PARI), Japan and his former colleagues in PARI, whose assistance in the test execution were extremely valuable. The technical supports from Ibaraki University, Hitachi, Japan, Bridgestone Corporation, Tokyo, Japan and Toa Corporation, Yokohama, Japan are also gratefully acknowledged.

REFERENCES

Abichou, T., Tawfiq, K., Edil, T.B., and Benson, C.H. (2004): Behavior of a Soil-Tire Shreds Backfill for Modular Block-Wall, *ASCE Geotechnical Special Publication*, No. 127, pp.162–172.

Bosscher, P.J., Edil, T.B., and Kuraoka, S. (1997): Design of highway embankments using tire chips, *Journal of Geotechnical and Geoenvironmental Engineering, ASCE*, Vol. 123, No. 4, pp. 295–304.

Edil, T.B. (2006): Green Highways: Strategy for Recycling Materials for Sustainable Construction Practices, *Seventh International Congress on Advances in Civil Engineering*, Istanbul, Turkey, pp. 1–20.

Hazarika, H., Sugano, T., Kikuchi, Y., Yasuhara, K., Murakami, S., Takeichi, H., Karmokar, A.K., Kishida, T., and Mitarai, Y. 2006a. Evaluation of Recycled Waste as Smart Geomaterial for Earthquake Resistant of Structures, *41st Annual Conference of Japanese Geotechnical Society, Kagoshima*, pp. 591–592.

Hazarika, H., Kohama, E., and Sugano, T. 2007. Underwater Shake Table Tests on Seismic Performance of Waterfront Structures Protected with Tire Chips Cushion, *Journal of Geotechnical and Environmental Engineering*, ASCE (In Press).

Http://www.bousai.go.jp 2006. Cabinet Office, Government of Japan: Central Disaster Management Council, *Available: http://www.bousai.go.jp/jishin/chubou*, 2006 (in Japanese).

http://unfccc.int 2005. Kyoto Protocol to the United Nations Framework Convention on Climate Change, *http://unfccc.int/resource/docs/convkp/kpeng.html*.

Humphrey D.N. (1998): Civil Engineering Applications of Tire Shreds, *Manuscript Prepared for Asphalt Rubber Technology Service*, SC, USA.

Humphrey, D.N., and Tweedie, J.J. (2002): Tire Shreds as Lightweight Fill for Retaining Walls – Results of Full Scale Field Trials, *Proceeding of the International Workshop on Lightweight Geomaterials*, Tokyo, Japan, pp. 261–268.

Iai, S. (1989): Similitude for Shaking Table Tests on Soil-structure-fluid Model in 1g Gravitational Field, *Soils and Foundations*, Japanese Geotechnical Society, Vol. 29, No. 1, pp. 105–118.

Iai, S., and Sugano, T. (2000): Shake Table testing on Seismic Performance of Gravity Quay Walls, *12th World Conference on Earthquake Engineering, WCEE*, Paper No. 2680.

Inagaki, H., Iai, S., Sugano, T., Yamazaki, H., and Inatomi, T. (1996): Performance of Caisson Type Quay Walls at Kobe Port, *Special Issue of Soils and Foundations*, Japanese Geotechnical Society, Vol. 1, pp. 119–136.

Ishihara, K. (1997): Geotechnical aspects of the 1995 Kobe earthquake, Terzaghi Orientation, *14th International Conference of International Society of Soil Mechanics and Geotechnical Engineering*, Hamburg, Germany, Vol. 4, pp. 2047–2073.

Japanese Geotechnical Society (JGS) and Japan Society for Civil Engineers (JSCE) (1996): Joint Report on the Hanshin-Awaji Earthquake Disaster (in Japanese).

JATMA/JTRA (2003): Tire Recycling Handbook, Japan Automobile Tire Manufacturers Association & Japan Tire Recycle Association, *Report No. 105-0001*, Tokyo, Japan (in Japanese).

Kamon, M., Wako, T., Isemura, K., Sawa, K., Mimura, M., Tateyama, K., and Kobayashi, S. (1996): Geotechnical Disasters on the Waterfront, *Special Issue of Soils and Foundations*, Japanese Geotechnical Society, Vol. 1, pp. 137–147.

Miura, K., Otsuka, N., Kohama, E., Supachawarote, C., and Hirabayashi, T. (2003): The Size Effect of Earth Pressure Cells on Measurement in Granular Materials, *Soils and Foundations*, Japanese Geotechnical Society, Vol. 43, No. 5, pp. 133–147.

PIANC (International Navigation Association) (2001): *Seismic Design Guidelines for Port Structures*, Balkema Publishers, Rotterdam.

RMA (2004): US Scarp Tire Markets 2003 Edition, *7th Biannual Report of the Rubber Manufacturers Association*, USA, pp. 1–51.

SEAOC, (1995): Performance based seismic engineering of buildings. Structural Engineers Association of California, Sacramento, California, USA.

Towhata, I., Ghalandarzadeh, A., Sundarraj, K.P., and Vargas-Monge, W. (1996). Dynamic Failures of Subsoils Observed in Waterfront Area, *Soils and Foundations*, *Special Issue,* Japanese Geotechnical Society, pp. 149–160.

Tweedie, J.J., Humphrey, D.N., and Standford, T.C. 1998. Tire Shreds as Retaining Wall backfill, Active Conditions, *Journal of Geotechnical and Geoenvironmental Engineering*, ASCE, Vol. 124, No. 11, pp. 1061–1070.

Yasuhara, K., Hazarika, H., and Fukutake, K. 2006. Applications of Scrap Tires in Civil Engineering, *Doboku Gijutsu*, Vol: 61, No. 10, pp. 79–86 (in Japanese).

Use of scrap tire derived shredded geomaterials in drainage application

A.K. Karmokar

Central Research, Bridgestone Corporation, Tokyo, Japan

ABSTRACT: Permeability tests on scrap tire shreds have been carried out using a prototype apparatus designed indigenously for evaluating drainage/filtration potentiality upon considering compression phenomena of tire shreds mass. Permeability coefficient of tire shreds has been found to be in a level ($\times 10^0$ cm/s) comparable with the conventionally used permeable geomaterials e.g., sand, gravels, etc., in spite of inducing high level of compressions that may be incurred in the field due to surcharge on tire shreds. A full-scale field trial on the use of tire shreds as drainage layer under high embankment has also been undertaken in Hokkaido, Japan. *In-situ* performance of tire shreds drainage layer has been found to be highly satisfactory.

1 INTRODUCTION

Tires from end-of-life vehicles, known as used or scrap tires, constitute a large volume of solid waste. It is estimated that 88% of about 106 millions of scrap tires generated in Japan for 2006 are reused and/or recycled [JATMA (2007)]. Japan scrap tire management trends in the recent years are shown in Fig. 1. As may be seen, the overall share recorded for material recycling sector is almost constantly limited to about 15%. With the aim of increasing the share of scrap tires in material recycling sector, a research project has been initiated by our group to explore the possibility of their use as geomaterials in civil engineering applications [Karmokar et al. (2005, 2006), Kawaida et al. (2005), Yasuhara et. al. (2004, 2006)].

The development of highly potential tire recycling technologies for civil engineering applications is believed to reduce unaccountable tires (12% of total scrap tire volume as shown in Fig. 1). Moreover, the increase of material recycle is directly involved with the lowering of CO_2 emission in the environment. It is reported that scrap tires have essential engineering properties viz., resilient, water

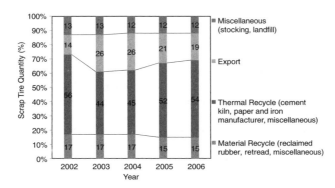

Figure 1. Japan scrap tire management trends.

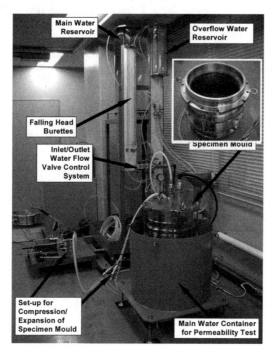

Figure 2. Test apparatus.

proof, insulation, bondable, durable, and such properties are commonly recognized as beneficial to many civil engineering applications [(Humphrey & Manion (1992)].

As a part of the project, the present research is framed to explore the drainage/filtration applications of shredded scrap tire materials. Permeability tests on scrap tire derived rubber chips and tire shreds were carried out in laboratory using a prototype apparatus which is designed to evaluate drainage behavior upon taking into consideration of compression phenomena of the tire shreds mass. With the view of undertaking a field trial on drainage application using tire shreds, model permeability tests were conducted on composite field soil-tire shred system for evaluating clogging potentiality of the tire shreds by the surrounding soil. Based on test results, a full-scale field trial on the use of tire shreds as drainage layer under high embankment had been undertaken in Hokkaido, Japan. *In-situ* measurements of the post-construction consolidation behavior of tire shreds layers were also carried out.

2 EXPERIMENTAL PROCEDURE

2.1 *Test apparatus*

Permeability tests were carried out in laboratory using a large sized test apparatus as shown in Fig. 2. It is comprised of a specimen mold (shown in the inset of Fig. 2) of 300 mm in diameter which is designed to work in the vertical water-flow condition. It may be argued that rubber chips/tire shreds are compressive in nature under vertical load, and drainage potentiality would change according to the levels of compression. In turn, evaluating drainage behavior of tire rubber chips/shreds essentially needs the compression phenomena of the materials to be taken into account. In order to evaluate the drainage behavior of tire rubber chips/shreds at different compression, the length of the cylindrical specimen mold is made mechanically adjustable. Consequently the tire rubber chips/shreds in the mold is able to compress and expand upon squeezing and expansion of the cylinder mold. Therefore, within the mechanical limits of this apparatus and/or sample used, the

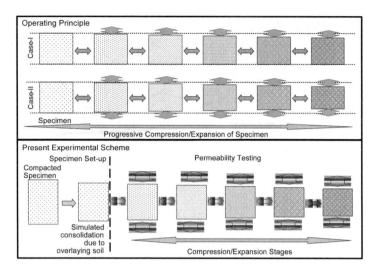

Figure 3.   Operating principle.

permeability tests could be performed at various intermittent stages, on the same specimen but under different compressive conditions.

2.2   *Operating principle*

The main part of this permeability test apparatus is a cylindrical specimen mold which is 400 mm high in expanded form, and could be shortened to a length of 300 mm in its shortened form. The lengthwise adjustments of this mold could be made possible at either and/or both ends of the specimen as shown in case-I and case-II of Fig. 3, respectively.

The process of progressive compression/expansion could be made repeatable (i.e., cyclic experiments) for a number of times, if necessary, during the experiment. The experiments in this study are limited to one compression/expansion cycle only, and conducted in accordance with the process of case-II shown in Fig. 3. The specimen is pre-compressed up to an intermediate stage for simulating the *in-situ* consolidation due to soil capping over tire shreds, followed by experimental compression/expansion stages for simulating different levels of surcharge loading.

3   RESULTS AND DISCUSSION

3.1   *Basic tests*

Two kind of samples, viz., scrap tire derived rubber chips (TC20) and tire shreds (TS20) are included in this study. Both the samples are composed of mainly non-spherical particles, and contain chips/shreds of size ranging 4.75–19 mm. The cumulative particle size distribution curves of these two samples are quite identical. As shown in Fig. 4, tire shreds contain steel and textile cords embedded inside the rubber as it was in the tire, while tire chips sample is composed of rubber only materials i.e., steel/textile cords are separated out from rubber. Consequently, tire shreds behave less compressive than tire chips. The specific gravity of rubber chips and tire shreds are 1.15 and 1.20, respectively.

Tire derived specimens are compacted into the cylindrical mold. Compacted density maintained for TC20 and TS20 are 0.66 and 0.71 kN/m^3, respectively. In order to simulate the field consolidation of specimens by overlaying soil, the specimen in the mold is then compressed for up to a level of normal pressure. The specimen so set is used for permeability testing.

(a) Scrap tire rubber chips (TC20)   (b) Scrap tire shreds (TS20)

Figure 4.   Scrap tire derived test samples.

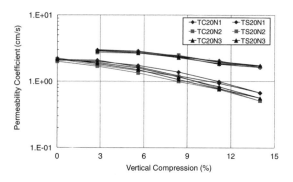

Figure 5.   Effect of compression on permeability.

The permeability coefficients of TC20 and TS20 samples for different compressions (and expansions) are shown in Fig. 5. Three repetitions for each tests conducted show quite identical permeability behavior. As evident, the levels of permeability coefficients of tire derived materials are in the high ($\times 10^0$) level in absolute term which is comparable with the highly permeable geomaterials e.g., coarse grained sand and gravels. In spite of high level of compression in rubber chips/tire shreds materials, which may be incurred in the field due to surcharge, the permeability coefficient still remains in the higher level.

As compared to rubber chips sample, tire shreds sample shows slightly higher permeability which may be related to material compressibility. Rubber chips in TC20 are more deformable under the normal stress, and therefore, the chances of better packing of rubber chips take place that leads to a lower void ratio as compared to tire shreds (TS20).

In general, the permeability coefficients obtained in the compression stage and expansion stage are in the same order which may indicate that the rubber chips and tire shreds specimens are almost recovered to its previous form even after undergoing a sizable degree of compression. Such a kind of material property may be very useful in many field applications.

The relationships between void ratio and permeability coefficients of the samples are shown in Fig. 6. Due to easy packing of rubber only chips into the mold, TC20 shows void ratio in the lower range (0.45–0.25) while TS20 shows the range of 0.6–0.4. However, similar to other permeable geomaterials, both rubber chips and tire shreds samples show overall linear behavior.

3.2   Applied tests

As for a guideline in selecting materials for drainage/filtration application, preliminary screening rules followed in Japan are shown in Table 1. Accordingly, grain size distribution curve for soil collected form a probable site in Hokkaido is compared with those of the two tire shreds samples

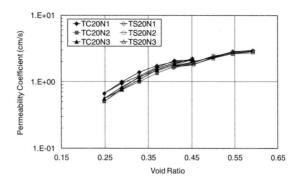

Figure 6.   Effect of void ratio on permeability.

Table 1.   Drainage/filtration guidelines.

Category	Guidelines
Permeability function	$\dfrac{D_{15}\ (\textit{filter material})}{D_{15}\ (\textit{field soil})} > 5$
Non-clogging potential	$\dfrac{D_{15}\ (\textit{filter material})}{D_{85}\ (\textit{field soil})} < 5$

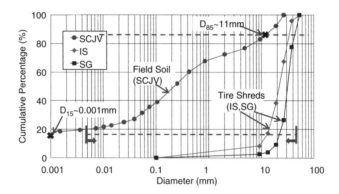

Figure 7.   Grain size distribution of field soil and tire shreds.

proposed for drainage application under high embankment. As shown in Fig. 7, both the guidelines were well satisfied with the tire shreds materials intended for use in this case.

In spite of general satisfaction of the guidelines stated above, permeability tests on field soil (SCJV), conventionally used coarse gravel drainage material (GRVL) and one of the newly proposed tire shreds samples (SG) are conducted in laboratory using the prototype test apparatus. The photographs of the respective materials and their permeability coefficients are shown in Fig. 8. Bulk densities of the specimens in the permeability mold are given in Table 2.

Field soil compacted to different specimen densities show quite different permeability values. Densely compacted soil (SCJV/J2) shows permeability coefficient of a level of 10^{-07}–10^{-06} which may indicate that the field soil has very low permeability characteristics, and thus additional drainage/filtration measures must be taken for maintaining stability in the structure constructed using such clayey soil. Conventionally used gravel (GRVL/H) drainage material shows permeability value of 10^{-1}–10^{0}. Tire shreds specimen is compressed (SG/Com) and return (SG/Ret) gradually in the mold during testing. Permeability coefficients corresponding to respective compression (also return) of tire shreds are obtained almost flat, with only a marginal overall declination for

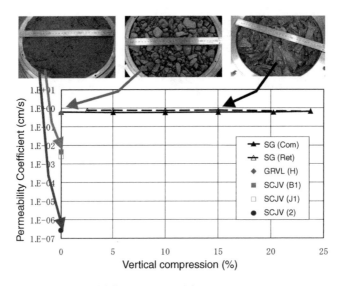

Figure 8.   Permeability coefficient of different geomaterials.

Table 2.   Bulk density of specimens.

Materials	Code	Density (g/cc)
Tire shreads	SG (Com)	0.65–0.85
(compression)	SG (Ret)	0.85–0.65
and extension)		
Gravel	GRVL (H)	1.77
Field soil	SCJV (B1)	1.72
	SCJV (J1)	1.71
	SCJV (J2)	1.80

higher compression. This may be indicative of maintaining equal (if not high) level of permeability coefficients of tire shreds to that of gravel even after imparting as high as 23–24% compression level (equivalent to about 15 m high soil layer above tire shreds).

3.3 *Soil clogging test*

In course of time, drainage/filter layers may get clogged (or blocked) due to soil particles carried along with water from the surrounding soil. The water-borne soil particles get entrapped into the voids in the drainage layer, causing deterioration of permeability characteristics of the layered material. Therefore, to obtain a steady water flow through the drainage/filtration layer, it is wise that the layer do not get clogged.

In spite of physical agreement of the guidelines for non-clogging potentiality of tire shreds with field soil as stated in Table 1 and Fig. 7, model permeability tests were conducted on composite soil-tire shred system for evaluating the clogging potentiality of the tire shreds by the surrounding soil. As for specimen, tire shreds (SG) were placed in the mold for a height of 35 cm, above which a layer of field soil of 5 cm thick was placed. The composite system is then compressed for up to a level of about 20% (equivalent to soil pressure of about 15 m high embankment) before doing the permeability test, comparatively for a long stretch, on the specimen.

132

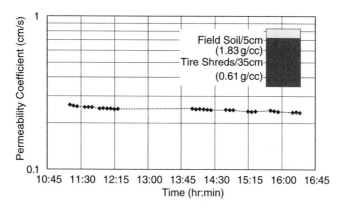

Figure 9.   Permeability coefficient of soil-tire shreds system.

Figure 10.   Clogging phenomenon is very less in tire shreds.

Permeability coefficients of tire shreds are calculated from the composite soil-tire shreds system flow rate, and are shown in Fig. 9. As may be seen, the permeability coefficient of composite soil-tire system is still in the higher level of 10^{-1}–10^{0}, and remain in the range even after high pre-compression of the composite specimen. The effect of time on permeability coefficient is also negligible, which means that the tire shreds layer did not get clogged by the soil placed on top of tire shreds layer in the mold.

A few of the photographs taken at various depth levels during dismantling of the composite specimen from the specimen mold are shown in Fig. 10. As may be seen, traces of soil in the tire shreds layer is almost negligible at or onward the depth level beyond 6 cm. (soil placed for up to a depth of 5 cm in the specimen mold). Also, the filter paper placed at the bottom of tire shreds layer show no soil deposition/strain on it which could ascertain that soil did not get penetrated or percolated through tire shreds layer during the permeability test.

3.4   Consolidation test

One-dimensional consolidation tests were carried out in laboratory to evaluate the compressive strains of tire shreds under incremental loading conditions. A large sized test apparatus comprised of a specimen mold of 300 mm in diameter and 400 mm in height is used. A piston-head of size 297 mm in diameter is used for loading the specimen in the mold.

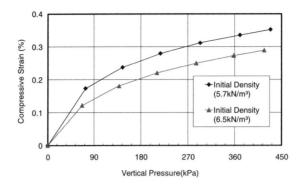

Figure 11. Compressive strain of tire shreds specimens.

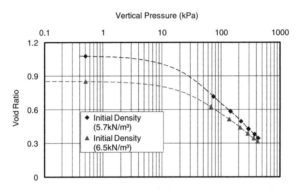

Figure 12. Relationships of vertical pressure and void ratio.

Tire shreds sample (SG) was used in this study. It may be argued that tire shreds mass is highly compressive in nature [Kawaida et al. (2005)], and consequently, their consolidation behavior may change according to the levels of compaction. Therefore, two kinds of specimen settings, viz., tire shreds compacted with initial bulk densities of 5.7 kN/m^3 (loosely filled) and 6.5 kN/m^3 (quite heavily compacted) in the mold are studied.

Compressive strains obtained for the tire shreds specimens under different vertical pressures, and the corresponding e-log(p) relationships are shown in Fig. 11 and Fig. 12, respectively. The tire shreds specimen with higher initial bulk density (6.5 kN/m^3) still exerts a high level of compressive strain under loading though the same is quite heavily compacted in the mold.

Inherently, tire shreds mainly made up of rubber material are susceptible to elastic deformation under loading during compaction, which however, rebounds back to its almost original shape when loading is released (such a material property may be useful in many civil engineering applications). Consequently, a high level of void ratio still remains in the tire shreds specimen, as also may be seen in Fig. 12, even after imparting quite heavy compaction.

4 FIELD TRIAL

Based on laboratory test results a full-scale field trial had been undertaken in Hobetsu, Hokkaido, Japan. A line diagram of the high embankment where tire shreds drainage layers (two) are to be incorporated is shown in Fig. 13. As depicted, the final thickness of drainage layer is planned to be 400 mm each with an inclination of 5% down toward the embankment face. Consequently, upon consideration of the relevant consolidation percentage of tire shreds, the laying thickness for respective drainage layers were adjusted during construction.

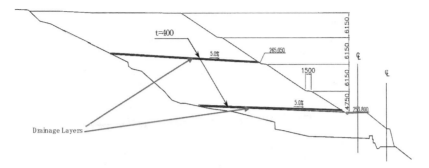

Figure 13.   A line diagram of high embankment for tire shreds drainage layers.

Figure 14.   Laying of tire shreds at the site.

Figure 15.   Dressing of tire shreds.

Length and width of each drainage layer was 40 m and 10 m, respectively. Altogether, 240 ton scrap tire shreds (SG) were used in the form of blanketed drainage/filtration layer under 20 m high embankment. Conventionally available constructional equipments and machinery were used for construction in this field trial. A few snaps on the constructional process e.g., laying, dressing and compaction of tire shreds are shown in Fig. 14 to Fig. 16, respectively.

A view of the completed embankment after one year of construction is shown in Fig. 17. It is essential to mention that *in-situ* consolidation of tire shreds layers placed under this embankment is also measured upon setting up of consolidation plates during construction (Fig. 18). Consolidations measured for about one year after the end of construction are shown in Fig. 19.

Figure 16.   Compaction of tire shreds.

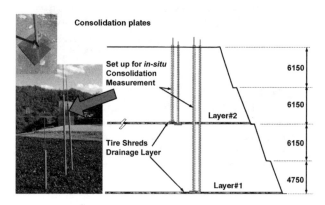

Figure 17.   A view of completed embankment (after 1 year).

Figure 18.   A set-up for measuring *in-situ* consolidation.

As evident, total consolidation of tire shreds layer#1 is 139 mm with respect to the original layer thickness of 520 mm (tire shreds layer#2 was placed a little bit thick than estimated design thickness due to amount of tire shreds supplied at the field in excess with a safety margin). From above, the level of *in-situ* consolidation of bottom tire shreds layer is in the level of 27% which is almost in parity with the compressive strain obtained in laboratory under equivalent vertical pressure of about 20 m high soil capping (compressive strain corresponding to about 360 kPa normal pressure in Fig. 11) on the tire shreds.

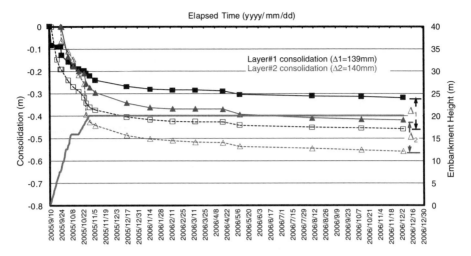

Figure 19. *In-situ* consolidation of tire shreds used as drainage layers under high embankment.

5 CONCLUSION

The theme of this research is to evaluate permeability behavior of tire chips/shreds for their use as geomaterials. In this direction, permeability of tire rubber chips and tire shreds at different levels of compression are evaluated using a newly fabricated large scale test apparatus.

Permeability coefficient of tire chips/shreds has been found to be in the high ($\times 10^0$ cm/s) level which is comparable with the conventionally used highly permeable geomaterials e.g., sand and gravels. Permeability coefficients still remain in the higher level in spite of incurring high level of compressions that may be induced in the field due to surcharge on tire shreds. Permeability coefficients obtained in the compression stage and return stage are in the same order which may indicate that tire shreds specimen is almost recovered to its previous forms even after undergoing a sizable degree of compression. Permeability coefficient of the composite soil-tire system is still in the level of 10^{-1}–10^0 cm/s (almost same as virgin tire shreds or gravel), and remains in the range even after compression of the specimen.

The high levels of permeability characteristics, along with the non-clogging phenomenon of tire shreds had encouraged us to undertake a full scale field trial at Hokkaido on the use of tire shreds as drainage layer under embankment. Altogether, 240 ton scrap tire shreds were used in the blanket formed drainage layer placed under about 20 m high embankment. Almost all the constructional processes were conventionally used machinery controlled, and finished quite smoothly. The performance of tire shreds drainage layer is found to be highly satisfactory till to date. *In-situ* consolidations of tire shreds layers placed under this embankment are measured for about one year after the end of construction. Consolidation test results obtained in laboratory are found to be well in parity with the *in-situ* consolidation. Comparatively, higher level of compressive strain of tire shreds materials may be useful in many civil engineering applications.

ACKNOWLEDGEMENTS

The content of this report is based on results of a joint research conducted by a group of members comprising Takeichi H., Bridgestone Corporation, Tokyo, Japan; Hirotaka K., Shimizu Corporation, Tokyo, Japan; Kawaida M., Central Nippon Expressway Company, Tokyo, Japan; Moki H., Expressway Technology Center, Tokyo, Japan; and Yasuhara K., Ibaraki University, Ibaraki, Japan. The author is deeply grateful, and acknowledges the contributions made by each member.

REFERENCES

JATMA. 2007 –. Tyre Recycling Handbook , Tokyo 105-0001, Japan, p. 25.

Humphrey, D.N. and Manion, W.P. 1992 – Properties of tire chips for lightweight fill. *Grouting, Soil Improvement, and Geosynthetics, American Soc. Civil Engineers,* Vol. 2, pp. 1344–1355.

Karmokar, A.K., Hazarika, H., Takeichi, H. and Yasuhara, K. 2005 – Direct shear behavior of tire chips for their use as lightweight geomaterials. *Proc. 40th Geotech.Res. Conf., Japan Geotech. Soc. Japan,* pp. 641–642.

Karmokar, A.K., Takeichi, H., Kawaida, M., Kato, Y., Mogi, H. and Yasuhara, K. 2006 – Study on thermal insulation behavior of scrap tire materials for their use in cold region civil engineering applications. *Proc. 60th Annual Meeting, Japan Soc. Civil Engineers, Japan,* pp. 851–852.

Kawaida, M., Hamazaki, T., Sano, Y., Fujioka, K., Karmokar, A.K. and Kawasaki, H. 2005. Compaction and consolidation behavior of tire shreds geomaterials. *Proc. the 6th Natl. Conf. on Env. Geotech., Sapporo,* May, pp. 187–192.

Yasuhara, K., Kishida, T., Mitarai, Y., Kawai, H., Karmokar, A. K. and Kikuchi, Y. 2004 – Case study on the application of rubber chips in coastal area. *J. Found. Engg. & Equipment; General Civil Engg. Res. Center,* Vol. 32, No. 7, pp. 79–83.

Yasuhara, K., Karmokar, A.K., Kato, Y., Mogi, H. and Fukutake, K. 2006. Technology for the application of used tires in soil foundation and soil structures. *J. Found. Engg. & Equipment; General Civil Engg. Res. Center,* Vol. 34, No. 2, pp. 58–63.

Scrap Tire Derived Geomaterials – Opportunities and Challenges – Hazarika & Yasuhara (eds)
© 2008 Taylor & Francis Group, London, ISBN 978-0-415-46070-5

Change of engineering properties of cement treated clay by mixed with tire chips

Y. Kikuchi
Port & Airport Research Institute, Yokosuka, Japan

T. Nagatome & Y. Mitarai
Toa corporation, Yokohama, Japan

ABSTRACT: The purpose of this study is to evaluate both the failure and permeability properties of cement treated clay with tire chips using a micro-focus X-ray CT scanner. First, the effect of fraction of tire chips on the physical properties of cement treated clay is discussed. Next, a series of unconfined compression tests was conducted, and the failure mechanism was investigated by CT scanning the specimens during unconfined compression. A series of permeability tests using a triaxial apparatus was also performed during the process of unconfined compression, and their permeability change was examined. Based on the results, the engineering properties of cement treated clay with tire chips were evaluated.

1 INTRODUCTION

In Japan, almost 90% of used tires are recycled, but nearly half of this amount is used in thermal recycling, which produces heat for power generation but also generates CO_2. On the other hand, material recycling, which reduces consumption of virgin resources and does not cause the CO_2 emissions associated with combustion, is a far more effective recycling method from the viewpoint of environmental sustainability. Development of a material recycling method for reusing waste tires in civil works would make it possible to recycle a large number of tires at even a single construction site. Thus, a reuse method of this type would have great potential for recycling.

Many studies have been conducted on barrier design for marine disposal fills in recent years. One important point in designing a barrier is to minimize barrier breakage in the event of serious deformation of the revetments. Cement treated clay (CTC) is a promising sealing material for disposal fills due to its low permeability, but because it is a brittle material, it cannot follow revetment deformation. The authors considered that addition of tire chips to CTC offers an effective means of improving its ductility and maintaining its low permeability, enabling use in structures where deformation is anticipated (Mitarai et al. 2004).

The X-ray computed tomography (CT) scanner, which is used in medical diagnosis, has also been used for engineering purposes as a nondestructive observation method (Otani et al. 2002, Otani and Obara 2003, Kikuchi et al. 2005a, Kikuchi et al. 2005b), and a micro-focus X-ray CT scanner (SMX-225CT) was installed at the Port and Airport Research Institute, Japan in 2004.

The objective of this study is to evaluate both the failure and permeability properties of cement treated clay with tire chips (CTCT) using a micro-focus X-ray CT scanner. The specimens used in this study were CTCT and CTC. First, the effect of fraction of tire chips on the physical properties of the cement treated clay was investigated based on the results of CT scanning with image processing analysis. A series of unconfined compression tests was then conducted with the same specimens. In this paper, the failure mechanism under unconfined compression were discussed as a part of an evaluation of the failure properties of the specimens using the CT data. A series of permeability tests

Table 1. Specifications of micro-focus X-ray CT in PARI.

	Type	Non-enclosed
X-ray unit	Voltage	30–225 kV
	Current	10–1000 μA
	Maximum output	135 W
	Minimum focus size	4 μm
	X-ray irradiation angle	60 deg. in cone shape
Image intensifier	Field-of-view diameter for input window	9/7.5/6/4.5 inches (selectable)
Test samples	Maximum mounting dimensions	φ250 × 800 mm
	Maximum weight	600 N
X-ray shield box	Dimensions	2150 × 1310 × 2400 mm
	Weight	45100 N

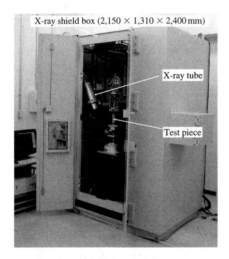

Figure 1. Overview of micro-focus X-ray CT scanner in PARI.

using a triaxial apparatus was also performed with the specimens during the process of unconfined compression, and permeability change was examined. The effect of tire chips on the engineering properties of cement treated clay was then evaluated.

2 X-RAY CT DATA

The X-ray CT scanner introduced in the Port and Airport Research Institute (PARI) in 2004 is the micro-focus type. The basic specifications of this apparatus are shown in Table 1. Figure 1 shows an overview of the apparatus. The shield box is designed such that triaxial tests can be performed in the box: water tubes, air tubes and electric cables are inserted from outside, and the object mounting table can be moved about 50 cm in the vertical direction.

CT image data are evaluated quantitatively using the so-called "gray level," which is a numerical number and proportional to the material density. The gray level was used as a parameter for evaluating density distribution of the specimen. In CT images as shown in Figure 4, darker regions represent lower density zones, while lighter regions represent higher density zones.

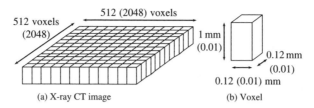

(a) X-ray CT image　　　　　　　　(b) Voxel

Figure 2. Concept of X-ray CT image and voxel.

Figure 3. Tire chips.

Figure 2 illustrates an X-ray CT image of the object with CT scanning. The voxel (volumetric pixel) size can be changed as required by the purpose. For example, in this study, the scanning space was divided into 512×512 or 2048×2048 voxels. In scans of the entire cross section of specimens, 512×512 voxels were used, and the size of one voxel was 0.12 mm $\times$ 0.12 mm $\times$ 1 mm. Microscopic scans of the cross section were performed using 2048×2048 voxels, giving a voxel size of 0.01 mm $\times$ 0.01 mm $\times$ 0.01 mm. Three-dimensional images are reconstructed by superposing 2-dimensional images.

3 PHYSICAL PROPERTIES OF CEMENT TREATED CLAY WITH TIRE CHIPS

3.1 Specimens

Dredged slurry and cement and tire chips were used to make CTC and CTCT. The dredged slurry was Tokyo Bay clay ($\rho_s = 2.716$ g/cm^3, $w_L = 100\%$). The water content of it was controlled to 2.5 w_L with adding sea water. Cement was normal portland cement, the particle density of which was 3.16 g/cm^3. The tire chips were cut in pieces from used automobile tires to a separate size of approximately 2 mm as shown in Figure 3, the particle density of which was 1.15 g/cm^3.

The CTC (Specimen-A) and CTCT (Specimen-B and C) were made by mixing three materials in designated mixing condition as shown in Table 2. The target of unconfined compressive strength of CTC at 28 days of curing time was 300 kN/m^2. Molds of 5 cm in diameter and 10 cm in height were filled up with the mixture. The molds were placed in a high moisture closed box (more than 95% of relative humidity).

The specimens were trimmed to 50 mmϕ × 100 mm. The measured wet densities of cured specimen were 1.284 g/cm^3, 1.257 g/cm^3 and 1.239 g/cm^3 for Sample-A, Sample-B and Sample-C, respectively.

3.2 Investigation of internal structure

CT images of each cross section are obtained based on the distribution of the gray level after scanning the specimen. Figure 4 shows an example of cross-sectional images of each specimen. These were chosen around the middle height in each specimen (50 mm). As described above, the CT

Table 2. Mixing condition.

	Specimen-A (Tire chips: 0%)		Specimen-B (Tire chips: 16.7%)		Specimen-C (Tire chips: 28.6%)	
	Mass (kg/m^3)	Volume (L/m^3)	Mass (kg/m^3)	Volume (L/m^3)	Mass (kg/m^3)	Volume (L/m^3)
Controlled clay	1225.7	980.5	1021.4	817.1	875.5	700.4
Cement	61.5	19.5	51.2	16.2	43.9	13.9
Tire chips	–	–	191.7	166.7	328.6	285.7
Total	1287.2	1000.0	1264.3	1000.0	1248.0	1000.0

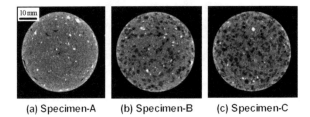

(a) Specimen-A   (b) Specimen-B   (c) Specimen-C

Figure 4. Cross-sectional images.

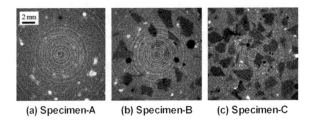

(a) Specimen-A   (b) Specimen-B   (c) Specimen-C

Figure 5. Cross-sectional high magnification images.

images were drawn darker for lower densities and lighter for higher densities. The parts shown in black are air voids which formed in the material during specimen preparation, pure white indicates the presence of a cement lump or piece of shell, and gray parts approximately 1–3 mm in size are tire chips. Figure 5 shows high magnification images of the cross section of each specimen around the middle height. The multiple concentric circles in these images are attributed to image noise. In the images, isolation of the tire chips can be observed in both the macro-range and micro-range. From this, it is considered that the tire chips are suspended in the matrix of the treated clay.

Here, fluctuations in specimen density were evaluated quantitatively using the CT scanning results. As described above, the CT images are constructed using the spatial distribution of gray levels and the gray level can be converted to material density.

Figure 6 shows the estimated density distributions of each specimen from gray levels. Dotted straight lines in this figure are the average wet density of the specimens from volume and mass measurements. The density fluctuations of Specimen-A, Specimen-B and Specimen-C were 1.272–1.292 g/cm³, 1.247–1.270 g/cm³ and 1.227–1.244 g/cm³, respectively. From profile of the estimated densities, density variances are not changed even if mixing the tire chips and it means the tire chips are uniformly mixed in the specimens.

142

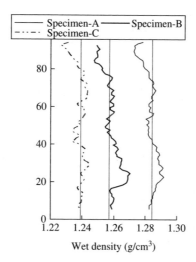

Figure 6. Density distributions.

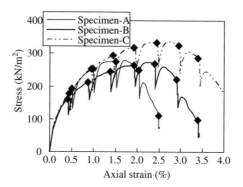

Figure 7. Relationship between the stress and axial strain.

4 FAILURE PROPERTIES

4.1 *Specimens and test procedure*

The specimens used in this study were the same specimens as those used in the investigation of physical properties (Sample-A, Sample-B and Sample-C). A series of unconfined compression tests was conducted with these specimens using a triaxial compression apparatus for the X-ray CT scanner. This apparatus was designed in such a way that the triaxial tests could be performed in the X-ray shield box (Fig. 1). CT scanning was conducted before compression and at 0.5% increments of axial strains. When nondestructive inspection was conducted with the X-ray CT scanner during unconfined compression tests, compression was interrupted momentarily while scanning was performed.

4.2 *Results of unconfined compression test*

Figure 7 shows the relationship between the stress and axial strain with Sample-A, Sample-B and Sample-C. The results show that the axial strain required to reach peak strength increases with increment of the fraction of the tire chips. The points marked by ♦ in the figure show the points where CT scanning were conducted. Compressive stress shows a clear drop in each scanning point, it is due to the relaxation of the specimen at this time.

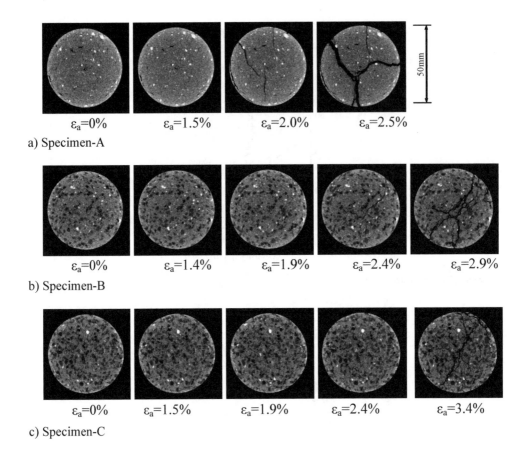

a) Specimen-A

$\varepsilon_a=0\%$ $\varepsilon_a=1.5\%$ $\varepsilon_a=2.0\%$ $\varepsilon_a=2.5\%$

b) Specimen-B

$\varepsilon_a=0\%$ $\varepsilon_a=1.4\%$ $\varepsilon_a=1.9\%$ $\varepsilon_a=2.4\%$ $\varepsilon_a=2.9\%$

c) Specimen-C

$\varepsilon_a=0\%$ $\varepsilon_a=1.5\%$ $\varepsilon_a=1.9\%$ $\varepsilon_a=2.4\%$ $\varepsilon_a=3.4\%$

Figure 8. Cross-sectional images.

Figure 8 shows the cross-sectional images of each specimen at initial state and axial strain of approximately 1.5%, 2.0%, 2.5%, 3.0% and 3.5%. These images were chosen at the middle height of each specimen (about 50 mm height). As a whole, there were no large changes before and at the stress peak condition (the axial strain at failure of Specimen-A, Specimen-B and Specimen-C was 1.5%, 1.9% and 2.4%, respectively), but cracks occurred in each specimen after the stress peak. However, the appearance of the cracks in the respective cross sections was different in CTC (Specimen-A) and in CTCT (Specimen-B and Specimen-C). The cracks in CTC were almost straight and their thickness widened as axial strain increased. In contrast, the cracks in CTCT were thin and developed successively in a reticulated pattern as axial strain increased.

Figure 9 shows 3-dimensional images of the cracks after the peak condition from outside and inside in all specimens. The cracks in CTC (Specimen-A) were straight, while those in CTCT (Specimen-B and Specimen-C) were not straight. Thus, the mode of crack formation was changed by the presence of tire chips.

5 PERMEABILITY PROPERTIES

5.1 *Specimens and test procedure*

Table 3 shows the mixing conditions of these specimens. CTCT used in this series of tests included 16.7% of tire chips in volume. The materials used and procedure prepared were the same as

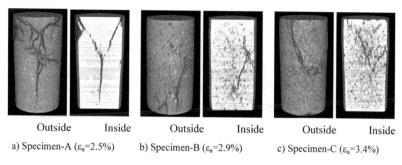

| Outside | Inside | Outside | Inside | Outside | Inside |

a) Specimen-A (ε_a=2.5%) b) Specimen-B (ε_a=2.9%) c) Specimen-C (ε_a=3.4%)

Figure 9. 3-dimensional images.

Table 3. Mixing condition.

	Specimen-D (Tire chips: 0%)		Specimen-E (Tire chips: 16.7%)	
	Mass (kg/m^3)	Volume (L/m^3)	Mass (kg/m^3)	Volume (L/m^3)
Controlled clay	1199.0	974.8	999.2	812.3
Cement	79.6	25.2	66.4	21.0
Tire chips	–	–	191.7	166.7
Total	1278.6	1000.0	1257.3	1000.0

described in 3.1. The water content ratio was 2.8 w_L and the cement content were slightly higher than Specimen-A, Specimen-B and Specimen-C.

The specimens were trimmed to 50 mmϕ × 100 mm. The measured wet densities of cured specimen were 1.270 g/cm^3 and 1.251 g/cm^3 for Sample-D (CTC) and Sample-E (CTCT), respectively.

A series of permeability tests using the triaxial compression apparatus was performed some points during the process of unconfined compression. The compression tests were conducted under the unconfined condition. The permeability tests were conducted under confined condition, because of water pressure of 20 kPa was applied to the specimen bottom to create a water pressure gradient. That is, first, the specimen was scanned with X-ray CT scanner. Then the permeability test was performed during the specimen was consolidated with confining pressure of 40 kPa. After the permeability test the specimen was released from consolidation pressure and compressed in an unconfined condition until the planned axial strain was achieved. This procedure was repeated at the designed axial strain.

5.2 *Results of permeability tests during process of unconfined compression*

Figure 10 shows the relationship between the stress and axial strain of specimen D and E. The points marked by ◆ in the figure show the points where CT scanning and permeability test were performed. The results show that the stress – strain relationships are rather similar in both cases. But there are some differences in around failure strain and in post failure strain level. As mentioned above, compression was interrupted when each CT scanning and permeability test were conducted. Compressive stress shows a clear drop in each scanning and permeability test. It is due to the relaxation of the specimen at this time.

Figure 11 shows the cross-sectional images of each specimen at initial condition, stress pre-peak condition, stress peak condition and axial strain of approximately 4.0%, 6.0% and 15.0%. These images were chosen at the middle height of each specimen. Specimen-D displays no large change

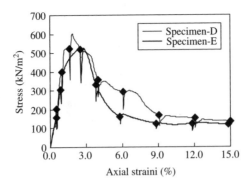

Figure 10.   Relationship between the stress and axial strain.

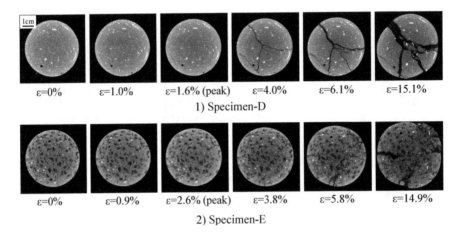

1) Specimen-D

2) Specimen-E

Figure 11.   Cross-sectional images.

from initial condition to stress peak condition. The cracks occurred in the specimen after the peak condition. The cracks in Specimen-D were almost straight and they were widened as axial strain increased. On the other hand, Specimen-E exhibits no large change from initial condition to pre-peak condition. Then the thin cracks occurred in the specimen at around the stress peak condition. And their cracks in Specimen-E were developed one after another in a reticulated pattern as axial strain increased. At the 15% axial strain, a few part of their cracks were widened. As mentioned, both the appearance and growth of the cracks in the respective cross sections was different in CTCT from in CTC.

Figure 12 shows high magnification images of the cross section of each specimen under uncon-fined compression around the middle height. In the image of Specimen-E at the initial condition, isolation of the tire chips can be observed in micro-range. The images of Specimen-E shows the cracks appeared around tire chips before stress peak condition. It is attributed to the fact that the Poisson's ratio of both the tire chips and cement treated clay are different. That is, Poisson's ratio of CTC is about 0.15 and that of tire chips is 0.5 (Kikuchi et al. 2006). If compressive stress is applied to CTCT, the part of CTC is more compressive than tire chips. It makes small thin cracks in CTC part. These thin cracks developed one after another around the tire chips, and their growth into wide cracks was prevented.

Figure 13 shows 3-dimensional images under the process of unconfined compression from inside in each specimen. The cracks in Specimen-D were straight, while in Specimen-E were not straight

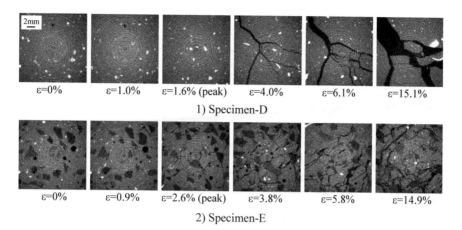

2mm ε=0% ε=1.0% ε=1.6% (peak) ε=4.0% ε=6.1% ε=15.1%

1) Specimen-D

ε=0% ε=0.9% ε=2.6% (peak) ε=3.8% ε=5.8% ε=14.9%

2) Specimen-E

Figure 12. High magnification images of the cross section.

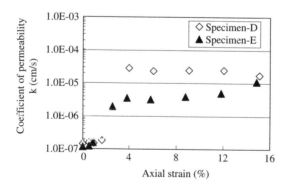

ε=0% ε=1.6% ε=4.0% ε=6.0% ε=0% ε=2.6% ε=3.8% ε=5.8%

1) Specimen-D 2) Specimen-E

Figure 13. Three-dimensional images.

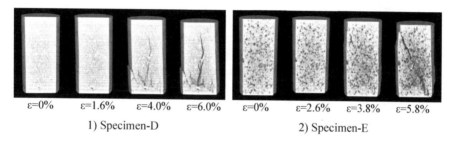

Figure 14. Relationship between the coefficient of permeability and axial strain.

and connected little by little. Thus, the mode of crack formation was changed by the presence of tire chips.

Figure 14 shows the relationship between the coefficient of permeability and axial strain for Specimen-D and Specimen-E. With Specimen-D, the coefficient of permeability increased readily after crack initiation. In contrast, in Specimen-E, the coefficient of permeability is about one order smaller than that of Specimen-D at this time. It is considered that large cracks formed in the Specimen-D created large water-bleeding channels, and they made the coefficient of permeability high. On the other hand, the cracks in Specimen-E were thin and did not create water-bleeding channels until large axial strain as 15%, and maintained the low coefficient of permeability.

147

6 CONCLUSION

In this study, the failure and permeability properties of cement treated clay with tire chips are discussed based on unconfined compression tests and permeability tests using a micro-focus X-ray CT scanner. The internal structure and failure mechanism during shear deformation were investigated visually and also evaluated quantitatively. The following conclusions were drawn from this study:

1. It was observed that tire chips are isolated and suspended in the matrix of the cement treated clay. Tire chips are found to be well mixed in cement treated clay from the observation of the density distribution by using CT scanner.
2. The failure mechanism of cement treated clay with tire chips under unconfined compression is different from that of cement treated clay without tire chips.
3. The water interception capability of cement treated clay was significantly degraded by the appearance of cracks after peak strength, while crack growth after peak strength was inhibited in cement treated clay with tire chips and the coefficient of permeability is one order smaller than that of without tire chips. This is attributed to differences in crack formation and development under shear deformation.

REFERENCES

Kikuchi, Y., Nagatome, T., Otani, J. and Mitarai, Y. (2005a): Engineering property of cement treated soil with tire-chips using X-ray CT scanner, *Proc. of the Sixth National Symposium on Environmental Geotechnology*, 343–350 (in Japanese).

Kikuchi, Y., Nagatome, T., Mitarai, Y. and Otani, J. (2005b): Failure properties of cement treated clay with added tire chips using X-ray CT scanner, *Proc. of the 2nd China-Japan Joint Symposium*.

Kikuchi, Y., Nagatome, T. and Mitarai, Y. (2006): Failure mechanism of cement treated clay with tire chips and change of its permeability under shear deformation, *Japanese Geotechnical Journal*, **1** (2), 19–32 (in Japanese).

Otani, J., Mukunoki, T. and Kikuchi, Y. (2002): Visualization for engineering property of in-situ light weight soils with air foams, *Soils and Foundations*, **42** (3), 93–105.

Otani, J. and Obara, Y. (2003): *International Workshop on X-ray CT for Geomaterials – GeoX2003 –*

Mitarai, Y., Kawai, H., Yasuhara, K., Kikuchi, Y. and Karrmokar, A.K. (2004): The mechanical properties of composite geo-material mixing with scraped tire-chips and application example to harbor construction works, *Proc. of the 49th Symposium on Geotechnical Engineering*, 77–84 (in Japanese).

Technical papers

Part 1 Mechanical properties, modeling and novel applications

Size effect on tire derived aggregate mechanical properties

M. Arroyo, J. Estaire,[1] I. Sanmartin & A. Lloret
Department of Geotechnical Engineering and Geosciences, UPC, Barcelona, Spain
[1]*Laboratorio de Geotecnia, CEDEX, Madrid, Spain*

ABSTRACT: Tire derived aggregate might substitute for other granular materials in civil engineering structures. Processing the tires to obtain granulate materials has a cost that increases rapidly as the aggregate size decreases; this reason favors the use of large aggregate sizes. Minimizing the risk of self-combustion is another reason why large aggregate sizes are favored in civil engineering applications. The use of relatively large aggregate sizes (30 to 3 cm) poses a characterization problem. Geotechnical test equipment is classically designed to deal with smaller granulate sizes. Tire derived aggregates of smaller sizes are available and may be employed as an analogue, but the doubt remains about the validity of the properties thus derived for the larger aggregate sizes employed in civil engineering structures. A testing program has been designed to explore possible size effects in shear resistance and compressibility of tire derived aggregates. This communication presents some results of that testing program, with an emphasis on shear resistance.

1 INTRODUCTION

During the last decade disposal of used tires has been a subject of major regulatory activity in Europe in general and Spain in particular. The EU Landfill Directive (EU, 1999) forbade the landfill disposal of whole tires since 2003 and of shredded tires since 2006. In 2001 Spain Environment ministry presented the first national plan on waste tire (PNNFU, MMA 2001). A hierarchy of waste management strategies was established: prevention, re-use, recycle, energy production and disposal, in order of priority. The plan established a goal of 20% of total waste tire production recycled by 2006.

The last available statistics (MMA, 2007) indicate that currently about 300,000 tons of waste tires are generated per year in Spain. In 2005 (last year for which data is available) only 13,5% of the total production was recycled and still 50% of the residue was disposed of in landfills. Therefore there is still a large scope for improving the situation.

The use of shredded tires as tire derived aggregate (TDA, Humphrey, 2007) in civil engineering works has a number of benefits from the environmental viewpoint. First, it is a low-cost recycle product, because large sizes (up to 30 cm) might be employed and the production costs are lower than for finer materials. Second, it has a large consumption potential: thousands of tons of residue may be employed in a single project. Thirdly, exploiting its properties (mechanical, hydraulic, acoustic, thermal) may offer substantial improvements on construction costs in a number of circumstances.

The mechanical property of TDA fill that has attracted more interest up to now is its being lightweight. TDA fill may substitute advantageously for other construction lightweight materials (expanded clay, polystyrene). A lightweight fill has advantages when building over low-bearing capacity soils or when fill loads over or onto structures.

TDA fill has been already employed successfully in a number of civil engineering projects. The vast majority of these projects have been embankments; there is a more limited experience in other applications, like retaining walls. To consolidate this experience and expand into newer applications, some fundamental knowledge of the TDA mechanical properties seems advantageous.

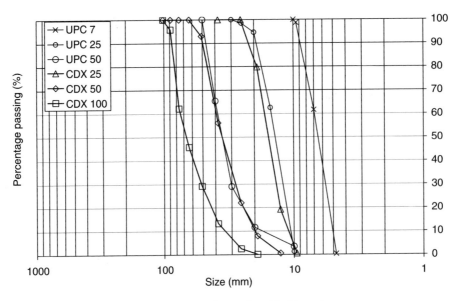

Figure 1. Granulometric curves of the TDA samples employed in this study.

Laboratory studies of TDA have been already numerous (Humphrey et al., 1993; Ahmed, 1993; Drescher et al., 1999; Lee et al., 1999; Moo-Young et al., 2003). However, there are still a number of open issues with respect to TDA mechanical behaviour. A prominent one is how best to model this material: a comprehensive constitutive model is still lacking. With direct bearing on this main problem a second one appears: how best to test this material on the laboratory? The sizes of most TDA employed as fill are larger than what is easily tested with conventional soils laboratory equipment. If the same material employed in field applications has to be tested large and cumbersome specimens are required. However, if smaller TDA sizes are tested there might be doubts about how representative are the results obtained for field applications.

This last question can be summarized as follows: is there any significative size effect on TDA mechanical response? Some results from an experimental campaign specifically designed to answer that question are presented in this communication.

2 MATERIALS AND METHODS

Several samples of TDA were sent from a commercial shredding operation to the geotechnical laboratories at UPC and CEDEX. The granulometric curves of these samples are presented here in Figure 1. The main identification characteristics of the samples are summarized in Table 1, including median grain diameter, D_{50}, specific weight, G, initial water content, w_{ini}, maximum adsorbed water w_{max}. The smaller sized samples have a more uniform granulometric curve, a consequence of their being obtained at a later stage in the shredding process. These smaller fragments do also show a higher ability to absorb water.

Table 1 does also include several metallic content measures, determined according to ASTM D6270; the weight percentage of free metallic fragments, M_f, the weight percentage of TDA with exposed metal of length above 25 mm, M_{e25}, and the weight percentage of TDA with exposed metal of length above 50 mm, M_{e50}. Samples with small sizes do not have any metal fragments, since they are obtained at a later stage in the shredding process. The larger sized samples have a relatively large amount of metal, free and/or exposed, enough to make them fail the material criteria for Type II fill specified by ASTM D6270 (ASTM, 1998).

152

Table 1. Characteristics of the TDA samples employed in this study.

Sample	D_{max} (mm)	D_{50} (mm)	C_u	G (%)	w_{ini} (%)	w_{max} (%)	M_f (%)	M_{e25} (%)	M_{e50}
UPC 7	10	6.6	1.35	1.11	0.43	6.92	0	0	0
UPC 25	30	14	1.39	1.22	1.45	6.19	0	0	0
UPC 50	50	36	2.11	1.28	0.70	3.18	2.2	55.4	8.4
CDX 25	30	16	1.57	1.16			0	0	0
CDX 50	65	36	1.96	1.44			1.47	18.1	1.78
CDX 100	100	66	2.25	1.16			0	24.5	5.77

Table 2. Test program.

	Apparatus characteristic length (mm)					
	Oedometer			Direct shear		
Nominal tire size (mm)	50	300	1000	50	300	1000
7	X	X	p	X	X	p
25		p	X		X	p
50		X	p		X	X
100			p			X

X: Results available at the time of writing, p: pending.

On these materials a series of shear box and oedometric tests has been planned. As specified in Table 2, the experimental program is ongoing and not all tests have been yet performed. This communication focuses almost exclusively on direct shear results.

Testing at CEDEX took place on a large scale box of $1m^3$ capacity, as well as in a 300 mm square shear box. Initially designed for rockfill testing (Estaire & Olalla, 2006), the $1m^3$ apparatus can act both as a shear box and as an oedometer. When employed as an oedometer, grease was applied on the box sides and a load cell was placed at the bottom of the box, taking care of embedding it in a layer of fine sand. In direct shear, a dummy was located at the bottom of the box, restricting the sample height to 750 mm and thus avoiding compression below the shear plane. Shear was always applied at a rate equal to 1.5 mm/min.

Testing at UPC employed medium scale apparatuses of 300 mm size as well as standard sized apparatuses of 50 mm diameter. The 300 mm oedometer had three load cells located below the base plate, and silicone grease was applied to the walls. Direct shear tests proceeded generally at a rate equal to 1 mm/min. For comparative purposes, a different shear rate of 0.027 mm/min was applied twice on the 50 mm apparatus. No significant change in the results was observed.

At UPC the material for all the tests was placed without compaction. In the direct shear tests the same material was tested at all pressures, after unloading and reloading. At CEDEX, part of the material was replaced after each shear test, and the rest was removed with a fork. The material states (density, normal stress) before shear was applied are shown in Figure 2.

3 DIRECT SHEAR RESULTS

The results in Figure 3 and Figure 4 are representative of the direct shear results obtained on the campaign, across all materials and equipments. Shear strength does not show a peak, but most tests attain a clearly defined shear stress plateau at large horizontal displacements, clearer, at least, than

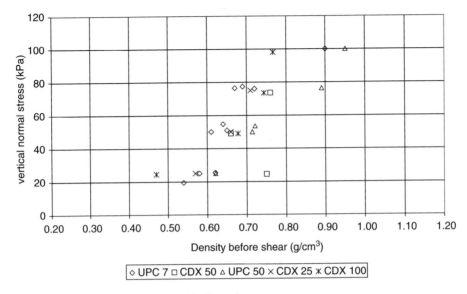

Figure 2. Initial state of the samples tested in direct shear.

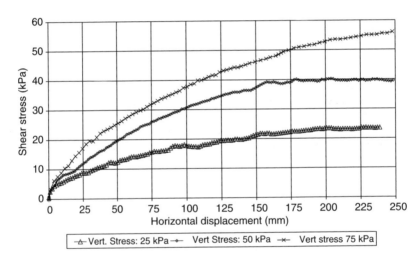

Figure 3. Shear stress vs horizontal displacement for CDX 50 material tested in the 1000 mm shear box.

that in most previously reported results (Humphrey et al. 1993, Yang et al. 2002, Moo-Young et al. 2004). All tests, without exception, showed contractive behaviour during the initial phase of shear, a result in contrast with that of Yang et al. (2002). The later stages of the vertical displacement measurements seem less consistent between tests at different stress levels and might have been affected by test artifacts (frame or load plate rotation, load plate jamming).

A number of comparisons are possible across material and apparatus scales. Here the representative scales are denoted, respectively, by Lm (taken equal to D_{50}) and Ld. In Figure 5 and Figure 6 a single device size is employed to test three different sized TDA, placed under similar confining stress at similar densities. There is good agreement between the tests on the stress – displacement curve, although the intermediate material shows a slightly different start. The vertical displacement pattern disagrees at large displacements, but agrees at the more reliable moderate displacements.

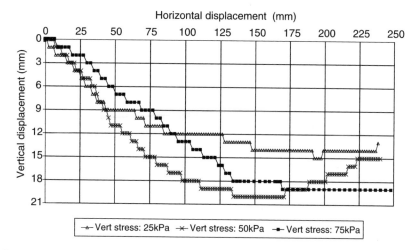

Figure 4. Vertical vs horizontal displacement for CDX 50 material tested in the 1000 mm shear box.

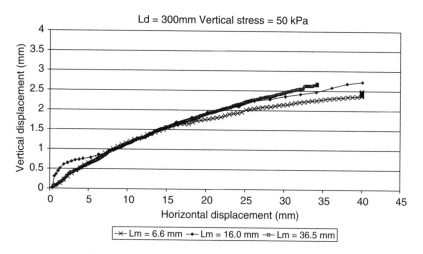

Figure 5. Shear stress vs horizontal displacement for three TDA material sizes. Device length 300 mm.

To compare results from different apparatus some kind of normalization is necessary. For a similar direct shear testing series on sand, Cerato (2005) employed as normalizing parameter either the thickness of the sample or that (presumed) of the shear zone. There is very limited evidence of what might be the shear zone thickness for TDA. On the other hand, and since many previously reported curves have not show a clear strength plateau, it has become a standard practice of TDA direct shear reporting to quote as failure strength that corresponding to a displacement equal to some fraction of the apparatus shear length (fraction reported here as xf in Table 3). Therefore, to make comparisons across apparatus length, it seemed appropriate to normalize the horizontal displacement by sample length. A typical result is that shown in Figure 7, showing a fair degree of similitude between the normalized displacement vs shear curves.

The use of a standard mobilised friction is, therefore, advantageous to make comparisons valid across all apparatus and material scales. The standard mobilised friction is taken as that corresponding to a horizontal displacement equal to Ld/10. When that value is plotted against the characteristic

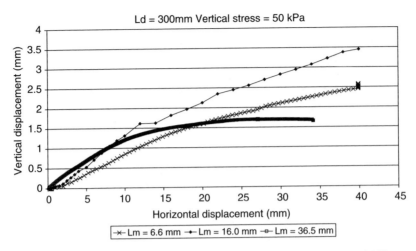

Figure 6. Vertical vs horizontal displacement for three TDA material sizes. Device length 300 mm.

Table 3. Studies of TDA on the direct shear apparatus. All lengths in mm and stresses in kPa.

Reference	Material	Device	Tests
Humphrey et al., 1993	Dmax = 76/51/38 Compacted: Yes Protruding steel: Y/N	Square box Ld = 305/406 H = 228	xf = Ld/10 dx = 7.6 mm/min σ = 17–68
Foose et al., 1996	Dmax = 150/100/150 Dmin = 100/50/6 Not compacted Protruding steel: Yes	Circular box Ld = 279 H = 314	xf = Ld/10 dx = 1.3 mm/min σ = 0–80
Yang et al., 2002	Dmax = 10 Dmin = 2 Compacted: Yes Protruding steel: No	Circular box D = 63.5	xf = Ld/30 to Ld/10 dx = 1 mm/min σ = 0–83
Moo-Young et al., 2003	Dmax = 300/200/100/50 Dmin = 200/100/50/2 Compacted: Yes Protruding steel: Yes	Square box Ld = 610 H = 305	dx = 0.05 mm/min σ = 4–40
This study	Dmax=100/65/50/30/10 Compacted: Y/N Protruding steel: Y/N	Square/circular box Ld = 50/1000 H = 228	xf = Ld/10 dx = 0.03/1/1.5 mm/min σ = 20–100

material length, Figure 8, a considerable dispersion is observed, but unrelated to the material length. A similar effect was observed when the results where plotted against device length or against a normalized length ratio, Ld/Lm, as in Figure 9.

It is therefore clear that all sizes of TDA tested here behaved similarly in direct shear across all the direct shear devices employed. It seems therefore legitimate to plot all the standard mobilised friction results together and try to infer something about the direct shear resistance of the material.

In Figure 10 all the shear strength results obtained are represented, alongside two possible interpretations. The first is a simple linear regression, with a line forced through the origin. Such linear regression, of course, corresponds to a shear resistance defined by a 29.2 friction angle and

156

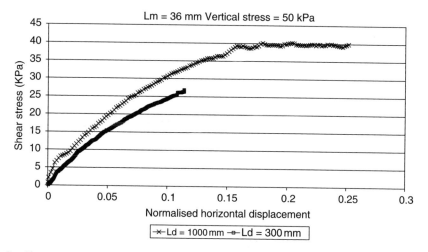

Figure 7. Shear stress vs normalized horizontal displacement for two apparatus. Material length 36 mm.

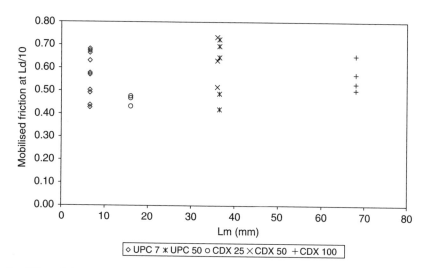

Figure 8. Effect of characteristic material length on TDA standard mobilized friction.

no cohesion. The second is an exponential relation, suggested by Yang et al (2002), namely

$$\tau = 1.6\sigma^{0.75} \tag{1}$$

It is visually apparent that the nonlinear relation does not improve the fit of the linear one. A more precise comparison might be obtained using the sum of squared residuals: those of the linear relation add up to just 7% of those of the exponential relation.

Note that, for sands (Lings & Dietz, 2004), the relation between the direct shear friction angle at large displacements, φ_{ld}, and the critical state friction angle, φ_{crit}, is given by $\tan \varphi_{ld} = \sin \varphi_{crit}$. Large displacements are roughly equivalent to the Ld/10 criteria previously selected here. Therefore, applying that relation here results on a critical state angle for TDA of 34.1 degrees.

A summary of the main characteristics of previous direct shear testing campaigns on TDA is presented in Table 3. The testing campaign here described is more comprehensive than the precedent campaigns. It does also maintains a higher ratio of apparatus size to characteristic material

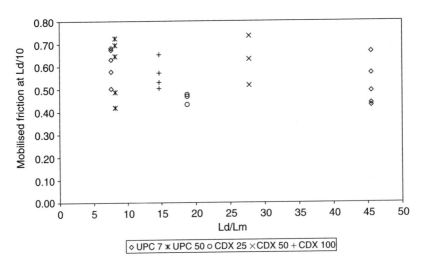

Figure 9. Effect of normalized length ratio on TDA standard mobilized friction.

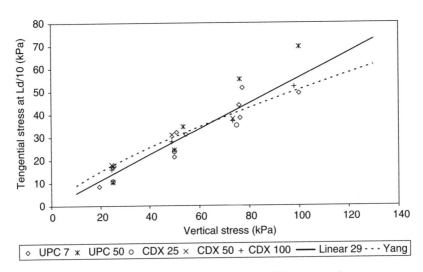

Figure 10. Shear strength results for all the available tests and possible interpretations.

size (7.6 being here the minimum) than some previous research. It is believed that some erratic results previously obtained might be explained by the use of very low ratios of apparatus size to characteristic material size (less than 3 in some cases).

4 OEDOMETRIC RESULTS

The number of oedometric tests performed at the time of writing was still limited. The noval or virgin loading branch of the available results is shown in Figure 11. For the 300 and 1000 mm devices, the equivalent load there indicated corresponds to that obtained as the average of the load applied on top of the apparatus and that registered at the base. Therefore side friction effects, amounting roughly to 20% of the applied load, but variable across different apparatuses, have been already discounted.

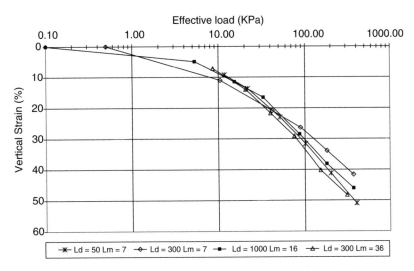

Figure 11. Noval branch of oedometric tests on TDA. Ld, device length. Lm, matrial length.

5 SUMMARY AND CONCLUSIONS

The results of a direct shear campaign here described show no evidence of size effect on the shear resistance of tire derived aggregate. The campaign covers relative sizes Ld/Lm ranging from 7.6 to 45. Within the range explored, the presence or not of protruding steel on the TDA had no clearly identifiable effect on the direct shear results. Despite some dispersion, the results obtained for all materials and apparatus support a linear relation between confining pressure and shear resistance. The direct shear friction angle thus derived is 29.2 degrees, equivalent to a 34.1 degrees critical state friction angle. The absence of size effect for shear resistance has been also pointed at by Humphrey (2007).

In oedometric conditions, the limited amount results here explored seem to indicate again that no important size effect is present while testing TDA. This result needs to be confirmed by further testing.

That no size effect appears in the mechanical response of TDA might seem surprising. However, a possible reason for this difference may be that, as opposed to what is observed in other granular materials (e.g rockfill, Oldecop & Alonso, 2000), the tire fragments that compose TDA are not prone to fracturing in the TDA operative stress range.

ACKNOWLEDGEMENTS

The research described in this paper was partly supported by the research funds of the Spanish ministry of the Environment (Ministerio de Medio Ambiente) through grant 197/2006 2.5-2 Samples of TDA were procured by SIGNUS Ecovalor.

REFERENCES

Ahmed, I (1993) Laboratory study on the properties of rubber soils, PhD thesis, Purdue University
ASTM (1998) Standard practice for use of scrap tires in civil engineering applications. ASTM D6270-98
Cerato, A.B. (2005) Scale effects of shallow foundation bearing capacity on granular material, PhD, Department of Civil and Environmental Engineering, University of Massachusetts in Amherst

Drescher, A., Newcomb, D. & Heimdahl, T. (1999) Deformability of shredded tires, Report, University of Minessota, Department of Civil Engineering

Estaire, J. & Olalla, C. (2006) Analysis of the strength of rockfills based on direct shear tests made in 1 m³ shear box, 22nd ICOLD, Barcelona, Q.86-R.36, 529–540

Foose, G.J., Benson, C.H. & Bosscher, P.J. (1996) Sand reinforced with shredded waste tires, ASCE, J. of Geotechnical and Geoenvironmental Eng. 122, 9, 760–767

Humphrey, D.N. (2007) Tire derived aggregate as lightweight fill for embankments and retaining walls, IW-TDGM, Yokosuka, Japan, Preprint Proceedings, 56–79

Humphrey, D.N., Sandford, T.C., Cribbs, M.M. & Manion, W.P. (1993) Shear strength and compressibility of tire chips for use as retaining wall backfill, Transportation Research Record, 1422, 29–35

Lee, J.H., Salgado R., Bernal A. & Novell, C.W. (1999). Shredded tires and rubber sand as lightweight backfill. J. of Geotech. and Geoenvir. Engrg., ASCE, pp. 132–141.

Lings, M. L. & Dietz, M. S. (2004). An improved direct shear apparatus for sand. Geotechnique, 54,4, 245–256

Moo-Young, H., Sellasie, K., Zeroka, D. & Sabnis, G. (2003) Physical and Chemical Properties of Recycled Tire Shreds for Use in Construction. J. of Envir. Engrg., ASCE, pp. 921–929.

MMA, (2001) Plan Nacional de Neumáticos Fuera de Uso, Ministerio de Medio Ambiente (in Spanish)

MMA, (2007) Borrador del Plan Nacional de Residuos, Ministerio de Medio Ambiente (in Spanish, draft available at http: //www.mma.es/portal/secciones/calidad_contaminacion/residuos/planificacion_residuos/borrador_pnir.htm)

Oldecop, L.A. & Alonso, E.E. (2000) A model for rockfill compressibility Géotechnique, 51, 2, 127–139

Yang, S., Lohnes, R. A. & Kjartanson, B. H. (2002) Mechanical Properties of Shredded Tires, Geotechnical Testing Journal, GTJODJ, Vol. 25, No. 1, pp. 44–52.

Warith, M.A. & Rao, S.M. (2005) Predicting the compressibility behaviour of tire shred samples for landfill applications. Waste Management 26, 268–276.

Wu, W.Y., Benda, C.C. & Cauley, R.F. (1997) Triaxial determination of shear strength of tire chips J. of Geotech. and Geoenvir. Engrg., ASCE, pp. 479–482.

Compressibility and liquefaction potential of rubber composite soils

P. Promputthangkoon & A.F.L. Hyde
University of Sheffield, UK

ABSTRACT: Used tyres have become an increasingly problematic global problem whose disposal is posing dangers to the environment. Disposal of whole used tyres has been prohibited by a new EU Landfill Directive since July 2003. Thus, recycling needs to be considered imaginatively and the solutions must be sustainable. One method of disposing of tyres is by mixing tyre chips with soil and using them for landfill purposes. The first consideration is how this will affect the compressibility of the landfill materials. Initial results show that a sand mixture containing 5% tyre chips by solid volume has a similar compressibility to that of pure sand. The compressibility of the mixtures, however, begins to change noticeably when the percentage of tyre chips is increased to 10% and composites with greater than 20% may not be tolerable in geotechnical design, suggesting that the limiting percentage of tyre chips mixed sand should not exceed 20%. A major consideration for fill materials in seismic zones is their susceptibility to liquefaction. Initial cyclic triaxial load testing of sand rubber composites with similar grain sizes for each fraction has demonstrated a more gradual build up of pore water pressures and an overall reduction in the liquefaction resistance with increasing rubber chip content.

1 INTRODUCTION

Geotechnical engineers have paid attention to the occurrence and consequences of soil liquefaction since the 1964 earthquake in Niigata and Alaska. They now largely understand the phenomenon in terms of how it happens. However, many aspects of liquefied soil are still not fully understood due to the complexity of earthquake patterns as well as the interaction between soil particles and generated pore water pressures during seismic loading. When soil liquefies, it completely loses its shear strength and stiffness and behaves like a viscous fluid having a unit weight equal to a saturated soil. In such conditions, if superstructures have a unit weight greater than that of the liquefied soil, will sink. On the other hand, buried structures of which the unit weight is smaller, will float (JGS, 1998). Damage is then inevitable. If a structure is to be constructed in seismic zones containing liquefiable soil, countermeasures are needed to maintain the function of the structure (PHRI, 1997). The basic principles for reducing the risk are increasing the shear strength and stiffness of the liquefiable soil. This could involve replacement of fills, removal of water, soil compaction, chemical grouting and deep mixing, and thermal treatment (Sitharam, 2003).

In element tests under undrained conditions, a soil is considered to have liquefied when the generated pore water pressure is approximately equal to the initial effective stress. In the cyclic triaxial test, it is customary to use either 5% double-amplitude axial strain in case of isotropic consolidation or 5% axial plastic strain in case of anisotropic consolidation as a failure criterion (Hyodo et al., 1996; Hyde and Higuchi 2005). The liquefaction strength of a soil in laboratories is defined as a stress level required to cause a specified level of strain after a specific number of load cycles (Ishihara, 1993).

Recently, discarded tyres have become an increasing problem around the world because disposing of them in open areas is a danger to the environment. They are vulnerable to fire and subsequently may contaminate ground water. As a result, disposing whole used tyres has been prohibited by a

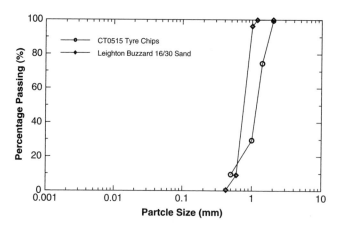

Figure 1. Grain size distribution curves of Leighton Buzzard sand and CT0515 tyre chips.

New EU Landfill Directive since July 2003. Thus, making use of them needs to be considered imaginatively and solutions must be sustainable. Used tyres have been recycled and used in civil engineering applications such as road construction and it has been suggested that they might be used for back fill for retaining walls. An important property of waste tyres is that they are much lighter compared to soil, helping geotechnical engineers to overcome the settlement and stability problems of embankments constructed on very soft clay. However, the high compressibility and exothermic reactions must be taken into account.

This paper focuses on the compressibility and liquefaction potential of coarse, liquefiable sand mixed with tyre chips. The compressibility and the liquefaction strength of the mixtures were determined using a conventional oedometer cell and a computer-controlled cyclic triaxial testing system, respectively.

2 TEST MATERIALS

The sand used was Leighton Buzzard 16/30 sand obtained from WBB Minerals, UK. It is a silica sand having D_{50} of 0.7 mm, brown colour, sub-rounded to sub-angular grains. the maximum and minimum void ratios were 0.821 and 0.460 and $G_s = 2.67$. The tyre chips CT0515 were obtained commercially from Charles Lawrence International. They had an average particle size D_{50} of 1.2 mm, uncompacted bulk density of 0.432 Mg/m^3, and $G_s = 1.17$ compared to the ASTM (1998) average value of 1.02–1.27, and 1.13–1.36 from Edil and Bosscher (1994). The grain size distributions of the tyre chips and sand were chosen to be approximately similar as is shown in Fig. 1.

3 TEST EQUIPMENT AND PROCEDURES

3.1 Oedometer cell

Normally sand is relatively incompressible and thus settlements are not a major concern in geotechnical design. The compressibility characteristics of sand mixed with tyre chips, of course, are somewhat different from those of plain sand and it is predictable that the mixture will be more compressible. Hence, it was first necessary to determine a threshold percentage of tyre chips below which settlements might be acceptable. Thus one-dimensional compression tests were used to determine the compressibility of the mixtures. A conventional oedometer cell having a diameter of 75 mm and a height of 20 mm was used to determine the compressibility of sand and sand mixed

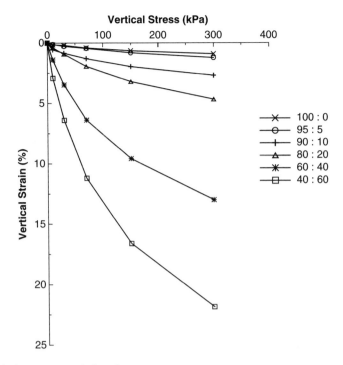

Figure 2. Vertical stress vs. vertical strain.

with tyre chips. The sand and tyre chip mixtures were expressed in terms of solid volumes and prepared with an initial void ratio of 0.83. The ratios of sand to tyre chips were 100:0, 95:5, 90:10, 80:20, 60:40, and 40:60.

3.2 Cyclic triaxial testing system

A computer-controlled closed-loop cyclic triaxial testing system manufactured by ELE International and later modified by Higuchi (2001) was used to perform the undrained cyclic triaxial tests. It basically comprises a triaxial cell, a loading frame fitted with a double acting actuator connected to a pneumatic servo-valve and a Control and Data Acquisition System (CDAS) linked to a computer. The system was modified by adding a vacuum gauge and a CO_2 facility for aiding sample preparation and saturation.

Cyclic triaxial tests were performed on cylindrical specimens 100 mm in diameter and 200 mm high. The mixtures for each test were separately mixed in eight portions and were reconstituted in a mould by means of dry funnel deposition with zero drop height in eight layers. The initial void ratio for the triaxial samples was 0.70; the ratios of sand to tyre chips were 100:0, 99:1, 97.5:2.5, 95:5, and 85:15, by solid volume.

An initial cell pressure of 50 kPa was applied and the specimen was percolated with CO_2 for 10 minutes, and subsequently was flushed with de-ionised-deaired water for 5 minutes. An initial back pressure of 40 kPa was then applied. At the end of the saturation process, the initial effective stress was 10 kPa. The specimen was consolidated by increasing the effective cell pressure to 100 kPa. A B value (Skempton, 1954) of 0.99–1.00 was easily achieved at back pressures of 90 to 140 kPa. However, the back pressure was raised to 190 kPa for all tests to prevent cavitation of the pore water (Head, 1986) and to ensure full saturation which is crucial in liquefaction studies. The specimen was consolidated for one hour. The desired cyclic deviator stress q_{cyc} then was applied using a 0.1 Hz sinusoidal wave-form.

Table 1. Variation of vertical stress and vertical strain of mixtures.

Sand:Rubber	Vertical strain (%) at a vertical stress of		
	70 kPa	150 kPa	300 kPa
100:0	0.41	0.65	0.91
95:5	0.45	0.82	1.23
90:10	1.30	1.96	2.70
80:20	1.94	3.21	4.68
60:40	6.39	9.58	12.99
40:60	11.20	16.61	21.86

4 ONE-DIMENSIONAL COMPRESSION TEST RESULTS

The relationships between vertical stresses and associated strains for the mixtures are shown in Fig. 2 and is summarized in Table 1.

At a vertical stress of 300 kPa, it can be observed that the vertical strains for pure sand (100:0) and the 95:5 mixture are relatively small and quite similar, but there is a noticeable increase as the tyre chip content is increased to 10%. However even with 20% tyre chips the vertical strain is still less than 5% at $\sigma_v = 300$ kPa.

When the percentage of tyre chips is greater than 20%, there is a marked increase in the compressibility which may not be tolerable in geotechnical design. These results suggest that the limiting percentage of tyre chips to be mixed with sand should not exceed 20%. It should be noted however that a higher percentage of tyre chips could be used if the excessive settlements were to be prevented. One example might be to use higher proportions of rubber in lower layer pre-loaded and pre-compressed by upper layers.

5 CYCLIC TRIAXIAL TEST RESULTS

The cyclic triaxial test results are summarised in Table 2 and typical cyclic triaxial test results are shown in Figs. 3–7 for $\sigma'_{3c} = 100$ kPa and $q_{cyc} \cong 30$ kPa, while the cyclic strength curves of all mixtures are illustrated in Fig. 8.

In Table 2 it can be seen that void ratios of the tyre sand mixtures consolidated to 100 kPa are slightly lower than those of the pure sand specimens. For example comparison of 100:0 and 85:15 shows average void ratios of 0.670 and 0.656, respectively. This conforms with the one-dimensional compression test results where as more tyre chips were added, the greater was the compressibility. The cyclic deviator stress range used to determine the liquefaction strength is shown for each mixture. It was necessary to decrease the maximum and minimum values as the proportion of rubber increased. It was also found that the liquefaction criterion, $\Delta u / \sigma'_{3c} = 1.0$, (where $\Delta u =$ excess pore water pressure and $\sigma'_{3c} =$ initial effective stress) was almost always reached before the 5% double-amplitude axial strain, which was reached after a few more cycles of loading. Thus separate data is presented for these two criteria.

Figures 3 to 7 show a comparison of the characteristics of the range of mixtures tested under similar amplitudes of cyclic deviator stress $q_{cyc} \cong 30$ kPa. During cyclic loading, the specimens strained largely in extension as shown in Figs. 3(a)–7(a). Larger compressive strains were observed as the excess pore pressure ratio approached 1.0, with a resulting increase in cyclic strain amplitude, as can be observed in Figs. 3(b)–7(b), however they were still much smaller than the extension strains. As a result samples always failed in extension indicating perhaps that some anisotropy occurred

Table 2. Cyclic triaxial test results.

Sand: Rubber	Initial void ratio	Consolidated void ratio	Deviator stress, kPa	Cyclic stress ratio	Skempton's B value	N at initial liquefaction	N at 5% axial strain
100:0	0.675	0.667	41.00	0.2050	1.00	5.02	8.00
	0.678	0.671	36.00	0.1800	0.99	9.02	11.00
	0.678	0.672	32.00	0.1600	0.99	55.00	57.00
	0.678	0.671	29.00	0.1450	0.99	109.00	111.00
99:1	0.682	0.675	42.00	0.2100	1.00	6.04	8.00
	0.681	0.675	38.00	0.1900	0.99	8.02	10.00
	0.679	0.673	35.00	0.1750	0.99	31.00	34.00
	0.678	0.671	30.00	0.1500	1.00	36.00	38.00
	0.681	0.674	23.00	0.1150	1.00	264.00	266.00
97.5:2.5	0.675	0.661	43.00	0.2150	1.00	2.02	3.00
	0.675	0.668	33.00	0.1650	1.00	24.00	27.00
	0.674	0.666	30.00	0.1500	1.00	35.00	37.00
	0.680	0.673	27.00	0.1350	1.00	38.00	40.00
	0.676	0.669	25.00	0.1250	0.99	113.00	117.00
95:5	0.683	0.674	37.00	0.1850	1.00	5.02	6.00
	0.676	0.666	30.00	0.1500	0.99	10.04	12.00
	0.685	0.677	28.00	0.1400	0.99	23.04	25.00
	0.690	0.683	20.00	0.1000	0.99	213.02	219.00
85:15	0.677	0.662	32.50	0.1625	0.99	3.04	3.00
	0.675	0.657	25.00	0.1250	0.99	13.04	20.00
	0.669	0.651	20.00	0.1000	1.00	55.04	60.00
	0.669	0.653	16.00	0.0800	1.00	219.00	226.00

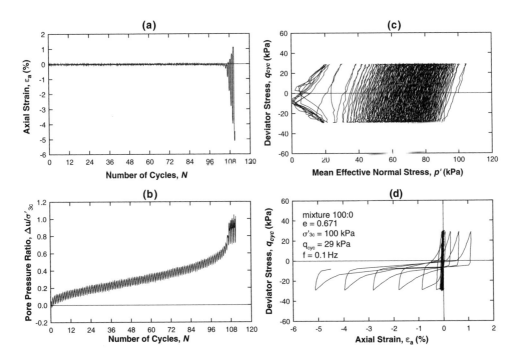

Figure 3. Typical cyclic triaxial test result of pure sand.

165

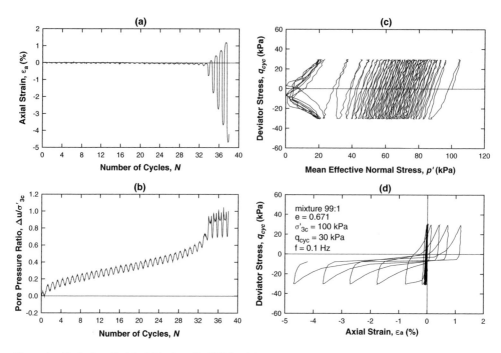

Figure 4.   Typical cyclic triaxial test result of 99:1 mixture.

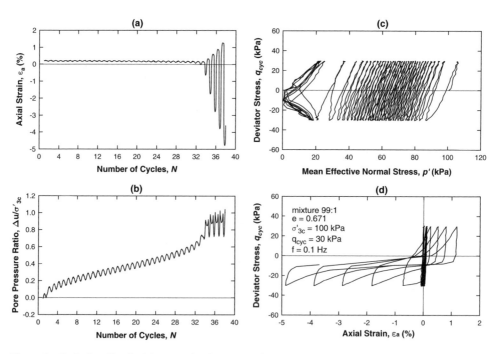

Figure 5.   Typical cyclic triaxial test result of 97.5:2.5 mixture.

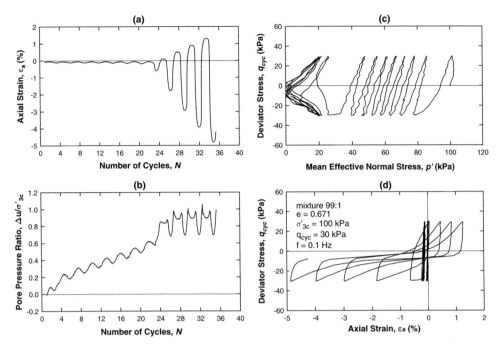

Figure 6.   Typical cyclic triaxial test result of 95:5 mixture.

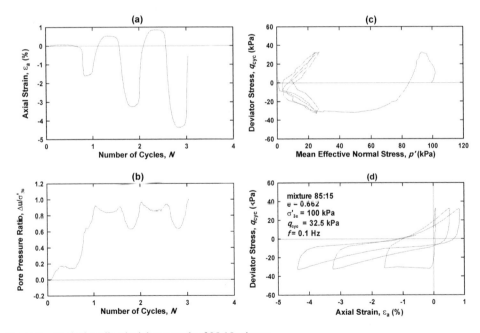

Figure 7.   Typical cyclic triaxial test result of 85:15 mixture.

during sample preparation. The development of the cyclic stress paths is shown in Figures 3(c)–7(c). It can be observed that the number of cycles to achieve liquefaction failure decreases in each case as the proportion of rubber particles increases. The change from relatively stable cyclic behaviour to liquefaction occurs over 2 or 3 cycles for pure sand and mixtures up to 2.5% rubber. At rubber

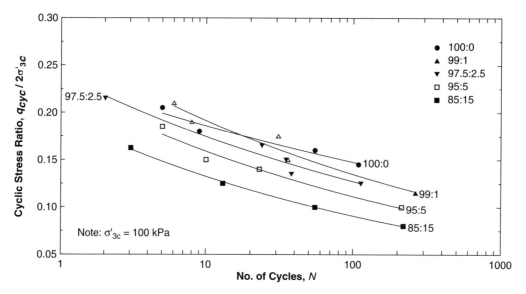

Figure 8.   Cyclic strength curves.

Table 3.   CSR at 10, 15, and 20 cycles.

Sand:Rubber	CSR at 10 cycles	CSR at 15 cycles	CSR at 20 cycles
100:0	0.186	0.178	0.173
99:1	0.191	0.180	0.172
97.5:2.5	0.175	0.165	0.159
95:5	0.159	0.149	0.143
85:15	0.129	0.121	0.116

contents above 2.5% the onset of liquefaction occurs in one or two cycles and this behaviour is particularly marked at 15% rubber content.

The stress-strain hysteresis curve of pure sand shown in Fig. 3(d) indicates a sharp increase in deviator stress on the compression side at constant peak strain; however, when tyre chips are added a more gradual response was observed, as can be seen in Fig. 4(d)–7(d).

The cyclic strength curves for sand mixed with tyre chips, shown in Fig. 8, have a steeper gradient than that for pure sand specimens. Apart from 99:1 mixtures, the cyclic strengths of the mixtures are lower than those for pure sand. However, there are two tests for the 99:1 mixtures that give a higher liquefaction strength and suggest that in this case the use of rubber tyre chip admixtures is beneficial.

At 20 cycles, CSRs of all sand mixed with tyre chips specimens are lower than those of pure sand specimens. At 10 and 15 cycles, however, the CSRs of 99:1 mixture are 0.191 and 0.180 which are slightly higher than those for pure sand suggesting there is a marginal benefit from small quantities of rubber for shorter seismic events. However overall the addition of tyre chips with a similar D_{50} to the sandy matrix material results in a dramatic decrease in cyclic strength. Figure 9 illustrates the change in CSR with increasing rubber content.

6   CONCLUSIONS

Two materials having a similar D_{50}, Leighton Buzzard 16/30 sand and CT0515 tyre chips, were mixed to investigate the compressibility characteristics and liquefaction potential of the composite

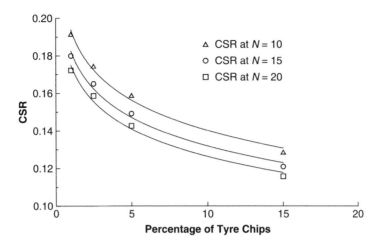

Figure 9. CSRs at 10, 15, and 20 cycles.

material. The compressibility of the mixtures was obtained by means of one-dimensional compression tests using a conventional oedometer cell. A computer-controlled cyclic triaxial testing system was used to investigate the liquefaction of the mixtures. From the test results, the following conclusions can be drawn:

1) One-dimensional compression test results show that the pure sand specimen and the sand mixed with 5% tyre chips specimen have a similar compressibility. The compressibility increased noticeably when up to 10% of tyre chips were added. When the tyre chip content is increased to 20% or more, the compressibility is so significant that it may be intolerable.
2) Initial liquefaction defined as the pore water pressure becoming equal to the initial effective stress occurred a few cycles before 5% double-amplitude axial strain which is often used as a liquefaction criterion.
3) During cyclic loading, the strains occurred largely in extension and specimens always failed in extension indicating the possibility of induced anisotropy caused by sample preparation.
4) From cyclic stress paths, the change from relatively stable conditions to cyclic liquefaction occurred over 2 or 3 cycles for pure sand and mixtures up to 2.5% rubber. This reduced to only 1 or 2 cycles when the rubber content was greater than 2.5%, and was particularly marked at 15% rubber content.
5) The hysteresis curve for pure sand shows a sharp increase of deviator stress on the compression side at constant peak strain. When tyre chips were added, however a more gradual response is observed.
6) All cyclic strength curves for sand mixed with tyre chip specimens have a steeper gradient than those of pure sand.
7) The CSRs at 20 cycles for all sand tyre chip specimens were lower than those for pure sand apart from the 99:1 mixture which is slightly higher at 10 and 15 cycles suggesting there is a marginal benefit from small quantities of rubber for shorter seismic events.
8) Overall, the mixtures show no sign of increasing the liquefaction strength except for the 99:1 mixture at 10 and 15 cycles. However only one size of tyre chips was tested and further studies considering other factors such as size, shape, and materials properties are required.

169

ACKNOWLEDGEMENTS

The work presented herein has been funded by the Commission of Higher Education, Thailand, and the University of Sheffield, UK., to whom the authors are gratefully indebted. Support and contribution to the development of the test equipment and laboratory work from the technicians in the Civil and Structural Engineering Department of the University of Sheffield are kindly acknowledged.

REFERENCES

ASTM (1998) Standard practice for use of scrap tires in civil engineering, ASTM D6270-98, American Society of Testing and Materials, W. Conshohocken, PA.

Edil, T. B., and Bosscher, P. J. (1994) Engineering properties of tire chips and soil mixtures, Geotechnical Testing J., ASTM, Vol. 17, No. 4, 453–464.

Head, K. H. (1986) Manual of soil laboratory testing: Volume 3 effective stress tests, Pentech Press, London.

Higuchi, T. (2001) Liquefaction and cyclic failure of low plasticity silt, PhD Thesis, University of Sheffield, U.K.

Hyde, A. F. L., and Higuchi, T. (2005) Post liquefaction characteristics of low plasticity silt, 16th International Conference on Soil Mechanics and Geotechnical Engineering, 2659–2662.

Hyodo, M., Aramaki, N., Itoh, M., and Hyde, A. F. L. (1996) Cyclic strength and deformation of crushable carbonate sand, Soil Dynamics and Earthquake Engineering, Vol. 15, 331–336.

Ishihara, K. (1993) Liquefaction and flow failure during earthquakes, Geotechnique, Vol. 43, No. 3, 351–415.

JGS (1998) Remedial measures against soil liquefaction: From investigation and design to implementation, A. A. Balkema, Rotterdam, Netherlands.

PHRI (1997) Handbook on liquefaction remediation of reclaimed land, A.A. Balkema, Rotterdam, Netherlands.

Sitharam, T. G. (2003) Earthquake geotechnical engineering: An overview, Workshop on Current Practices and Future Trends in Earthquake Geotechnical Engineering, Golden Jubilee Seminar Hall, Civil Engineering Department, Indian Institute of Science, Bangalore, India.

Skempton, A. W. (1954) The pore-pressure coefficients A and B, Geotechnique, Vol. 4, 143–147.

Developing design variants while strengthening roadbed with geomaterials and scrap tires on weak soils

S.A. Kudryavtsev, L.A. Arshinskaya & T.U. Valtseva
Far Eastern State Transport University (FESTU), Khabarovsk, Russia

U.B. Berestyanyy
Scientifically-Introducing Company <<DV-GEOSYNTHETICS>>, Khabarovsk, Russia

A. Zhusupbekov
Geotechnical Institute, Astana, Kazakhstan

ABSTRACT: The institute of transport construction of Far Eastern State Transport University (FESTU) together with company <<DV-GEOSYNTETICS>> has developed modern geotechnical decisions with use of properties of geosynthetic materials for many nonstandard building objects of the Far East. One of the problems of modern geotechnologies is considered to be the opportunity of design changing and technology of construction of roadbed strengthening on especially weak from fluid slimes bases.

1 ENGINEERING-GEOLOGICAL STRUCTURE OF A CONSIDERED SITE

The construction of a highway site Far Eastern, Russia began in 2001 as a result of the engineering-geological researches done, basically concerning places of water-streams crossings.

Geologic-lithologic cut of a river valley is presented by lake-alluvial and alluvial adjournment.

In lake-alluvial adjournment there are loams, sandy loams, the clay, that have mainly fluid consistence slimes. The river-bed consists of gravel-pebble and sandy adjournment.

Lie in the basis of an embankment there are over humidified soils up to depth of 10–13 m: slimes clay, loamy with prevalence fluid and flow-plastic consistence and of natural humidity of 38–66 %. For all the soils is typical the presence of high porosity (e is characteristic. $= 0,97–1,39$), the low structural communications predetermining character of their mechanical properties, high compressibility under loading.

1.1 *Deformations of an embankment*

After the construction of bridge transitions and filling the first layer of roadbed construction works were suspended. The further construction was continued only in 2004.

In August – September, 2005 after filling of an roadbed up to height above 3,0 m on a construction site deformations of the basis began to occur locally accompanied by local destructions of fining of roadbed and countervailing berms (figure 1).

According to results of additional engineering-geological inspection, strengthening of the bottom part of roadbed is by soil holders from high-strength women geotextiles is recommended.

In August, 2006 after the performance of constructional actions and land filling of roadbed up to 4–5 m on height on separate motorway sites have deformations of destruction of an roadbed with whistler of a ground basis, and destruction of constructions of holders with breaks of high-strength geotextiles occurred.

Figure 1. Deformations of an embankment.

For detailed studying engineering-geological and hydro-geological conditions, and also for definition of physic mechanical properties of soils, lying in the basis of an embankment, field works and laboratory researches soils were carried out.

From additional researches was determined the followings:

1. The rivers-bed is combined with weak soils which form the basis of an existing embankment. Processes of deformation of a roadbed periodically. Destruction of the basis and a roadbed occurs on round-centric surface of sliding with выпорами of ground at the basis of an embankment.
2. Proceeding from the analysis of cross-section engineering-geological cuts it is obvious, that the greatest subsidence of soils occurs on an axis of an embankment accompanied by shaft whistler, countervailing prisms are insufficiently counted. Consolidation process of the basis occurs owing to filling of embankments up to maximal height.

Thus, technical decisions for the device of a "floating" embankment on especially weak basis, recomended earlier, did not justify the purpose already at the initial stage of construction of the given motorway.

Other constructive decisions are required to provide safety and uninterrupted operation of movement of transport not only during the construction period accompanied by work of building technics, but also during operation.

1.2 The constructive decisions offered for strengthening of a roadbed

For the salvation of such problems the application of Russian and foreign experience of designing and construction of a road-bed on the weak bases, on the basis of methods of geotechnical modelling with use in constructions of modern geosynthetic materials is the most expedient.

Geotechnical modelling of construction of strengthening of a roadbed and the weak basis for the construction of a highway site Far Eastern, Russia was carried out on a program complex <<FEM models>>. The basis of the given complex is the finite elements method, allowing to research thermal and intense-deformed condition of constructions and their bases in 3-demention modelling.

172

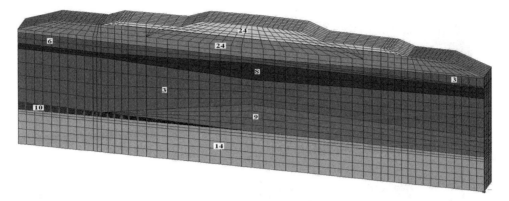

Figure 2. The settlement scheme of a road-bed and the basis No 3–24 – settlement soils layers: 3 – silt clay, fluid; 6,8 – silt loamy, fluid; 9 – silt loamy, flow-plastic; 10 – loam heavy weak-plastic; 14 – pebble with sandy loam firm; 24 – gravel ground with sand.

For the description of a road-bed performance of a highway Far Eastern, Russia on the weak bases of a fluid consistence during the given design stage is applied elasto-plastic model with the limiting surface described by criterion Coulomb -Moor. Properties of modern geosynthetic materials– integrated geogrid were included in the given model. Properties of integrated geomaterials are researched and confirmed by the results of large-model experimental researches of Russian and foreign experts.

The principle of performance of integrated geogrid consists in strengthening of non-connected layers of road clothes. When the granulated material condenses above a geogrid, its particles get through apertures of geogrid and are fixed, creating effect of "blocking". Possessing high rigidity, geogrid SS allows to maintain high loadings at very low deformations.

The used technique and program complex are realized by authors on construction objects in Russia. Application of methods and approaches for calculations and designing of geotechnical constructions by use of appendices of a program complex «FEM-models» has shown, that it allows most authentically and objectively to carry out selection and calculations of the most rational geotechnical constructions.

Geotechnical modelling is done for different loadings and the bases. In calculations it some variants are determined. Their key parameters of the intense and deformed conditions are determined.

One of the variants is the construction which is offered to be made in the shape of 2 rows of piles with of definite size set up along an embankment on berms. The step of piles and their parameters are defined by calculations by geotechnical modelling.

Between the headstalls piles the integrated geogrid of high durability with rigid fastening in headstalls pairs in the ranks of the friend opposite to the friend uniaxial is set up. The number of layers of a geolattice is defined by amount of loadings and geological features, durabilities of soils of bottom supporting of piles in each settlement task.

It is supposed for all settlement cases, irrespective of capacity of bases weak soils, the supporting of the ends of piles on strong soils with depth in them not less than down to 1,5 m.

In figures 2–7 as an example the settlement scheme of a road-bed and the basis on the picket 17+00, the scheme of destruction, a construction of strengthening and results of modelling is demonstrated (shown).

For decrease in deformation of a road-bed the design of strengthening is offered (fig. 4).

As a result of numerical modelling vertical deformations on an axis of a roadbed reach up to 0,182 m (fig. 5), and horizontal deformations develop in the basis up to 0,037 m (fig. 6).

Thus horizontal pressure makes in a body of a roadbed up to 114 kPa (fig. 7).

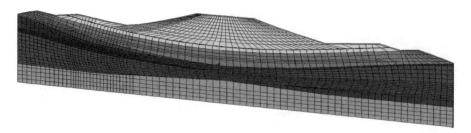

Figure 3.   The deformed scheme of a road-bed.

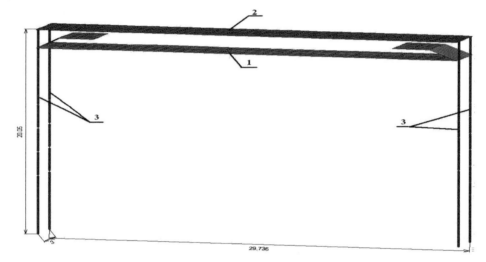

Figure 4.   The scheme of a strengthening construction:1 – woven geotextiles; 2 – uniaxial integrated geolattice 160RE in 2 layers; 3 – precast piles section 0,25 × 0,2 м, reinforcing standard.

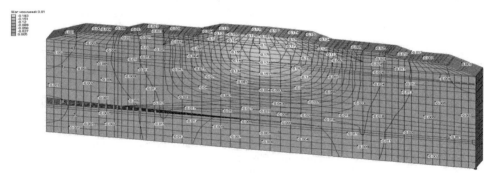

Figure 5.   Isocurves of vertical movings in a construction.

The results of modelling have shown, that for conditions where height of an embankment is more than 8,0 m, and capacity fluid slimes is in the basis more than 14,0 m, the construction of a road-bed is possible only on the continuous pile basis with a distributing platform and strengthening at slopes of an embankment made of integrated geogrid (fig. 8).

As a result of geotechnical modelling technical solutions of using considered designs are developed for various soil conditions of the basis and capacity of constructions. In figure 9 the schedule of settlement parameters of designs of strengthening of the basis, corresponding to limits of height of an embankment is presented.

174

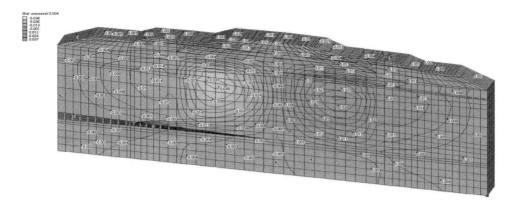

Figure 6.   Isocurves horizontal movings in a construction.

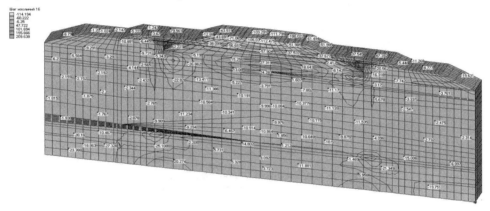

Figure 7.   Isocurves horizontal pressure (kPa) in a construction.

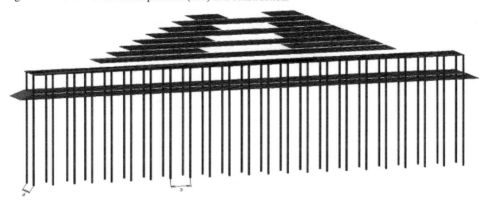

Figure 8.   The scheme of a strengthening design for small-sedimentary bases of a construction.

As is apparent from the schedule for the definition of design type, use of the facilitated design assuming deposits during operation, is admissible at loadings from weight of an embankment of height no more than 8,3 m.

At height of an embankment on a site of more than 8,3 m the maintenance of durability, reliability and durability of a construction for the set parameters is possible only by application as the basis of a pile field with ростверком from a geocomposite of a ground and an integrated biaxial geogrid.

175

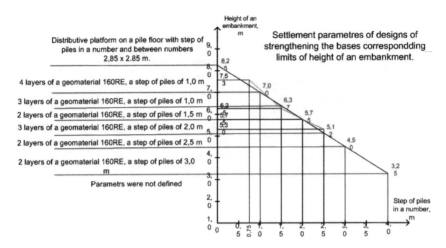

Figure 9. Settlement parameters of designs of strengthening of the basis heights of an embankment corresponding to limits.

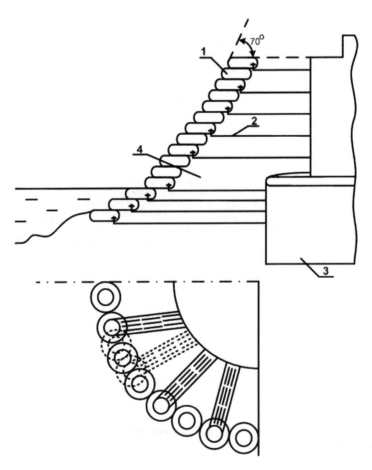

Figure 10. Facing of a cone of the bridge by trunks: 1 – defective automobile tires; 2 – uniaxial an integrated geogrid; 3 – body of a support of the bridge; 4 – ground of backfilling.

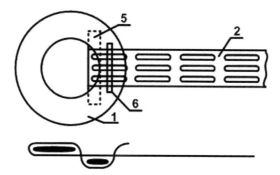

Figure 11.   Units of fastening: 5 – plastic anchor; 6 –the connecting probe.

The received results of modelling of a roadbed of a highway for various engineering-geological conditions allow to pick up a rational design with admissible sizes of deformation.

2   USAGE OF SCRAP TIRES

In parts of the world where automobile transport takes over the problem of scarp tire utilization has arisen and it demands significant in vestments. So, their usage in soil constructions has been a successive solution. Reinforced soil design to strengthen the abutments of bridges at a highway of Far East, Russia, serves an example of scrap tires utilization. Here scrap tires are used as a cladding material at wing wall abutments across the permanent water stream.

The stripes of a uniaxial in intergraded geogrid of certain length are used as anchor elements. The scheme of the design is shown on fig. 10. The developed units linking tries and anchor grid are simple, technological and reliable (fig. 11). Made this way, the stripes of an anchor geogrid working together with soil filling produce stable reinforced soil design.

Steepness of a bridge approaching embankment is 70 degrees that allows to save 27–30% of soil fill.

The optimum geometrical indexes (parameters of designing and reinforcing) are defined by calculation of a strain deforming state using the method of finite elements.

The stability of the design was tested on the supposition of possible distortion mechanism according to Coulomb.

3   CONCLUSIONS

1. It is got, that in designs of a highway on the set site with especially weak bases at height of con-structions up to more than 4,0 m without strengthening, already by initial loading, deformations not compatible to demanded parameters for roads of II category are realized, including initial stages of destruction.
2. The done geotechnical modelling of conditions of constructions has allowed establishing the most rational designs, capable to provide safe and uninterrupted movement of transport during the construction and performance periods.
3. The design of strengthening of the basis in the form of a pile field with flexible foundation grill from integrated biaxial geogrids of type SS to the deformability properties is capable to provide operational reliability of a construction without additional deformations already practically right after the concluding of construction.
4. At height of embankments of more 6,0 m stretching pressure in its top part and slopes are high enough and close on sizes to pressure in zones of loss of strength of bases. For maintenance of

reliability of constructions and traffic safety partial reinforcing the stretched zones of a body of embankments and batter parts is offered.

5. On sites of construction, where deposits of a roadbed during operation are structurally assumed, it is necessary to provide a building stock on width of an embankment. A prospective a deposit during operation will make 70–100% from settlement size for the construction period.

REFERENCES

Kudryavtsev S.A., Berestyanyy U.B.,Valtseva T.U., Barsukova N.V. Practice of use of positive properties of geosynthetic materials on building objects in severe climatic conditions of the Far East of Russia. 1st International conference on new developments in geoenvironmental and geotechnical engineering. November 9–11, 2006. University of Incheon. Korea. P. 423–427.

Kudryavtsev S.A. Proceedings of the international coastal geotechnical engineering in practice. 21–23 May 2002. Atyray, Kazakhstan. P. 145–147.

Kudryavtsev S.A. Permafrost engineering. Proceeding of the fifth International symposium on permafrost engineering (2–4 September 2002, Yakutsk, Russia). – Yakutsk. Permafrost Institute Press, 2002. – Vol. 1. P. 198–202.

Scrap Tire Derived Geomaterials – Opportunities and Challenges – Hazarika & Yasuhara (eds)
© 2008 Taylor & Francis Group, London, ISBN 978-0-415-46070-5

Shaking table tests on effect of tire chips and sand mixture in increasing liquefaction resistance and mitigating uplift of pipe

Taro Uchimura, Nguyen Anh Chi, Shanmugaratnam Nirmalan, Takuya Sato, Mehrashk Meidani & Ikuo Towhata
The University of Tokyo, Tokyo, Japan

ABSTRACT: The authors propose to use a mixture of tire chips and sand as backfill material for buried pipes. It was found in the authors' research that a backfill with a mixture of tire chips and sand has higher liquefaction resistance compared to sand-only backfill. This implies that use of such materials for backfilling buried pipes could mitigate floating up damages.

Herein, a series of shaking table tests on scaled model of buried pipes in sand ground backfilled with tire chips-sand mixture, and a series of undrained cyclic loading tests on triaxial specimens of the mixtures are reported to discuss on advantages of use of the mixture materials.

1 INTRODUCTION

More than one million ton of used tires are wasted annually in Japan. Of these tires, nearly 90% are reused or recycled for applications such as tire retreads, thermal recycling, material recycling (for cement, steel, paper and pulp, etc). The other about 10% of waste tires has to be used in any applications such a way that dumping of waste tires could be avoided.

Several ideas to use scrapped tire materials in civil engineering have been proposed. Tire chips have good engineering properties as backfill material, such as, long durability, low specific gravity, high flexibility to reduce stress concentration, less adverse effects the environment, and etc. Karmokar et al. (2003) reported results from leaching tests not showing any detrimental effects. Besides, they also proved the durability of tire chips.

The authors propose to use Tire Chips and sand mixture as backfill material around buried pipes. It was found in the authors' research that a backfill with a mixture of tire chips and sand has significantly higher liquefaction compared to sand-only backfill, which implies that use of such materials for backfilling buried pipes could results in mitigation of floating up damages. The low specific gravity of tire chips (1.17 for the tire rubber) will make the total density of liquefied tire chips-sand mixture, which also results in lower buoyant force to the buried pipe in the liquefied backfill. Another possible advantage is that a compacted mixture of tire chips and sand has lower stiffness compared to sand-only backfill, which reduces the risk of damaging buried pipes due to various situations, such as compaction work on the backfill, possible non-uniform ground displacement, and long-term static earth pressure, etc.

The schematic diagram of use of tire chips-sand mixture for backfilling buried pipes is shown in Figure 1. Tire chips are usually poorly graded with larger particle size compared to the sandy backfill materials. Therefore, if tire chips are used without mixing with sand, surrounding sand particle will flow into the large void among the tire chip particles, causing subsidence in road surface around it. Consequently, the authors propose to use tire chips and sand mixture.

Herein, a series of shaking table tests on scaled model of buried pipes in sand ground with backfill of tire chips-sand mixture, and a series of undrained cyclic loading tests on triaxial specimens of

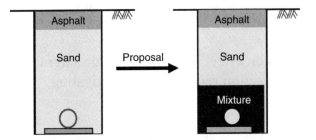

Figure 1. Proposal of use of tire chips-sand mixture for backfilling buried pipes.

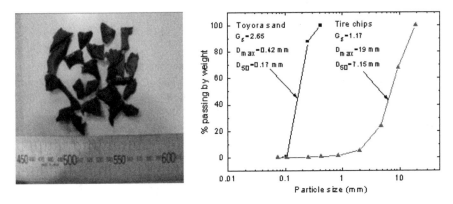

Figure 2. Photo of tire chips and particle size distribution of tire chips and Toyoura sand.

Table 1. Physical properties of tested tire chips material.

Properties	Value
D_{10} (mm)	2.48
D_{30} (mm)	5.23
D_{50} (mm)	7.15
D_{60} (mm)	8.38
D_{max} (mm)	19
Coefficient of uniformity, Cu	3.38
Coefficient of gradation, Cc	1.32
Specific gravity, Gs	1.17

tire chips-sand mixture are reported to discuss on the effect of use of the mixture materials on mitigation of damages due to liquefaction.

2 MATERIALS

The tire-chips sample used in this study was shredded tire rubber (Figure 2 and Table 1; Gs = 1.17, $D_{max} = 19$ mm, $D_{50} = 7.15$ mm, and $U_c = 2.5$), instead of real tire chips which contains other materials like steel, fiber, etc. As it is shredded mechanically, the particle shape was very irregular and angular, and some particles had very long shape. As general properties of rubber, the tire chips particles has negligible water absorption, and negligible volumetric compression due to isotropic pressure. Toyoura sand was used for the mixture, as- well as sand-only backfill, and sand ground part of the models.

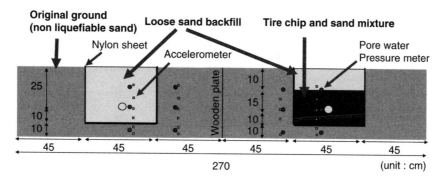

Figure 3. Cross-section of test model.

The mixture ratio of tire chips were evaluated by the dry weight of the tire chips relative to that of the total mixture material as the following formula:

$$TC_r = \frac{M_{TC}}{M_{TC} + M_{TS}}$$

(TC$_r$: Tire chips contents, M$_{TC}$: Weight of tire chips, M$_{TS}$: Weight of Toyoura sand).

TC$_r$ = 0% (sand-only), 15%, 30%, and 40% were selected as the ratio. If the tire chips content was larger than this, the sand cannot fill the entire void among tire chip particles, and the models and specimens became non-uniform.

The compaction density of the mixture was evaluated by the relative density of the Toyoura sand which fills the void among the tire chips particles, that is:

$$e_S = \frac{V_{Total} - V_{TS} - V_{TC}}{V_{TS}}$$

(e$_s$: Void ratio of Toyoura sand, V$_{total}$: Total volume of mixture, V$_{TC}$: Volume of tire chips particles, V$_{TS}$: Volume of Toyoura sand particles)

$$D_r = \frac{e_{max} - e_S}{e_{max} - e_{min}}$$

(D$_r$: Relative density of tire chips-sand mixture, e$_{min}$(=0.597), e$_{max}$(=0.977): Minimum/maximum void ratios of Toyoura sand, respectively).

As another index for amount of tire chips used in compacted mixture, volumetric contents of tire chips, θ_{TC} can be calculated as:

$$\theta_{TC} = \frac{V_{TC}}{V_{Total}} = \frac{1}{(1 + (1 + e_s)(\frac{1}{TC_r} - 1)\frac{GS_{TC}}{GS_{TS}})}$$

(GS$_{TC}$ (=1.17), GS$_{TS}$ (=2.65): Specific gravity of tire chips and Toyoura sand, respectively).

3 SHAKING TABLE TESTS

3.1 Experimental procedures

Figure 3 shows the structure of the models. Their preparation methods are mainly after Koseki et al. (1997 and 1998). A dense Toyoura sand ground (L = 2700 mm, H = 450 mm, W = 400 mm,

Table 2. Specification of models for shaking table tests. (17L & 19L: backfilled by sand only).

Case No.	D_r of original ground (%)	D_r of loose backfill (%)	D_r of tire chips-sand mixture (%)	Tire chips content, TC_r (by weight) (%)	Volumetric tire chips content, θ_{TC} (%)	Uplift of pipes after shaking (mm)
17L	88	30	–	–	–	31 mm
17R	90	30	27	40	44.6	0.2 mm
19L	100	0	–	–	–	260 mm (up to ground surface)
19R	100	4	0	40	43.3	6 mm
20R	100	5	0	42	45.3	2 mm

a)   b)

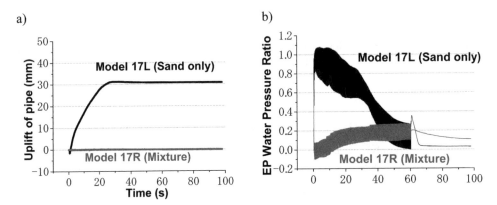

Figure 4. Test results with liquefiable backfill ($D_r = 30\%$): a) uplift of pipes, b) excessive pore water pressure ratio at the side of pipe.

$D_r = 90$ or 100%) was made as a non-liquefiable original ground in a sand box primarily. In order to make a uniform ground, the sand was pluviated in water, keeping constant falling height of 50 cm in air and 10 cm in water respectively. And the sand box was shaken in orthogonal two directions with various accelerations for several minutes until the prescribed density was achieved. A wooden plate was rigidly fixed at the center to separate the ground so that two models of buried pipes can be tested at once with the same input acceleration. Then, the water in the sand box was drained out, and a trench was excavated for 450 mm-long and 350 mm-deep at the center of each section. Nylon sheet was placed along the excavated surface to avoid rapid dissipation of water pressure from liquefied loose backfill part to surrounding original ground. A model of buried pipe (PVC, average specific gravity 0.4, length 35 cm, diameter 4.8 cm) was placed at the center of each trench with a depth of 250 mm from the ground surface. Then, the trench was backfilled with a 350 mm layer of loose Toyoura sand (left section in Figure 3), or two layers of the mixture for 250 mm and loose Toyoura sand for 10 mm (right in Figure 3). Finally, CO_2 gas was filled in the sand box for better saturation, and the model was filled with water slowly, taking around 3 hours, and the model was left quietly for 12 hours before shaking.

The detailed compaction densities of each part of the models are listed in Table 2. The mixture ratio TC_r of tire chips was 0% (sand only) or around 40% for the models reported in this paper. The shaking input to the sand box was 440 Gal, 5 Hz, 60 sec. The floating up displacement of the pipes and the pore water pressure around the pipes were observed.

3.2 Results from shaking model tests

Figure 4 shows the lift up displacement of the buried pipes and the excessive pore water ratio (excessive pore water divided by initial effective confining pressure) measured at the side of the

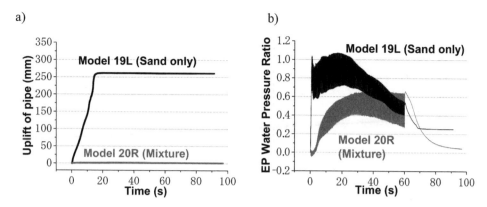

Figure 5. Test results with liquefiable backfill ($D_r = 0\%$, mixture was weekly compacted): a) uplift of pipes, b) excessive pore water pressure ratio at the side of pipe.

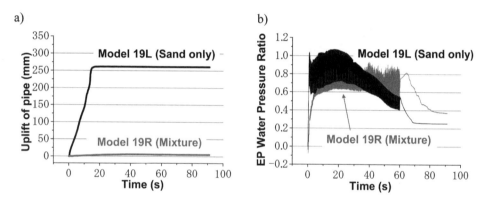

Figure 6. Test results with liquefiable backfill ($D_r = 0\%$, mixture was not compacted): a) uplift of pipes, b) excessive pore water pressure ratio at the side of pipe.

pipe (250 mm under the ground surface) obtained from the models 17L and 17R. The relative densities of the backfills are around 30%, with which pure Toyoura sand is expected to liquefy due to shaking. The pipe buried in sand-only backfill (17L) lifted up by 31 mm, while the pipe in mixture backfill (17R) showed an ignorable displacement. Correspondingly, the pore water pressure in the sand-only backfill increased showing liquefaction immediately after starting shaking, while that in the mixture backfill increased gradually to a much lower extent.

Figures 5 and 6 shows the test results from the models with very loose backfill, $D_r = 0\%$. Model 19R and 20R had similar backfill density to each other, but the backfill of 20R was slightly compacted to adjust the density, while 19R was just placed in the trench without compaction. As a result, the pipe in a very loose sand-only backfill (19L) lifted up to the ground surface, while the pipes in mixture backfill lifted up by less than 1 cm. The excessive pore water pressure due to shaking with the slightly compacted mixture backfill (20R) was larger than that of denser backfill (17R), but still it did not indicate liquefaction in the backfill. In the non-compacted mixture backfill (19R), the excessive pore water pressure was much larger, but still it was lower than liquefaction level, and the floating up of the pipe was small.

These results mean that the mixture backfill of tire chips and sand showed higher liquefaction resistance.

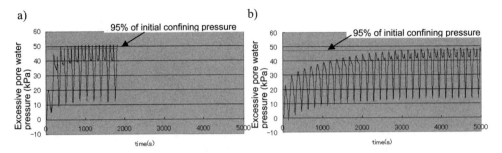

Figure 7. Excessive pore water pressures ($D_r = 65\%$): a) $TC_r = 0\%$ (sand-only), b) $TC_r = 30\%$.

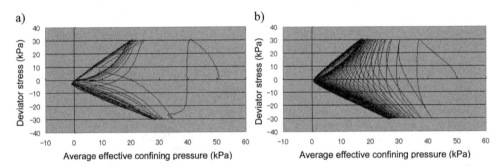

Figure 8. Effective stress paths ($D_r = 65\%$): a) $TC_r = 0\%$ (sand-only), b) $TC_r = 30\%$.

4   LIQUEFACTION TESTS ON TRIAXIAL SPECIMENS

4.1   *Experimental procedures*

A series of undrained cyclic loading tests were conducted on cylindrical triaxial specimens of the same materials as the backfill for the shaking model tests. The specimens were 100 mm in diameter, and 200 mm in height, and prepared with various combinations of mixture ratio ($TC_r = 0\%$ (sand only), 15%, and 30%), and relative density ($D_r = 35\%$, 50%, and 65%). Double vacuum method was employed for better saturation, and it was confirmed that the B-values of each specimen was more than 95%. The specimens were consolidated with an isotropic effective confining pressure of 50 kPa, and then loaded in an undrained condition with cyclic axial stress with a single-amplitude of 30 kPa and a frequency of 0.005 Hz. The axial strain and pore water pressure of the specimens were observed, and the specimens were considered to be liquefied when the excessive pore water pressure became 95% of the initial effective confining pressure.

4.2   *Results from liquefaction tests on triaxial specimens*

Figures 7 to 10 show the results from the tests on sand-only and tire chips-sand mixture ($TC_r = 30\%$) specimens with relative density of $D_r = 65\%$. In Figures 7 and 8, the behaviours of the excessive pore water pressure shows that the sand-only specimen liquefied at the 3rd cycle of loading, while the mixture specimen liquefied at the 18th cycle with gradual increment in the pore water pressure. In Figure 9, the amplitude of cyclic deformation of the mixture specimen was very stable within a limited range compared to the sand-only specimen, even after liquefaction. Figure 10 show that the mixture specimen had strong characteristics of cyclic mobility, that is, the effective confining pressure in the specimen recovered periodically corresponding to small shear strain. These comparisons suggest that a compacted mixture of tire chips and sand has significantly

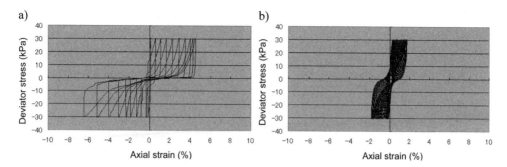

Figure 9.   Stress-strain relationships ($D_r = 65\%$): a) $TC_r = 0\%$ (sand-only), b) $TC_r = 30\%$.

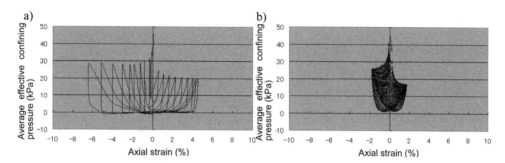

Figure 10.   Effective confining pressure and axial strain ($D_r = 65\%$): a) $TC_r = 0\%$ (sand-only), b) $TC_r = 30\%$.

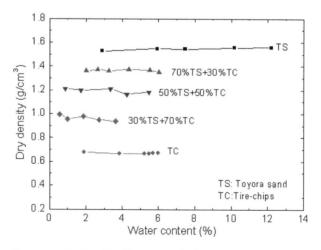

Figure 11.   Compaction curves for tire chips-Toyoura sand mixture.

higher liquefaction resistance compared to compacted pure Toyoura sand, consistently with the results from the model tests.

Figure 11 shows compaction curves of tire chips, Toyoura sand, and their mixtures with $TC_r = 30\%$, 50%, and 70%, obtained with the standard compaction energy of 552 kJ/m^3 (Shanmugaratnam, 2006). The compaction curves are flat, nearly independent of water contents. The higher tire chips content TC_r resulted in lower compaction density. Taking account of the difference in specific gravities between tire chips ($GS_{TC} = 1.17$) and Toyoura sand ($GS_{TS} = 2.65$), the void ratio

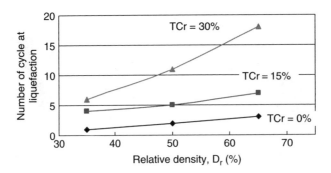

Figure 12. Number of loading cycles required for liquefaction (Liquefaction is defined as loss of 95% of the initial effective confining pressure).

e_s of the mixture can be obtained as below:

$$e_s = \frac{GS_{TS}}{(1-TC_r)}(\frac{1}{\gamma_t} - \frac{TC_r}{GS_{TC}}) - 1 \quad (\gamma_t: \text{Total density of the mixture})$$

By this calculation, the relative density of pure Toyoura sand was around $D_r = 70\%$, while that of mixture with $TC_r = 30\%$ was around $D_r = 50\%$. Thus, assuming the same compaction procedures are taken in actual construction work, the results on the sand-only specimen with $D_r = 65\%$ shown in Figures 7 to 10 should be compared with those on the mixture with $D_r = 50\%$ or lower.

The numbers of loading cycles required before the specimens were liquefied are summarized in Figure 12. The mixture specimen with $TC_r = 30\%$ and $D_r = 50\%$ took 12 loading cycles for liquefaction, still showing much higher liquefaction resistance than the sand-only specimen.

Such characteristics may be due to the large particle size of the tire chips. Another possible reason may be the flexibility of the tire chip particles. The particles are bended and contact to each other by compaction. The contact force, as well as the friction, between particles becomes more stable than the case with rigid sand particles, because the contacts are not easily lost when the pore water pressure increase, although the deformation in the particles may recover slightly.

5 SUMMARY AND CONCLUSIONS

A series of shaking table tests on scaled models of buried pipe backfilled with various mixture of tire chips and Toyoura sand was conducted. The backfill with mixture showed significantly higher liquefaction resistance compared to the backfill with pure Toyoura sand, resulting in very little floating up of the pipes. A series of undrained cyclic loading tests on triaxial specimens were also conducted, and it was found that the mixture of tire chips and sand has higher liquefaction resistance, consistently with the results from the model tests.

Such mixture materials may be useful to construct soil structures in which high liquefaction resistance is required.

REFERENCES

Koseki et al. (1997): Uplift behavior of underground structures caused by liquefaction of surrounding soil during earthquake, Soils and Foundations, vol. 37, no.1, pp. 97–108.
Koseki et al. (1998): Uplift of sewer pipes caused by earthquake-induced liquefaction of surrounding soil, Soils and Foundations, vol. 38, no. 3, pp. 75–87.
Nirmalan Shanmugaratnam (2006): Strength and Deformation characteristics of compacted scrap tire chips, Master Thesis, University of Tokyo.

Scrap Tire Derived Geomaterials – Opportunities and Challenges – Hazarika & Yasuhara (eds)
© 2008 Taylor & Francis Group, London, ISBN 978-0-415-46070-5

Undrained cyclic shear properties of tire chip-sand mixtures

M. Hyodo, S. Yamada, R.P. Orense & M. Okamoto
Dept. of Civil and Environmental Engineering, Yamaguchi University, Ube, Japan

H. Hazarika
Graduate School of System Science and Technology, Akita Prefectural University, Akita, Japan

ABSTRACT: Recycling of scrap tires has been gaining attention recently because of its potential economic and environmental benefits. Recently, tire chips derived from old tires are mixed with sand to produce new geomaterials that can be used in engineering structures. In order to understand the undrained cyclic shear properties and strength characteristics of such soil mixtures, a series of undrained cyclic triaxial tests was conducted on specimens with varying mix ratios of tire chips and sand. Test results showed that since the tire chips in the soil mixture control the build-up of excess pore water pressure, the undrained cyclic shear behavior and cyclic strength are affected by the amount of sand comprising the mixture.

1 INTRODUCTION

The recycling of used tires has been made compulsory with the enactment of the Automobile Recycling Law in 2002. As a result, searching for cheap methods to recycle old tires in large quantities became a top priority. In Japan, the recycling rate for old tires was about 87.8% in fiscal year 2005, unchanged from the rate in 2001, and it is expected to remain at the same level in the years to come. Moreover, it is known that tire is a better quality fuel than coal, and because the unit price is also cheaper by about half of that of coal, its demand as fuel in the cement industry is high. However, the demand for cement has recently reached its peak, and a decrease in thermal recycling rate is anticipated in the future. Therefore, it is expected that old tires will be used as new geomaterials. For the effective utilization of old tires as geomaterials, tires currently being examined are crushed into smaller pieces according to usage, and they are classified as either whole tire, tire shreds, tire chips or rubber powder, depending on the particle size.

In the United States, research on the effective use of old tires as geomaterials has advanced greatly during the first half of the 1990s, with several cases involving their use as road embankments (Bosscher et al. 1997) and as lightweight backfill in retaining walls (Lee et al., 1999). Furthermore, a standard has been provided on the use of old tires for engineering works through the ASTM Standards (ASTM, 1998). Although technological advancement in Japan regarding the effective use of old tires as geomaterials is still in its early stage, researches on their application as tire chip-mixed solidified soil (Kikuchi et al. 2006), earthquake-resistant reinforcement (Hazarika et al. 2006) and fill improvement (Mitarai et al. 2006) have been actively conducted. In the future, old tires will be effectively used as materials for soil structures and to address stability concerns, it is necessary to accumulate information on the mechanical characteristics of tire chips as well as tire chip-sand mixtures.

Considering this background, a series of undrained cyclic triaxial tests was conducted to understand the undrained shear behavior and strength characteristics of soil mixtures consisting of sand and tire chips. From the test results obtained, the undrained cyclic shear behavior and strength characteristics were examined in detail while paying attention on the mix ratio by volume of tire chip and sand.

Table 1. Physical properties of soil samples.

Sand fraction	ρ_s (g/cm^3)	ρ_{dmin} (g/cm^3)	ρ_{dmax} (g/cm^3)	e_{max}	e_{min}	D_{50} (mm)	U_c
sf = 41 (Soma sand)	2.645	1.273	1.574	1.077	0.680	0.395	1.65
sf = 0.9	2.576	–	–	–	–	0.399	1.67
sf = 0.8	2.498	–	–	–	–	0.403	1.67
sf = 0.7	2.410	0.939	1.234	1.565	0.953	0.407	1.69
sf = 0.6	2.309	–	–	–	–	0.414	1.72
sf = 0.5	2.192	0.744	0.988	1.948	1.218	0.423	1.75
sf = 0.3	1.892	0.563	0.735	2.361	1.576	0.453	1.91
sf = 0 (Tire chips)	1.150	0.347	0.442	2.318	1.600	0.655	2.72

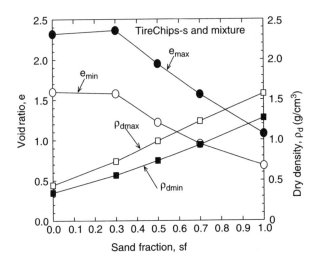

Figure 1. Relations between void ratio, dry density and sand fraction.

2 MATERIAL USED AND EXPERIMENTAL METHOD

2.1 *Physical properties of materials*

In this research, tire chips, Soma silica sand No. 5 and a mixture of these two materials were used. The tire chips were derived from used tires, with metals and fibers removed beforehand, by crushing them with a machine into smaller pieces measuring 1mm in diameter. The tire chips and Soma silica sand were mixed in various proportions, i.e., the mix ratios of sand to tire chips by volume were set at 10:0, 9:1, 8:2, 7:3, 6:4, 5:5, 3:7 and 0:10.

Table 1 summarizes the physical properties of the soil mixture, including density of soil particles (ρ_s), minimum and maximum dry densities (ρ_{dmin} and ρ_{dmax}), maximum and minimum void ratios (e_{max} and e_{min}), mean diameter (D_{50}) and coefficient of curvature (U_c), respectively, of the samples used in the experiments. In the table, *sf* (sand fraction) indicates the proportion by volume occupied by Soma silica sand in the tire chip-sand mixture. Thus, *sf* = 1 indicates samples consisting of sand only, while *sf* = 0 represents sample with tire chips only.

The density of the particles of tire chips is 1.15 g/cm^3, which is relatively light compared to conventional geomaterials and represents only 2/5 of the particle density of the Soma silica sand. The maximum and minimum dry densities and maximum and minimum void ratios for the composite materials shown in Table 1 are summarized in Figure 1 in terms of their relation with *sf*. It is evident from the figure that the values of e_{max} and e_{min} show almost the same values within the range of

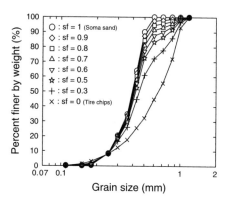

Figure 2.   Grain size distribution curves of various soil mixtures.

$sf = 0$–0.3, and when $sf = 0.3$–1, the values of e_{max} and e_{min} decrease with increase in sf. On the other hand, both ρ_{min} and ρ_{max} increase in value with increase in sand fraction from $sf = 0$ (tire chip only) to $sf = 1$ (Soma Silica Sand No. 5 only). It is apparent that the lower the dry density of the mixture, the higher is the void ratio because the densities of particles of tire chips and sand differ widely; and this is one of the features of this soil mixture.

Figure 2 illustrates the grain size distribution curve for each sample type. Due to the difference in particle density between Soma silica sand No. 5 and tire chips, the soil mixtures with sand : tire chip mix ratio by volume of 10:0, 9:1, 8:2, 7:3, 6:4, 5:5, 3:7 and 0:10, have the corresponding sand : tire chip mix ratio by dry unit weight as 100:0, 95:5, 90:0, 84:16, 76:24, 70:30, 50:50, and 0:100, respectively. Therefore, even if 70% of the entire volume of the sample with $sf = 0.3$ consists of tire chips, the particle size characteristics (D_{50}, U_c) of the soil mixtures as shown in Figure 2, with percent finer by weight in the vertical axis, are much closer to those of pure Soma silica sand than for pure tire chips.

2.2   Specimen preparation and experimental method

The tire chip-sand mixture specimen used for undrained cyclic triaxial tests were prepared by moist tamping method. First of all, the tire chips were washed with a detergent to remove impurities that adhered to their surfaces, and then exposed to warm air to dry for two days. Drying inside an oven with constant temperature of 50°C was attempted, but oil began to ooze out of the tire chips and as a result, warm air was selected to dry them instead. The dried tire chips and Soma silica sand No. 5 were mixed at the prescribed mix ratio by volume. Water was added to the mixture to obtain a sample with initial water content $w = 10\%$ after which the sample was thoroughly mixed again. Membrane was installed in the pedestal of the triaxial apparatus, and the mold 10cm high and 5cm in diameter was set up. The test specimen was prepared by placing the soil mixture inside the mold in five layers, with each layer compacted at a prescribed number of times by dropping a rammer from a prescribed height to control the compaction energy, Ec, which is given by the following expression.

$$Ec = \frac{W_R \cdot H \cdot N_L \cdot N_B}{V} \tag{1}$$

In the above expression, W_R is the rammer weight ($=0.00116$ kN), H is the drop height (m), N_L is the number of layers ($=5$), N_B is the number of drops per layer, and V is the volume of mold (m^3). In the experiment, the test specimens were formed by adjusting the height of drop H and the number of drops N_B in order to obtain two levels of compaction energy, $Ec = 51$ kJ/m^3 and 116 kJ/m^3. The compaction energy was chosen such that the relative density of Soma silica sand

No. 5 specimen ($sf = 1$) was $Dr = 15\%$ (for $Ec = 51\,\mathrm{kJ/cm^3}$) and $Dr = 50\%$ (for $Ec = 51\,\mathrm{kJ/cm^3}$). Because water content has a large effect on the compaction characteristics of soils, this experiment adopted constant initial water content and compaction energy, and test specimens of tire chip-sand mixtures were prepared with different sf. To saturate the specimens, the voids in the specimens were first filled with carbon dioxide and de-aired water was allowed to percolate, after which back pressure of 100 kPa was applied. Two hours after the application of back pressure, the Skempton's pore pressure coefficient was measured. As a result of this method, all test specimens were confirmed to have B-value ≥ 0.95. The saturated specimens prepared as outlined above were then isotropically consolidated at confining pressure of $\sigma_c' = 100\,\mathrm{kPa}$, and undrained cyclic triaxial tests were conducted at loading frequency of 0.1 Hz.

3 UNDRAINED CYCLIC SHEAR PROPERTIES

Typical results obtained from undrained cyclic triaxial tests for tire chips-sand mixtures are shown in Figures 3(a) and (b) in terms of the stress paths and the corresponding deviator stress q – axial strain ε_a relations, respectively. In the figures, test results corresponding to specimens prepared with $Ec = 166\,\mathrm{kJ/m^3}$ and subjected to cyclic stress ratio of about $\sigma_d/2\sigma_c' = 0.17$ are shown as representative results. Note that because similar experimental conditions were not performed for specimens with $sf = 0$ and $sf = 0.3$, the test results shown for these specimens are for conditions nearest the value of $\sigma_d/2\sigma_c' = 0.17$. For specimens with $sf = 1$ and 0.9, the excess pore water pressure accumulated as the cyclic shear loading progresses until the effective stress completely dropped to zero. Moreover, since the amplitude of axial strain suddenly increased at a particular time, it is understood that liquefaction took place. For specimens with $sf = 0.8$, 0.7 and 0.6, the effective stress did not disappear completely and reached a steady state at the end, with the amplitude of axial strain showing gradual increase. For specimens with $sf \leq 0.5$, the decrease in effective stress with the cyclic loading was controlled, while for specimen with $sf = 0$ (pure tire chips), the decrease in effective stress was very minimal. Moreover, the relation between deviator stress and axial strain for $sf = 0$ showed visco-elastic behavior, with large axial strain generated during the first cycle of loading followed by small increase in axial strain during subsequent loading. For specimens with $sf = 0.3$ and 0.5, the amplitude of axial strain appeared to show gradual increase, but in the end, the stiffness was maintained and followed a regular loop. Thus, the undrained cyclic shear behavior of specimens with $sf \leq 0.5$ clearly differed from that of specimens with $sf = 0.6$–0.9, with the characteristics of tire chips dominant when $sf \leq 0.5$.

From the deviator stress-axial relation depicted in Figure 3(b), the stress-strain relations corresponding to the compression side of the first cycle up to the peak value were obtained for each soil mixture and plotted in Figure 4. From the figure, it is evident that for specimen with $sf = 0$ (pure tire chips), the behavior is almost linear when the axial strain $\varepsilon_a < 1\%$. Moreover, the secant modulus at $\varepsilon_a = 1\%$ for specimen with $sf = 0$ is very low, about $E = 1.5\,\mathrm{MPa}$, and axial strain as large as 2% occurred during the first cycle. Comparing the stress-strain relations for various mixtures, it can be observed that as sf increases, the slope of the curve suddenly increases, indicating increasing initial stiffness. Furthermore, the relation between deviator stress and axial strain for $\varepsilon_a < 1\%$ shifts from linear type to non-linear behavior.

The relations between excess pore water pressure ratio, $\Delta u/\sigma_c'$, and double amplitude axial strain, ε_{DA}, versus the number of cycles, N, for each of the test case described in Fig. 3 are summarized in Figures 5 and 6, respectively. In the figures, data points for specimen with $sf = 1$ (pure Soma silica sand No. 5) with relative density $Dr = 50\%$ and prepared using air pluviation method were included, with the undrained cyclic shear test results corresponding to $\sigma_d/2\sigma_c' = 0.20$ plotted as •. Note that for specimen with $sf = 1$ and prepared by moist tamping method ($Ec = 166\,\mathrm{kJ/m^3}$), the relative density was also $Dr = 50\%$. Comparing the test results for both cases having $sf = 1$, it is clear that for air pluviated specimen which was subjected to smaller cyclic shear stress ratio, the condition of $\Delta u/\sigma_c' = 1$ was reached in fewer number of cycles and ε_{DA} was rapidly generated, indicating that liquefaction occurred. This clearly showed that the difference in sample preparation

190

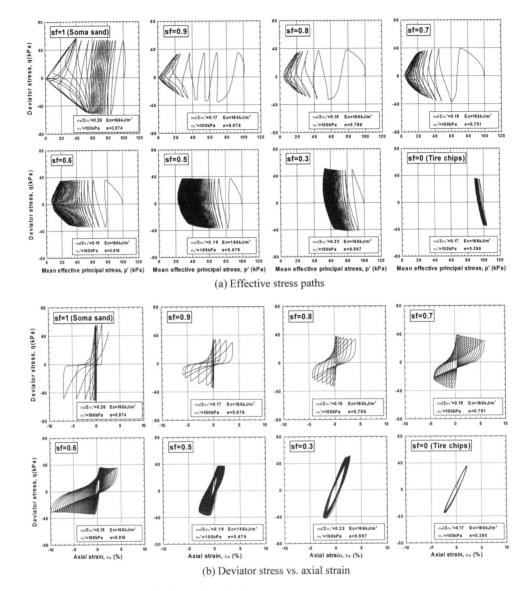

(a) Effective stress paths

(b) Deviator stress vs. axial strain

Figure 3   Typical results of undrained cyclic shear tests.

technique leads to the difference in cyclic behavior between the two specimens. Moreover, for specimens with $sf = 0.9$ and 0.8, excess pore water pressure, $\Delta u/\sigma_c$', as well as double amplitude axial stain, ε_{DA}, rapidly increased from the first cycle of load application, showing a behavior similar to specimen with $sf = 1$ and prepared by air pluviation method.

For $sf = 0$ (pure tire chips), because of the very low stiffness of the tire chips, the double amplitude axial strain, ε_{DA}, was about 4% during the first cycle of load application; however, the excess pore water pressure, $\Delta u/\sigma_c$', did not build up. In addition, no change in both ε_{DA} and $\Delta u/\sigma_c$' was observed even as the number of cycles of load application was increased. For specimens with $sf = 0.3$ and 0.5, $\Delta u/\sigma_c$' gradually increased with the number of cycles, but the condition of $\Delta u/\sigma_c$' = 1 was not reached. Similar to that observed for $sf = 0$, large ε_{DA} was noted during the first cycle, but afterward, the ε_{DA} did not increase even when the number of cycles was increased. Therefore, liquefaction of the soil mixture was not observed. For specimens with

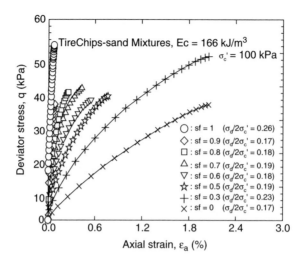

Figure 4. Stress-strain relation for the first cycle of loading.

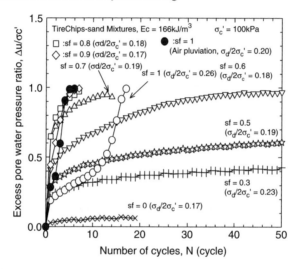

Figure 5. Relation between excess pore water pressure and number of cycles.

$sf = 0.6$ and 0.7, the excess pore water pressure $\Delta u/\sigma_c'$ built-up until near 1.0 and large ε_{DA} was observed at the end; however, such behavior showed gradual increase both in $\Delta u/\sigma_c'$ and ε_{DA} as the number of cycles was increased. Therefore, the characteristics of specimens with $sf = 0.6$ and 0.7 can be considered as intermediate or in-between those of specimens with $sf > 0.8$, which display liquefaction tendencies, and those of specimens with $sf < 0.5$, which are believed not to liquefy.

Figure 7 shows the relation between the maximum excess pore water pressure ratio, $\Delta u_{max}/\sigma_c'$, and sand fraction, sf, for all the test cases considered in this research. For each individual soil mixture, the results for tests involving different cyclic shear stress ratios were included, and the average values corresponding to each of the two compaction energy levels were connected. From the figure, although there was scattering of data points for each individual soil mixture, it can be said that $\Delta u_{max}/\sigma_c' > 0.9$ for soil mixtures with $sf = 1$–0.6, while for mixtures with $sf \leq 0.5$, a trend of lower $\Delta u_{max}/\sigma_c'$ for smaller sf was evident. In addition, the two lines connecting the average $\Delta u_{max}/\sigma_c'$ for each individual sf, which correspond to different compaction energies Ec,

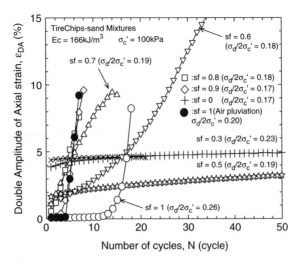

Figure 6.   Relation between double amplitude axial strain and number of cycles.

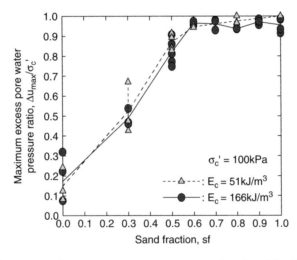

Figure 7.   Relation between maximum excess pore water pressure ratio and sand fraction.

were practically the same. Based on these results, the tire chips have the effect of controlling the accumulation of excess pore water pressure, and the effect was significant for specimens with lower *sf*. However, because the stiffness of the tire chip is very small, specimens with low *sf* have low stiffness and they deform easily for the prescribed loading.

Figure 8 shows the plots of cyclic shear stress ratio $\sigma_d/2\sigma_c'$ and number of cycles N required to produce double amplitude axial strain of $\varepsilon_{DA} = 1\%$, 2% and 5% for specimens with $sf = 1$, 0.8, 0.6, 0.5, 0.3, 0. In the figure, $e_{average}$ corresponds to the average void ratio of each specimen after consolidation at $\sigma_c' = 100\,kPa$. From the figure, the liquefaction strength of the specimen with $sf = 1$ and prepared at compaction energy $Ec = 166\,kJ/m^3$ is twice that of specimen prepared at $Ec = 51\,kJ/m^3$, confirming the effect of compaction on liquefaction strength. However, for specimens with $sf \leq 0.8$, the difference in liquefaction strengths between $Ec = 166\,kJ/m^3$ and $Ec = 51\,kJ/m^3$ decreases. Specially, when $sf = 0$ (pure tire chip), the relations be tween $\sigma_d/2\sigma_c' - N$

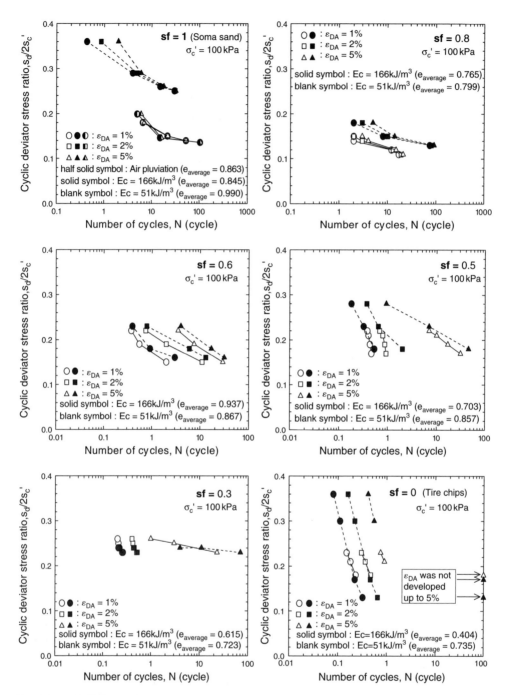

Figure 8. Relation between cyclic shear stress ratio and number of cycles required to produce the specified double amplitude axial strain.

for all values of ε_{DA} practically lie on the same curve, indicating that compaction energy has completely no contribution on the dynamic strength of specimens consisting of tire chips only.

Figure 9 shows the relation between *sf* and cyclic strength *R*, defined as the magnitude of cyclic stress ratio, $\sigma_d/2\sigma_c'$, required. to produce double amplitude axial strain $\varepsilon_{DA} = 5\%$ in $N = 20$ cycles

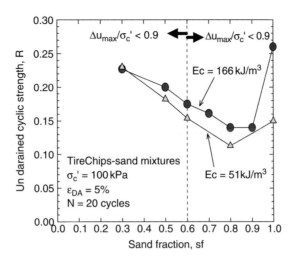

Figure 9. Relation between undrained cyclic shear strength and sand fraction.

of uniform load application. For specimen with $sf = 0$, the value of R was not defined because the condition of $\varepsilon_{DA} = 5\%$ was not achieved for any level of $\sigma_d/2\sigma_c'$ even when the number of cycles $N > 20$. In the figure, there is large difference between the cyclic strengths R of specimens with $sf = 1$ compacted with $Ec = 51\,kJ/m^3$ and $166\,kJ/m^3$, clearly showing the effect of compaction on the cyclic strength. For specimens with $sf \leq 0.9$, it is noted that although the cyclic strength corresponding to $Ec = 166\,kJ/m^3$ is larger for any value of sf, the effect of Ec on R is much smaller compared with the case of the specimen with $sf = 1.0$ (Soma silica sand only). In fact, as sf decreases, the difference in R between the two compaction energies become smaller, and when $sf = 0.3$, the two values of R are almost the same. Moreover, it can be observed that for $sf < 0.9$, there is a tendency for R to increase as the sf decreases, regardless of the compaction energy Ec.

Based on the above, it is can be said that the tire chips in the soil mixture control the build-up of excess pore water pressure, which is remarkable especially when sf is low. As the value of sf decreases, the behavior of the soil mixture transforms from one where the axial strain suddenly increases and shows liquefaction-type of response to one where the development of strain is slower as the cyclic loading progresses, gradually making the soil difficult to liquefy.

4 CONCLUDING REMARKS

In this research, undrained cyclic triaxial tests were performed on tire chip-sand mixtures in order to examine their cyclic shear behavior and strength characteristics. The main findings obtained from the test results are as follows.

- Pure tire chips have very small stiffness and displacements are easily generated during cyclic shear loading, resulting in visco-elastic stress-strain relation. Moreover, excess pore water pressure was not generated during cyclic loading and liquefaction did not occur.
- The occurrence of liquefaction was confirmed for specimens with $sf > 0.5$, but for specimens with $sf < 0.5$, liquefaction was not clearly observed.
- Tire chips appear to control the build-up of excess pore water pressure of the mixture during shearing, and for specimens with $sf < 0.7$, such effect was remarkable for low values of sf.
- When the volume of tire chips was about 10% ($sf = 0.9$), the effect of compaction on the cyclic shear strength decreases. For specimens with $sf < 0.3$, the effect of compaction on the cyclic shear strength was not observed.

REFERENCES

ASTM 1998. Standard practice for use of scrap tires in civil engineering applications, *Annual Book of ASTM Standards*, ASTM, 22 (6): 501–520.

Bosscher, P. J., Edil, T. B. and Kuraoka, S. 1997. Design of highway embankments using tire chips, *Journal of Geotechnical and Geoenvironmental Engineering*, 123 (4): 295–304.

Hazarika, H., Sugano, T., Kikuchi, Y., Yasuhara, K., Murakami. S., Takeichi, H., Ashoke, K.K., Kishida T., Mitarai, Y. 2006. Evaluation of a recycled waste as smart geomaterial for earthquake reinforcement of structure, *Proc., 41st Japan National Conference on Geotechnical Engineering*, 591–592, (in Japanese).

Kikuchi, Y., Nagatome, T. and Mitarai, Y. 2006. Failure mechanism during shear of rubber chip-mixed solidified soil, *Report of the Port and Airport Institute*, 45 (2): 87–103, (in Japanese).

Lee, J. H., Salgado, R., Bernal, A. and Lovell, C. W. 1999. Shredded tires and rubber-sand as lightweight backfill, *Journal of Geotechnical and Geoenvironmental Engineering*, 125 (2): 132–141.

Mitarai, Y., Kawai, H., Kishida, T., Nagatome, T., Yasuhara, K., Murakami, S., Sugano, T., Hazarika, H., Kikuchi, Y., Tatarazako, N., Takeichi, H. and Ashoke, K.K. 2006. "Damping capability of impact load by used tire chips," *Proc., 41st Japan National Conference on Geotechnical Engineering*, 595–596, (in Japanese).

Scrap Tire Derived Geomaterials – Opportunities and Challenges – Hazarika & Yasuhara (eds)
© 2008 Taylor & Francis Group, London, ISBN 978-0-415-46070-5

Basic properties of soil-cement slurry mixed with elastic material

Atsushi Shimamura
Engineering Development Department, Chemical Grouting Co., Ltd., Japan

Junichi Sakemoto
Marketing Department, T. I. C. Co., Ltd., Japan

ABSTRACT: Elastic geo-materials, such as rubber chips and EPS beads, are seldom used in soil structures by mixing with sandy soils to improve the dynamic properties. For the usages in a narrow space or underwater, slurry-type mixtures are more favorable; however, the properties of the mixtures have not been investigated. A series of lab-tests was conducted for soil-cement slurry containing fine elastic particles for their application as aseismic material. Effects on their workability and strength of cement and tire-chips content were checked to determine the appropriate composition of the slurry. High potentials for elastic geo-materials are confirmed on slurry workability and strength development, and the feasible components of the slurry are proposed based on this investigation.

1 INTRODUCTION

For increasing the dynamic durability of the structure, introducing elastic properties into the geo-material could be effective. Elastic geo-materials, such as rubber and EPS, are seldom used as insulators for structures. For example, elastic particle and soil mixtures have been applied in actual construction projects, in which elastic materials were added and mixed with soils and compacted at the construction sites. For this usage, the following precautions are important for the performance: (1) the materials should be accurately placed between ground and structure; and (2) a given homogeneity and density of the mixture should be kept. Slurries containing the elastic particles might be beneficial for the placement and are suitable for placement into narrow spaces or underwater spaces. Properties of the slurry-type mixtures, however, have not been investigated. From the point of the material recycling, waste tire-chips are the best choice for the elastic particles. A series of lab tests was conducted on soil-cement slurry and waste tire-chips mixtures. Effectiveness of the waste tire addition for aseismic usages is concluded on the test results.

2 LAB TESTS

2.1 *Material*

The physical properties of the materials used in this study are presented in Table 1. Cement and tire-chips were added to the basic soil slurry, collected from soil recycling plants. Tire-chips are made from old waste tires, and the diameter is from 1 to 5 millimeters. Slag cement*[4] was chosen for the cement because of its long-term stability and our experience.

2.2 *Sample preparation and test*

The compositions of the test samples are shown in Table 2. Thirteen types of the samples were prepared with three levels of the cement content of 50, 75, and 100 kgs in one cubic meter of

Table 1.   Material properties.

Materials	Properties
Soil slurry	Specific gravity $= 1.50 \pm 0.02$ Sand fraction[**1] $= 40 \pm 2.5\%$ Slump flow[*1] $= 400 \pm 50$ mm
Tire chips	Specific gravity $= 1.10$ Particle size $= 1$–5 mm
Cement	Slag cement[*4] Specific gravity $= 3.04$

[**1] The grain size is 74 μ m–2 mm.

Table 2.   Properties of soil-cement slurry mixed with tire-chips.

No.	Cement kg/m^3[***1]	Tire-chips kg/m^3[***2]	Tire-chips %[***3]	Unit weight g/cm^3
1	50	0	0	1.52
2	50	100	9	1.49
3	50	200	18	1.46
4	50	300	27	1.42
5	50	350	32	1.39
6	75	0	0	1.55
7	75	100	9	1.50
8	75	200	18	1.47
9	75	300	27	1.43
10	100	0	0	1.56
11	100	100	9	1.52
12	100	200	18	1.48
13	100	300	27	1.44

[***1] Cement / One cubic meter of slurry with soil and cement.
[***2] Tire-chips / One cubic meter of slurry with soil, cement, and tire-chips.
[***3] Tire-chips (volume) / Slurry with soil, cement, and tire-chips (volume).

soil-cement slurry, and five levels of tire-chips content of 0, 100, 200, and 350 kgs in one cubic meter of soil-cement slurry mixed with tire-chips.

In the preparation procedure, the specified amounts of cement and tire-chips were added into the readily prepared soil-slurry and mixed for 1.5 minutes by hand mixer.

Test samples were prepared by adding different amounts of tire-chips to the three basic slurries containing three different amounts of cement. In order to check the workability and segregation of the slurry, the slump flow test[*1] and bleeding test[*2] were conducted. Strength development was determined by the unconfined compression test[*3] at 7 and 28 days curing.

3   RESULTS

3.1   Workability

It was suspected that segregation could occur by the density difference between the materials; 1.1 g/cm^3 tire-chips in 1.6 g/cm^3 soil-cement slurry. But, from the observation during sample preparation, no segregation of the slurry was confirmed. The test results of the slump flow and bleeding are shown in Table 3. The bleeding on all samples was less than 0.5%. It was concluded that the cement and tire-chips did not affect the bleeding property in this test range.

Table 3. Properties of soil-cement slurry mixed with tire-chips.

No.	Cement kg/m^3	Tire-chips kg/m^3	Slump flow mm x mm	Bleeding (%) In 1 hr.	Bleeding (%) In 24 hrs.
1	50	0	380 × 380	Under 0.5	Under 0.5
2	50	100	330 × 330	Under 0.5	Under 0.5
3	50	200	260 × 270	Under 0.5	Under 0.5
4	50	300	175 × 175	Under 0.5	Under 0.5
5	50	350	135 × 135	Under 0.5	Under 0.5
6	75	0	360 × 360	Under 0.5	Under 0.5
7	75	100	320 × 320	Under 0.5	Under 0.5
8	75	200	260 × 260	Under 0.5	Under 0.5
9	75	300	170 × 175	Under 0.5	Under 0.5
10	100	0	350 × 350	Under 0.5	Under 0.5
11	100	100	300 × 300	Under 0.5	Under 0.5
12	100	200	260 × 260	Under 0.5	Under 0.5
13	100	300	180 × 180	Under 0.5	Under 0.5

Slurry properties:
Specific gravity = 1.50
    Sand fraction = 40%
    Slump flow = 410 × 415 mm
Mixing time: 1.5 min.
Mixing temperature: 10°C–14°C.

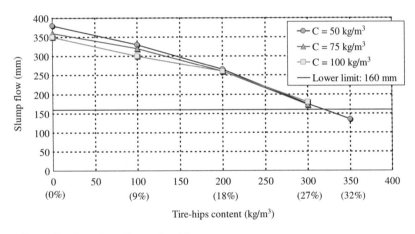

Figure 1. Slump flow test: slump flow – tire-chips content.

According to the guideline*[5] of the workability on soil-cement slurry, the criteria in the slump flow and the bleeding for the workability are more than 160 mm and less than 1%, respectively.

As shown in Figure 1, all samples, except sample number 5 (the amount of tire-chips is 350 kg/m^3), pass the criteria of 160 mm. The slump flow decreased (more viscous) with the increase in either cement and/or tire-chips content, but the tire-chips content appears to have a larger effect on the viscosity. From Figure 1, the upper limit of the tire-chips content is expected around 320 kg/m^3.

3.2 Strength development

The unconfined compression test was conducted on all the samples, except sample number 5, which did not meet the workability criteria. The test results of 7 days curing samples are shown in Table 4 and Figure 2, and the test results of 28 days curing samples are shown in Table 4 and Figure 3.

199

Table 4.   Unconfined compression test (Curing time: 7 days; 28 days).

No.	Cement kg/m^3	Tire-chips kg/m^3	Unconfined compressive strength kN/m^2		Rupture strain %	
			7 days	28 days	7 days	28 days
1	50	0	39	125	2.2	1.4
2	50	100	56	166	4.0	3.5
3	50	200	84	180	4.5	4.2
4	50	300	80	189	5.5	4.7
5	75	0	116	276	2.7	1.1
6	75	100	130	300	3.7	3.2
7	75	200	146	320	4.6	3.9
8	75	300	158	303	4.7	4.1
9	100	0	242	644	2.7	1.2
10	100	100	277	614	3.4	2.1
11	100	200	292	536	4.2	2.4
12	100	300	304	509	4.9	3.3

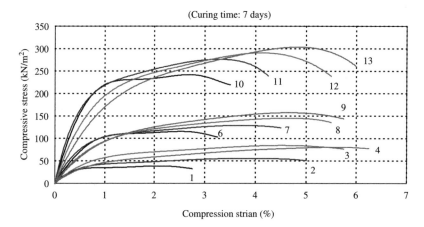

Figure 2.   Unconfined compression test (Curing time: 7 days) stress-strain curve.

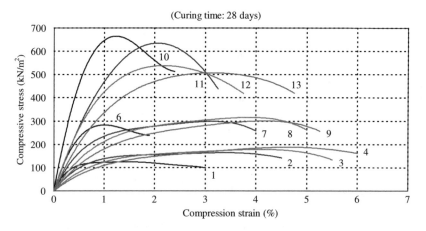

Figure 3.   Unconfined compression test (Curing time: 28 days) stress-strain curve.

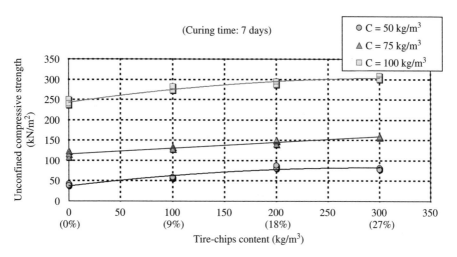

Figure 4. Unconfined compressive strength – tire-chips content (Curing time: 7 days).

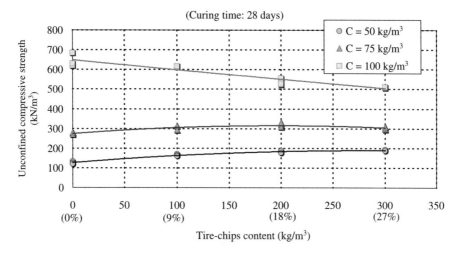

Figure 5. Unconfined compressive strength – tire-chips content (Curing time: 28 days)

At 7 days curing samples, compressive strength increases with the amount of cement, and rupture strain increases with the amount of tire-chips, as shown in Table 4 and Figure 2. The minimum and maximum unconfined compressive strength were 39 kN/m² (No. 1, C = 50 kg/m³, TC = 0 kg/m³) and 304 kN/m² (No. 13, C = 100 kg/m³, TC = 300 kg/m³), respectively. While the rupture strain of no tire-chips (No. 1) is the smallest of 2.2%, the largest rupture strain of 5.5% is achieved by tire-chips mixing. By the addition of 300 kg/m³ of tire-chips, rapture strain could be increased up to 2.5 times higher.

At 28 days curing samples, compressive strength increases with the amount of cement, and rupture strain increases with the amount of tire-chips, as shown in Table 4 and Figure 3. The minimum and maximum unconfined compressive strength were 125 kN/m² (No. 1, C = 50 kg/m³, TC = 0 kg/m³) and 644 kN/m² (No. 10, C = 100 kg/m³, TC = 0 kg/m³), respectively. While the rupture strain of no tire-chips (No. 6) is the smallest of 1.1%, the largest rupture strain of 4.7% is achieved by tire-chips mixing. At the same cement content, the addition of 300 kg/m³ of tire-chips increases the rapture strain to 2.8–3.7 times higher. The increase of the 28-day compressive strength by curing time in comparison with 7 days is 1.7–3.2 times.

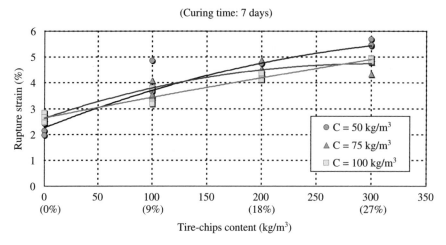

Figure 6.   Rupture strain – tire-chips content (Curing time: 7 days).

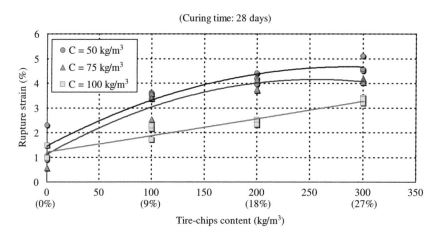

Figure 7.   Rupture strain – tire-chips content (Curing time: 28 days).

Figure 4 shows the relationships between the 7-day unconfined compressive strength and tire-chips content. In all three levels of cement contents, there is little difference in compressive strength, and the effect of tire-chips content is negligible. Therefore, the cement content is the major factor on the unconfined compressive strength.

Figure 5 shows the relationships between the 28-day unconfined compressive strength and tire-chips content. In the cement content of $50\,kg/m^3$, compressive strength increases with the tire-chips content. In the cement content of $75\,kg/m^3$, there is little difference in compressive strength, and the effect of tire-chips content is negligible; however, the opposite effect is confirmed that the increase of tire-chips content decreases compressive strength at $100\,kg/m^3$ tire-chips content.

The rupture strain against tire-chips content of 7-day curing samples is shown in Figure 6, where the rupture strain increases with the tire-chips content. Comparing the rupture strain at the same tire-chips content, the increase of cement content increases rupture strain; however, the opposite effect is confirmed that the increase of cement content decreases rupture strain at above $100\,kg/m^3$ tire-chips content.

The rupture strain against tire-chips content of 28-day curing samples is shown in Figure 7, where the rupture strain increases with the tire-chips content, but decreases with the cement content. Therefore, both cement content and tire-chips content are the major factors on the rupture strain.

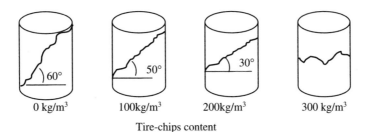

Tire-chips content

Figure 8.   Failure plane schematic (curing time: 7 days).

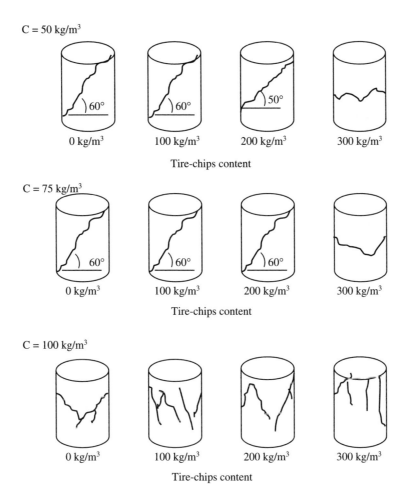

Figure 9.   Failure plane schematic (Curing time: 28 days).

The simple schematics of typical failure plane at 7-day curing samples are shown in Figure 8. Up to 200 kg/m³ of tire-chips content, the failure appeared to be sliding failure, but the 300 kg/m³ tire-chips content sample shows compression failure.

The chematics of typical failure plane at 28-day curing samples are shown in Figure 9. In the cement content of 50–75 kg/m³ with up to 200 kg/m³ of tire-chips content, the failure appeared to be sliding failure, but the samples with 300 kg/m³ tire-chips content show compression failure.

On the contrary, in the cement content of $100\,kg/m^3$, the samples of all levels of the tire-chips content show cracks like the ones shown in the unconfined compression test on concrete samples in general. It is concluded that the increase of the compression strength due to a long curing time made the characteristics of the sample similar to the ones of the concrete samples.

4 CONCLUSIONS/FUTURE WORK

Soil-cement slurry mixed with tire-chips is confirmed to have the good workability, segregation resistance, and strength comparable to improved soil, and it is planned to further investigate the strength changes with the curing time by performing long-term unconfined compression tests. It is also confirmed in the unconfined compression tests that the rupture strain of soil-cement is extended with the increase of the tire-chips content.

In the future, the dynamic deformation properties of the mixture with soil-cement tire-chips, such as 1) the relationship between the damping constant and shear strain ($h - \gamma$), and 2) the relationship between shear deformation modulus and shear strain ($G/G_0 - \gamma$), will be investigated through the cyclic shearing test in order to ensure the application of the mixture as aseismic material. From the point of view of the economy and recycling of waste, the influence to those properties will be tested by changing particle size of the tire-chips and the degree of the contents of the impurities other than rubber in the mixture. Furthermore, besides tire-chips, the addition of fibrous material for the purpose of increasing the elasticity of the mixture will also be included in the further research.

5 KEY WORDS

Tire-chips, Soil-cement slurry, Aseismic material, Elastic soil.

REFERENCES

*1 JHS A 313–1992, Cylinder method.
*2 JSCE-F 522–1994, Polyethylene bag method.
*3 JIS A 1216–1998, Unconfined compression test.
*4 JIS R 5211, Slag fraction is 30%–60%.
*5 Soil-cement slurry mixtures application technical manuals by Ministry of Land, Infrastructure & Transportation.

Stacks of tires for earth reinforcement using their resistance to hoop tension and land reclamation methods

K. Fukutake & S. Horiuchi
Shimizu Corporation, Tokyo, Japan

ABSTRACT: High bearing force and cost effective construction is possible if used tires are recycled in original form. But such reusing methods are seldom implemented. In this study, we made laminar bodies using tires which were packed with granular materials at the center. We validated the characteristics of the laminar bodies through compression tests. The experimental results revealed that laminar bodies had high strength especially under prestressed conditions. For land formation, we proposed easier and quicker construction methods. The construction methods can reuse large amounts of used tires and contribute to environmental conservation.

1 INTRODUCTION

Used tires have high recycling potential because they are available in large quantities at low cost. Used tires have, however, not been used effectively with less than 90% being recycled while illegal dumping of tires is causing problems (http://www.bridgestone...). Attempts have recently been made to develop technologies for recycling used tires and actually applying recycled tires in construction to reduce adverse effects of dumping on the environment (Yasuhara et al. 2006). For example, scrap tires have been shredded or chipped for reuse on a practical basis.

Crumbling tires requires efforts and fails to achieve the use of the benefits of whole tires. Tires have a complicated internal structure to prevent bursting and are highly durable. Thus It would seem more reasonable to find users for the tires in the original form. They are highly engineered structure. One engineer commented that "cutting tires up for various uses, was like tearing a good house down for the firewood" (Stuart 1999). Using tires whole instead of using them for fuel eliminates the labor for shredding and costs less. The method also helps use the strength of tires against hoop tension.

For using tires as a fill material or for land reclamation, (i) tires are crushed to form a bale or (ii) tires are used whole or cut for use as an earth reinforcement material. For type (ii) applications, numerous methods have been proposed such as slope protection, retaining wall construction and earth reinforcement (Yasuhara et al. 2006, Jones 1990, O'Shaughnessy 2000). Fill construction using cut tires stuffed with sand such as Ecoflex (http://www.ausimm...), a slope protection technology, have been implemented on a wide scale, but they have not made the best use of tire's resistance to hoop tension.

This paper describes the mechanical properties of multi-layered structures that use the mechanical properties of tires in order to adopt old tires as whole as possible. Also explained are the effectiveness and future developments of land reclamation technologies and methods using multi-layered structures.

2 BASIC PRINCIPLES OF METHOD

Applying loads on a cylindrical stack of tires filled with granules causes the granules to expand the tires horizontally as shown in Fig. 1 and photograph 1. Then, hoop tension acts on the tires and

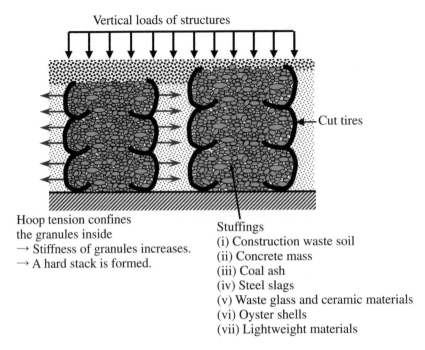

Vertical loads of structures

Cut tires

Hoop tension confines
the granules inside
→ Stiffness of granules increases.
→ A hard stack is formed.

Stuffings
(i) Construction waste soil
(ii) Concrete mass
(iii) Coal ash
(iv) Steel slags
(v) Waste glass and ceramic materials
(vi) Oyster shells
(vii) Lightweight materials

Figure 1.   Earth reinforcement using a stack of tires.

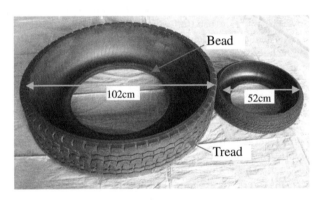

Bead

102cm

52cm

Tread

Photograph 1.   Cut old tires (truck and vehicle tires).

confines the granules. The stiffness of granules increases with the confining force. As a result, a hard cylinder is formed for stabilizing the ground. Tires have wires inside and are highly resistant to hoop tension, thus large confining stress acts on the granules. Granules can be made so hard that a stack of tires can resist large overburden loads.

Stacks are filled with stuffings by (i) cutting part of the tire to facilitate stuffing (Fukutake 2006) or (ii) placing stuffings into a stack of uncut whole tires (Fukutake & Horiuchi 2004, Fukutake et al. 2003). Method (i) uses bowl-shaped tires as shown in Photograph 1 to facilitate stuffing.

Waste materials may be used as granules as long as their density is expected to increase as a result of compaction. Lightweight aggregate, paper sludge ash or shredded tires can be used for constructing lightweight fills. In whole tires, materials have a relative density of approximately 90% as a result of compaction with a steel bar or by vibration (Fukutake & Horiuchi 2004, 2005).

3 STUFFED TIRE STACK COMPRESSION AND UNLOADING TESTS

Vehicle and truck tires were used in the tests either whole or with the top surface removed (Photograph 1). Tests were conducted in nine cases (Table 1).

3.1 Tests using whole vehicle tires

3.1.1 Test outline

Used vehicle tires were stacked in three layers. Tests were also conducted using five layers of tires (case ⑤, photograph 2(a)) and new tires (case ⑦) for comparison. Waste as well as soil materials were used as stuffings from environmental considerations.

Recycled aggregate is low-quality material with high adhesive mortar content. Paper sludge granules are obtained by granulating the mixture of ash produced from the incineration of paper sludge, cement and lime. They are slightly adhesive. Filling the opening of stacked tires was difficult because of the rough surface. Vibratory filling was made easier by connecting the beads of tires with clips (Fukutake & Horiuchi 2004). Photograph 2(b) shows fractured surface at the end of loading in case ② A. Tread was fractured in numerous cases as in this photograph, but bead was seldom fractured. The tread is the part of a tire that made contact with the road.

3.1.2 Test results and discussions

Figure 2 shows vertical stress-strain curves in all of the cases. Loads were removed once each at a stress of 1200 kPa and 4800 kPa, and four times at 2400 kPa (black triangles along the vertical axis). Unloading took place at other stress levels due to the severance of wires in the tread. Loads were held constant for three minutes at points of alternate loading and unloading to verify the effects of creep. The effects were small. Stress-strain curves at the time of loading tended to sag. This is because small voids left in the tire in the initial stage were crushed causing the stuffing to fit in with the tire. As the loads increased, the hoop tension fully acted on tires and created a highly confined state. The large confining force and the strength of tires were confirmed because stuffed tires had much higher stress than sandbags containing crushed stones (Fukutake et al. 2003). Several times of severance of wire in the tread ended in total fracture (Photograph 2 (b)). No fracture occurred suddenly but tires proved tough. In reality, the earth surrounding the tires confines them. Neither rapid fracture nor settlement is therefore unlikely to occur. Ultimate strength was lowest in the case of Toyoura sand filling five layers of tires (case ⑤). Fracture occurred at 4700 kPa. Wires were severed earliest at 2900 kPa, or 600 kN, in the case of compaction of Toyoura sand by watering (case ④). Thus, it was confirmed that whole vehicle tires provided a minimum strength of 600 kN, or 60 tf. Either used or new tires proved almost equally effective unless wires were severed. Variance between three- and five-layered stacks was small.

Figure 3 gives an enlarged diagram of stress-strain curves for stacks of tires filled with Toyoura sand, and Young's modulus E. The curves obtained as a result of unloading and re-loading are sloped sharply. Under the conditions, tires can provide high stiffness. Prestressed stacks of tires could work as very hard members. Voids in the tire not filled with stuffings are crushed by prestressing. Stacks of tires are available for a wide range of applications such as preloaded prestressed bridge piers using a rod (Fig. 4, Tatsuoka et al. 1998) and retaining walls because of their high stiffness and strength. Preloaded prestressed bridge piers aim at achieving high stiffness and elastic behavior in vertical direction, and aim at achieving flexible behavior like isolation system in horizontal direction.

Figure 5 shows relationships between vertical strain ε and Poisson's ratio v. Dilative materials including the Toyoura sand densely injected have high v. Loosely injected Toyoura sand or crushed granules have low v. Lateral displacement and the value of v increases when the material is about to fracture.

Table 1. List of compression tests.

Tire type	Number of tires, height of stack h	Stuffing material	Injection method	Density of stuffing material (t/m³)	Initial fracture stress*	Initial shear strain	Ultimate stress (maximum value)	Ultimate displacement (maximum value)	Remarks
Vehicle whole tire	① Used tires in three layers h: 36.5 cm	Mesalite	Steel bar	0.52			(Approximately 1100 kN)	(Approximately 40%)	Difficulty in compaction Crushing of particles No fracture
	② Used tires in three layers h: 37.8 cm	Toyoura sand	Vibro compactor	1.63	7150 kPa	8.2%	7150 kPa	8.2%	Fracture in the middle layer
	③ Used tires in three layers h: 37.0 cm	Paper sludge ash	Vibro compactor	1.24	7960 kPa	31.1%	8740 kPa	32.4%	Fracture in the middle layer
	④ Used tires in three layers h: 38.7 cm	Toyoura sand	Watering	1.37	2920 kPa	12.4%	6390 kPa	19.4%	Fracture in the middle layer
	⑤ Used tires in five layers h: 67.1 cm	Toyoura sand	Vibro compactor	1.76	3860 kPa	5.4%	4650 kPa	7.5%	Fracture in the middle layer
	⑥ Used tires in three layers h: 39.4 cm	Recycled aggregate	Vibro compactor	1.53	3220 kPa	10.9%	6410 kPa	19.3%	Difficulty in compaction Fracture in the upper layer
	⑦ New tires in three layers h: 37.9 cm	Toyoura sand	Vibro compactor	1.61	4440 kPa	5.5%	6840 kPa	9.5%	Fracture in the middle layer Loads kept constant for 15 minutes
Cut vehicle tire	⑧ Used tires in five layers	Toyoura sand	Vibro compactor	1.76	3050 kPa	6.5%	3550 kPa	8.3%	Stack made layer by layer Fracture in the middle layer
Cut truck tire	⑨ Used tires in three layers	Toyoura sand	Vibro compactor	1.76	6550 kPa	6.9%	6550 kPa	6.9%	Stack made layer by layer Fracture in the middle layer

*When the wire was cut for the first time.

(a) Cylindrical specimen composed of five layers of tires

(b) Fracture of a tire (ultimate state where Toyoura sand was used)

Photograph 2.   Stuffed used tire stack compression test (for whole vehicle tires).

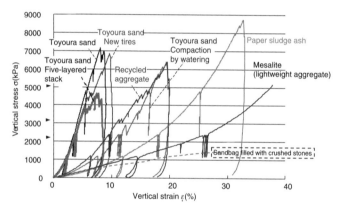

Figure 2.   Stress-strain curves in compression and unloading tests of stuffed whole tire stacks in cases ① through ⑦.

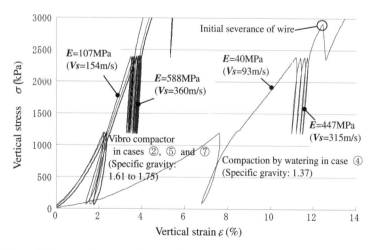

Figure 3.   Enlarged diagram of stress-strain curves for stacks of tires filled with Toyoura sand and stiffness (Colors of lines are the same as in Fig. 2).

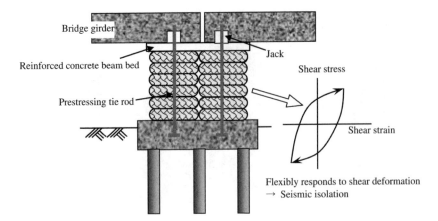

Figure 4.   Preloaded prestressed bridge pier.

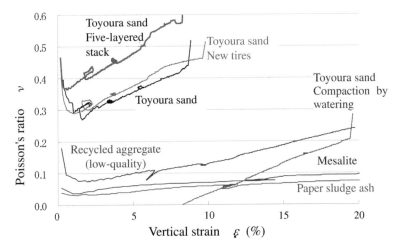

Figure 5.   Poisson's ratio and vertical strain.

3.2   *Tests using cut tires*

3.2.1   *Test outline*
Tires were cut by removing the top surface to use the bowl-shaped lower part in the tests (Photograph 1). Tires could be cut easily with existing cutting machinery. The upper part can be recycled as a weight for crop storage or chipped tires, or used for thermal recycle. Dry Toyoura sand was used for stuffing tires. The material was injected into cut tires consecutively in layers by a vibro compactor (Photograph 3).

Photograph 4 shows fractured surface of a truck tire at the end of loading. The bead on the top surface of the tire was removed when the tire was cut. Swelling was observed on the top surface as the load increased. When the load was increased to create an ultimate state, fracture occurred in the tread, but not in the bead. Strength deteriorated by a smaller margin than expected even with no bead on one side of the tire (Fig. 6(a)).

3.2.2   *Test results and discussions*
Figure 6 shows the relationships between vertical stress and strain, and Young's modulus E at a given point. In the initial stages of loading, curves sag. Loading involved unloading and

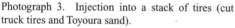
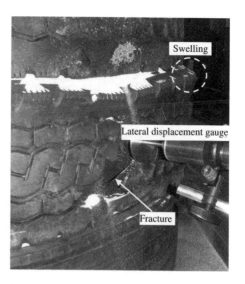

Photograph 3. Injection into a stack of tires (cut truck tires and Toyoura sand).

Photograph 4. Fracture and swelling of a truck tire.

re-loading. Wires were severed at points indicated by circles. Figure 6(a) compares the results for whole and cut vehicle tires. E was slightly smaller for cut tires than for whole tires. No other outstanding variances were found. Figure 6(b) provides the results for truck tires. Truck tires had larger E and much higher strength before fracture occurred. Truck tires had a cut surface four times as large as that of vehicle tires. A unit length of the truck tire therefore carried higher stress. The maximum vertical load was assumed to be smaller on the truck tire. Truck tires have double internal structures and are much more tough than vehicle tires. Stress σ_{max} and strength P_{max} at the time of initial severance of wire were 3800 kPa and 800 kN (80 tf) for whole vehicle tires, 3070 kPa and 650 kN (65 tf) for cut vehicle tires and 6600 kPa and 5300 kN (530 tf) for cut truck tires, respectively. Then, a strength of 60 tf was guaranteed in cut vehicle tires and a strength of 500 tf, 8.5 times the former, was guaranteed in cut truck tires. Truck tires had a very high strength.

Poisson's ratio v was 0.4 to 0.5 for five-layered stacks of either whole or cut vehicle tires. For three-layered stacks of cut truck tires, v was 0.2 to 0.3. Lateral displacement and v increased where fracture was about to occur also in cut tires. Swelling on the top surface of the tire was outstanding (Photograph 4). Lateral displacement was measured at the middle of the tire, so v indicates a mean value. Cut tires were assumed to exhibit higher v locally.

4 VARIOUS APPLICATIONS

Stacks of tires are applicable to actual structures as tire piles. Combining tire stacks with geotextile enhances total stability. Then, the friction between tires and the geotextile wrapping numerous tires increase the stability of earth reinforcement. Earth reinforcement can take on various shapes. Thus, greater freedom of design is provided.

An image of tire stacks applied to road embankment is given in Fig. 7. Securing traffic on at least one lane is sometimes required during a great earthquake with level-two ground motions. Stacks of tires may be used to provide for one-lane traffic (Fig. 8). As shown in Fig. 8, geotextile integrates tires into one and ensures safety even in the case of failure of the fill. A vertical embankment can be constructed with no slopes by applying geotextile densely.

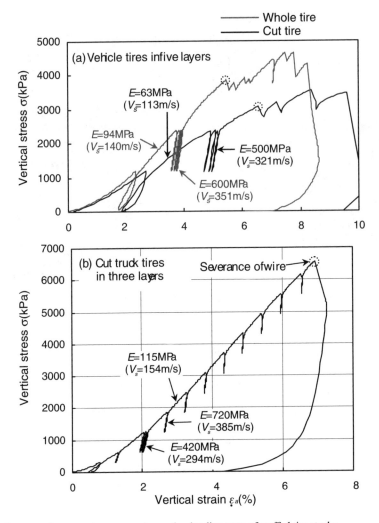

Figure 6.   Stress-strain curves in compression and unloading tests of stuffed tire stacks.

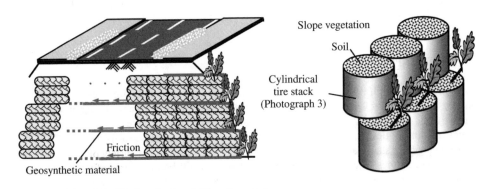

Figure 7.   Lightweight fill, sharply sloped fill and vegetation.

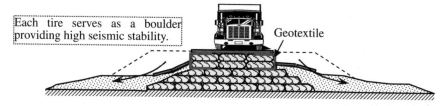

Figure 8.   Road fill for protection from level-two ground motions securing single-lane traffic at the minimum for emergency vehicles.

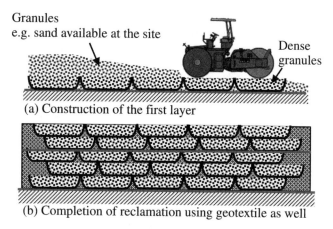

Figure 9.   Easy land reclamation method using cut tires.

In land reclamation in the field, soils are spread over the cut tires that are laid and compacted by rollers on a layer-by-layer basis as shown in Fig. 9. In the compacted soil, land can be reclaimed easily on a large scale. Cut truck tires have a thickness of 30 cm, fit for compaction by rollers. Stacking tires vertically, or in the shape of a cylinder, is not necessary. Geotextile is inserted between layers of tires whenever required for ensuring safety.

5   CLOSING REMARK

A method was presented for recycling used tires as whole as possible and using steel wires in tires. The method uses high strength of tires against hoop tension. Tests showed that earth reinforcements using stacks of tires filled with granules were mechanically strong. Prestressing stacks of tires was found to make stiff earth reinforcements. Then, it was described that stacks of tires could possibly be adopted not only for land reclamation or backfilling but also as structural members such as bridge piers. Finally, methods were suggested for constructing stacks of tires for land reclamation or backfilling on a large scale.

Used tires can be used effectively in construction in various forms. The value added of used tires can be increased by selecting appropriate methods. Tire stacks are globally applicable because used tires are available throughout the world. The proposed method also takes the environment into consideration. Burying tires underground eliminates such problems as fire and the growth of vermin. It is hoped that the proposed method will be adopted at a wide variety of sites.

The durability and service life of used tires are determined by the degree of wear. Hoenig has noted based on a 20-year monitoring of used tires buried in the soil that (i) no toxic materials leaked during the period and that (ii) used tires were assumed to have a service life of at least 200 years as long as they were not worn badly (Stuart 1999). There is little possibility of the rubber in buried tires being lost or the wires in tires exposed or corroded. The rate of tire deterioration varies according to the percentages of natural and synthesized* rubber, and soil properties. Thus, the rate is different from region to region. (*Vulcanized rubber is not biodegradable.) Studies will be required about the durability and quality control of used tires under more severe conditions.

Examinations should be made of the mechanical properties or seismic stability of stacks of tires that are staggered. Building distribution and management systems to facilitate the management and transport of used tires is also important.

ACKNOWLEDGEMENT

The authors would acknowledge with thanks the useful advice of Professor Kazuya Yasuhara of Ibaraki University, Mr. Hideo Takeichi and Dr. Ashoke K. Kamokar of Bridgestone Corporation, Mr. Minoru Kawaida of Central Nippon Expressway Company Limited and Mr. Hiroki Kawasaki of Shimizu Corporation.

REFERENCES

Fukutake, K. and Horiuchi, S. 2006. "Forming Method of Geostructure using Recycled Tires and Granular Materials" (in Japanese), *Proc. of the 41st Japan national conference on geotechnical engineering*, 597–598.
Fukutake, K. and Horiuchi, S. 2004. "Forming method of geostructure using recycled tires and granular materials" (in Japanese), *Proc. of the 39th Japan national conference on geotechnical engineering*, 653–654.
Fukutake, K., Horiuchi, S., Matsuoka, H. Liu, S. and Kawasaki, H. 2003. "Stable geostructure using recycled tires and properties of improved body using tires with granular materials" (in Japanese), *Proc. of the 5th Japan national conference on environmental geotechnical engineering*, 189–194.
Fukutake, K. and Horiuchi, S. 2005. "Stress-strain relationship of recycled tires compacted with granular materials" (in Japanese), *Proc. of the 40th Japan national conference on geotechnical engineering*, 247–248.
http://www.bridgestone.co.jp/eco/report/
http://www.ausimm.com/green2006/presentations/waste_tyre_recycling_opportunities_mining_industry.pdf.
Jones, C.J.F.P. 1990. "Construction influences on the performance of reinforced soil structures", STATE OF THE ART REPORT, *British Geotechnical Society*, 97–116.
O'Shaughnessy,V. and Garga, V.K. 2000. "Tire-reinforced earthfill, Part1-3", *Canadian Geotech. J.* Vol. 37, 75–131.
Stuart A. Hoenig: 1999. "The use of whole tires for erosion control, fences, cattle feed lots houses and water saving on grassy areas", *International Erosion Control Assoc.* Nashville, TN.
Tatsuoka, F., Uchimura, T., Tateyama, M. and Kojima, K. 1998. "Performance of Preloaded-Prestressed Geogrid-Reinforced Soil Railway Bridge Pier" (in Japanese), *Tsuchi-to-Kiso*, The Japanese Geotechnical Society, Vol. 46, No. 8, 13–15.
Yasuhara, K., Ashoke K Karmokar, Kato, Y., Mogi, H. and Fukutake, K. 2006. "Effective utilization technique of used tires for foundations and earth structures" (in Japanese), *Kisoko*, Vol. 34, No. 2, 58–63.

Scrap Tire Derived Geomaterials – Opportunities and Challenges – Hazarika & Yasuhara (eds)
© 2008 Taylor & Francis Group, London, ISBN 978-0-415-46070-5

Shaking table test on liquefaction prevention using tire chips and sand mixture

H. Hazarika
Akita Prefectural University, Akita, Japan

K. Yasuhara
Ibaraki University, Hitachi, Japan

A.K. Karmokar
Bridgestone Corporation, Tokyo, Japan

Y. Mitarai
Toa Corporation, Yokohama, Japan

ABSTRACT: A series of small-scale model shaking table test on a caisson type quay wall was conducted on liquefaction prevention measures using tire chips and sand mixtures. In one series, the caisson had the conventional sandy backfill. In another series, the backfill consists of composite soils made of tire chips mixed sand. The mixing percentage of tire chips was 50% of the total volume of the backfill sand. Sinusoidal motions were imparted to the soil-structure system, and the dynamic increment of the load acting on the caisson, associated displacement of the caisson, and the excess pore water pressure in various locations of the backfill were measured. The test results have demonstrated that, despite the fact that the tire chips reinforced composite backfill has a very low relative density, there was no liquefaction in the backfill, resulting in a good performance of the soil-structure system during earthquake loading.

1 INTRODUCTION

Growing emphasis on cost, environment and structural performance, have led to an increasing attention to research on recycled materials, including the tire derived geomaterials. Cost-performance turns out to be the key factor not only in the design and construction of new structures, but also in the upgradation of important structures, and to enhance their performances during earthquakes.

The joint report (JSCE/JGS, 1996) of the Japanese Geotechnical Society (JGS) and the Japan Society of Civil Engineers (JSCE) after the 1995 Hyogo ken Nanbu earthquake, Kobe, Japan has concluded that there were mainly two aspects that led to the devastating damages to retaining structures during the earthquake. They are: (1) soil failures, which were due mainly to liquefaction and subsidence of the backfill as well as the foundation soils. (2) structural failures, which were due mainly to the strong inertial force. Since that earthquake, significant theoretical and experimental works have been done on geotechnical failures during earthquake (Dickenson et al., 1998; Ghalandarzadeh et al., 1998; Iai, 1998; Inagaki et al., 1996; Ishihara, 1997; Ishihara et al., 1996; Kamon et al., 1996; Towhata et al., 1996). These researches have given a new dimension to the subsequent design and construction of geotechnical structures. On the other hand, 1990s have also experienced many other devastating earthquakes worldwide. The lessons learned from these earthquakes have led to the emergence of a new design methodology called the performance-based design (PIANC, 2001; SEAOC, 1995; Steedman, 1998).

The fundamental difference between the conventional limit state design and the emerging performance based design is schematically explained in Fig. 1. An overview of the performance based

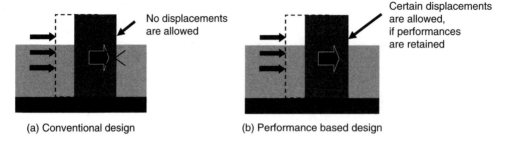

Figure 1. Underlying concept of performance based design.

approach for designing remediation of liquefiable soils is presented in Iai and Tobita (2005), in which the applicability and limitations of conventional approach are discussed. In the performance based design, the major shift is towards the emphasis on engineering judgment, and thus the design engineer has a significant and responsible role to play. The pressures on geotechnical engineering professionals, are, thus mounting towards development of cost-effective and environmentally friendly disaster mitigation techniques. Tire derived recycled geomaterials, can thus the materials of choice, due to potential economic and environmental benefits of these materials in developing disaster preventive technique that can enhance the structural performances during earthquakes. Due to many special characteristics, not present in other geomaterials, these emerging geomaterials of the millennium are coined *smart geomaterials* by Hazarika (2006).

Applications for utilizing the tire-derived geomaterials can be classified into two categories: (1) Stand alone application (use of tires and tire shreds/tire chips as it is) and (2) Composite application (tire chips mixed with soils and/or other materials). Both have many potential benefits. The Japanese experiences in both the category are described in Yasuhara et al. (2006). One of the later category applications is to mix tire chips with soils, and use them for backfilling purposes. Recent researches by Hyodo et al. (2007) have shown that such tire chips reinforced soil can prevent the liquefaction of loose sandy backfill if proper mixing percentage is selected. Various mechanical properties of sand tire chips mixers have been reported in Ghazavi (2004) and Youwai & Bergado (2003).

This research is an attempt towards achieving a better seismic performance of structures using the scrap tire derived geomaterials such as tire chips by mixing them with sand. The objective of this research is to evaluate the effect of tire chips mixed sand on preventing earthquake disaster mitigation such as liquefaction. Liquefaction related damages such as subsidence and lateral spreading are evident in almost every earthquakes affecting the performance of geotechnical structures (Hamada et al., 1996). The performances of many structures improved by various existing liquefaction prevention method during the 1995 Hyogoken Nanbu Earthquake have been reported in Yasuda et al. (1996). In this research, a series of small-scale model shaking table test on a caisson type quay wall was conducted to examine the soil-structure interaction effect and seismic performances of the soil-structure system, when tire chips are used as reinforcing material by mixing them with sand.

2 SHAKING TABLE TESTING PROGRAM

2.1 *Test model*

The shaking table assembly of soil dynamics laboratory of Port and Airport Research Institute (PARI), Japan was used in this study. Figure 2 shows the cross section and front view of the model caisson and the locations of various devices (load cells, pore water pressure cells, accelerometers and displacement gauges) used in the testing program. The soil box was made of a steel container 0.85 m long, 0.36 m wide and 0.55 m deep. The model caisson is 20 cm high and 10 cm wide as shown in the figure. All the monitoring devices were installed at the central caisson (20 cm long) to eliminate the effect of sidewall friction on the measurements.

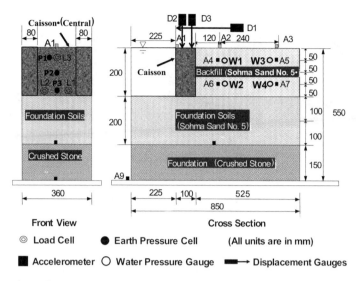

Front View   Cross Section

◎ Load Cell   ● Earth Pressure Cell   (All units are in mm)

■ Accelerometer   ○ Water Pressure Gauge   ■▶ Displacement Gauges

Figure 2. Experimental setup.

Table 1. Physical properties of the test materials

Test materials	ρ_s (t/m^3)	ρ_{dmax} (t/m^3)	ρ_{dmin} (t/m^3)	e_{max}	e_{min}
Sohma sand No. 5	2.645	1.574	1.273	1.077	0.680
Tire chips 1 mm	1.150	0.442	0.347	2.318	1.600

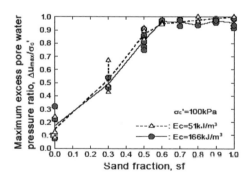

Figure 3. Relationship between excess pore water pressure and sand-tire chips mixing ratio (After Hyodo et al., 2007).

2.2 Test materials

For the conventional backfill, No. 5 Sohma sand was used. The tire chips of average grain size 1 mm were used as reinforcing materials. Table 1 summarizes the physical properties of the sand and the tire chips, including density (ρ_s), maximum and minimum dry densities (ρ_{dmax}, ρ_{dmin}), and maximum and minimum void ratios (e_{max}, e_{min}). The mixing percentage of tire chips was 50% of the total volume of the backfill sand. Hyodo et al. (2007) performed a series of undrained cyclic shear tests on tire chips to examine the effect of sand fraction on liquefaction. Their experimental results are shown in Figure 3, which shows that when tire chips mixing percentage (by volume)

217

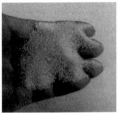

| (a) Sohma sand No. 5 | (b) Tire chips (1 mm size) | (c) Tire chips and sand mixture |

Figure 4. Test materials.

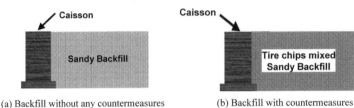

| (a) Backfill without any countermeasures | (b) Backfill with countermeasures |

Figure 5. Test conditions.

exceeds 50%, the mixture tends to liquefy. Thus, the adopted mixing percentage in this model test was reasonable from the liquefaction and practical point of view. Figure 4 shows the two materials that were used and the state of the mixture after mixing.

2.3 Model preparation and testing procedures

Both the backfill as well as the foundation were prepared using Sohma sand (No. 5). The dense saturated foundation sand was prepared in four layers using vibratory compaction technique. After preparing each layer, the whole assembly was shaken with 300 Gal of vibration of frequency 7 Hz for 4 sec. The process was repeated for 4 to 8 times depending on the requirements. Backfill (under submerged condition) was prepared in five layers using free falling technique. Compaction was achieved by shaking the soil box and the wall assembly 3 to 4 times (as needed) with 200 Gal of vibration of frequency 7 Hz for 4 sec. The achieved relative densities of the foundation soil were 95% to 100%.

Two sets of tests, that were conducted, are shown Fig. 5. They are: (1) a caisson with conventional sandy backfill, and (2) same caisson with backfill made from tire chips and sand mixture. In the case of tire chips mixed backfill, the mixing percentage were 50% by volume of the tire chips and the sand. The achieved backfill relative density, D_r was about 40%. In the case of conventional backfill without any liquefaction countermeasures, the backfill relative density, D_r was about 40%, implying that the backfill is liquefiable. The lower relative density used in the tire chips mixed backfill was to examine whether such low relative density could provide any resistance to backfill liquefaction.

3 TEST RESULTS

3.1 Liquefaction behavior

Figures 6–7 show the excess pore water pressure measured in the backfill at the locations, W2 and W4 shown in Fig. 2. Figure 6 is the case for the backfill without any countermeasures against liquefaction and Figure 7 is the case for the tire chips mixed sandy backfill. It can be seen that while the conventional backfill shows the increase of the excess pore water pressure with time, the

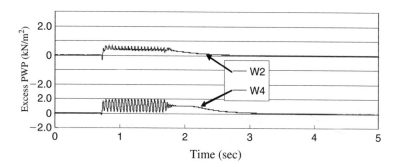

Figure 6. Liquefaction behavior (backfill without countermeasures).

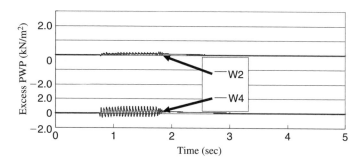

Figure 7. Liquefaction behavior (tire chips mixed sandy backfill).

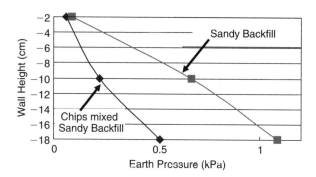

Figure 8. Incremental seismic earth pressure (Acceleration = 300 Gal).

tire chips mixed backfill shows almost no increase in the pore water pressure. This is, in spite of the fact that, the two backfills were having almost the same relative density (40%). The presence of tire chips in the sand could control the build up of the excess pore water pressure, and thus prevent any liquefaction related damages. Since liquefaction tends to increase the earth pressure against the wall, a look at the measured earth pressure on the wall will further substantiate this observation.

3.2 *Seismic incremental earth pressure*

The distribution of the incremental maximum earth pressure at a particular acceleration level (300 Gal) shows that (Fig. 8), as compared to the test case of without countermeasure (Fig. 5a), the tire chips reinforced backfill experiences significantly less earth pressures during the dynamic loading. The reason for this could be attributed to the liquefaction phenomenon and the related subsidence of

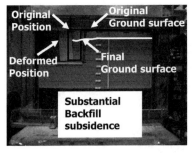 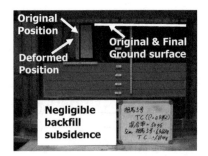

(a) Without liquefaction countermeasures   (b) Tire chips & sand mixed backfill

Figure 9. Structural performances while shaking at 400 Gal (Frequency = 7 Hz).

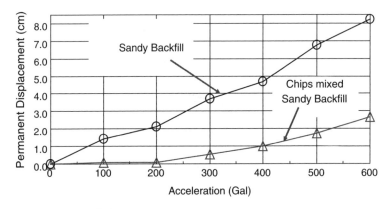

Figure 10. Residual displacement of the structure Vs. Acceleration amplitude.

the backfill. For loose backfill, such as the case here, the liquefaction increases the earth pressure against the structure. However, for the same relative density of the backfill, the presence of tire chips prevents the liquefaction of the backfill due to the earthquake resistant reinforcing effect of the tire chips.

3.3 Seismic performance of the structure

The effect of backfill liquefaction on structural performances is evident from the picture shown in Figures 9a and 9b. The state of deformation of the backfill at the acceleration amplitude of 400 Gal with a frequency of 7 Hz are shown in these figures. It can be observed that, in the case of the backfill with no countermeasures, the performance of the structure is completely lost with high structural deformation and substantial subsidence of the backfill. However, the tire chips reinforced backfill does not experience any subsidence in the backfill. In addition, the residual deformation experienced by the structure is within the limit that can preserve the structural performance. A quantitative discussion in the following subsection will further elaborate on the residual displacement and performance of the structure.

3.4 Residual deformation state

Figure 10 shows the relationship between the residual displacement of the structure and the input acceleration for the two test cases, at the end of each dynamic load. The figure indicates that with increasing acceleration amplitude, while conventional caisson experiences higher residual displacement, the caisson with the tire chips mixed backfill with same relative density, shows only a mild increase. The residual displacement could be reduced to more than half, even at higher level of accelerations. This implies that even at higher acceleration levels, the adopted countermeasure could enhance the performance of a soil-structure system.

4 CONCLUSIONS

The model shaking table test results on liquefaction prevention using sand-tire chips mixtures lead to the following conclusions:

(1) Tire chips mixed sand does not undergo liquefaction, if proper mixing percentage is selected.
(2) Prevention of liquefaction could reduce the incremental dynamic earth pressure on soil structures.
(3) A substantial reduction of the dynamically induced displacement of structures could be achieved using the liquefaction preventive measures. This implies a better structural performance during earthquake loading.

These conclusions imply that using a relatively cheaper and smarter geomaterials, such as tire chips, as performance enhancing reinforcing element, catastrophic failure of important structures could be prevented during earthquakes. The liquefaction prevention method developed, described and evaluated here is, thus, one of the options that satisfy the needs of the emerging and popular design methodology called the performance based design.

ACKNOWLEDGEMENTS

The authors gratefully acknowledge the financial support for this research provided by the Japan Society of Promotion of Science (JSPS) and Ministry of Education, Culture, Sports, Science and Technology (MEXT), Japan under the Grant-in-Aid for scientific research (Grant No. 18206052, Principal Investigator: Hemanta Hazarika). The authors also would like to express their gratitude to Mr. Hiroyuki Yamazaki, Head of the Soil Dynamics Division, PARI for his kind gesture in allowing to use the shaking table facilities of his laboratory. Special thanks go to Messrs Nagashima and Nagano of the Soil Dynamics Laboratory, PARI for their technical support during the testing.

REFERENCES

Dickenson, S., and Yang, D.S. (1998). Seismically-induced Deformations of Caisson Retaining Walls in Improved Soils, *Geotechnical Earthquake Engineering and Soil Dynamics III,* Geotechnical Special Publication, No. 75, ASCE Vol. 2, pp. 1071–1082.

Ghalandarzadeh, A., Orita, T., Towhata, I., and Yun, F. (1998). Shaking Table Tests on Seismic Deformation of Gravity Quay Walls, *Special Issue of Soils and Foundations,* Japanese Geotechnical Society, pp. 115–132.

Ghazavi, M. (2004): Shear Strength Characteristics of Sand Mixed with Granular Rubber, *Geotechnical and Geological Engineering,* Vol. 22, pp. 401–416.

Hamada, M., Isoyama, R., and Wakamatsu, K. (1996). Liquefaction Induced Ground Displacement and Its Related Damage to Lifeline Facilities, *Special Issue of Soils and Foundations,* Vol. 1, pp. 81–98.

Hazarika, H. (2006). "Lightweight Recycled Waste as an Earthquake Resistant Smart Geomaterial", International Conference on Civil Engineering in the New Millennium: Opportunities and Challenges (CENeM-2007), 11–14 January 2007, Shibpur, Howrah, India.

Hyodo, M., Yamada, S., Orense, R.P., Okamoto, M., and Hazarika, H. 2007. Undrained Cyclic Shear Properties of Tire Chip-Sand Mixtures, International Workshop on Tire Derived Geomaterials (IW-TDGM 2007), Hazarika and Yasuhara Edit., Taylor and Francis, The Netherlands.

Iai, S. (1998). Seismic Analysis and Performance of Retaining Structures, *Geotechnical Earthquake Engineering and Soil Dynamics III,* Geotechnical Special Publication No. 75, ASCE, pp. 1020–1044.

Iai, S. and Tobita, T. 2005. Performance-based Approach for Designing Remediation of Liquefiable Soils, *Proceedings of the Geotechnical Earthquake Engineering Satellite Conference on Performance Based Design*, Osaka, Japan, pp. 14–21.

Inagaki, H., Iai, S., Sugano, T., Yamazaki, H., and Inatomi, T. (1996). Performance of Caisson Type Quay Walls at Kobe Port, *Special Issue of Soils and Foundations*, Japanese Geotechnical Society, Vol. 1, pp. 119–136.

Ishihara, K. (1997): Geotechnical aspects of the 1995 Kobe earthquake, Terzaghi Orientation, 14th *International Conference of International Society of Soil Mechanics and Geotechnical Engineering*, Hamburg, Germany, Vol. 4, pp. 2047–2073.

Ishihara, K., Yasuda, S., and Nagase, H. (1996). Soil Characteristics and Ground Damage, *Special Issue of Soils and Foundations*, Japanese Geotechnical Society, Vol. 1, pp. 101–118.

Japanese Geotechnical Society (JGS) and Japan Society for Civil Engineers (JSCE) (1996). Joint Report on the Hanshin-Awaji Earthquake Disaster (in Japanese).

Kamon, M., Wako, T., Isemura, K., Sawa, K., Mimura, M., Tateyama, K., and Kobayashi, S. (1996). Geotechnical Disasters on the Waterfront, *Special Issue of Soils and Foundations*, Japanese Geotechnical Society, Vol. 1, pp. 137–147.

PIANC (International Navigation Association) (2001). *Seismic Design Guidelines for Port Structures*, Balkema Publishers, Rotterdam.

SEAOC, (1995). Performance based seismic engineering of buildings. Structural Engineers Association of California, Sacramento, California, USA.

Steedman, R.S., (1998). "Seismic design of retaining walls", *Geotechnical Engineering*, Institution of Civil Engineers, Vol. 131, pp. 12–22.

Towhata, I., Ghalandarzadeh, A, Sundarraj, K.P., and Vargas-Monge, W. (1996). Dynamic Failures of Subsoils Observed in Waterfront Area, *Special Issue of Soils and Foundations,* Japanese Geotechnical Society, pp. 149–160.

Yasuda, S., Ishihara, K., Harada, K., and Shinkawa, N. (1996). Effect of Soil Improvement on Ground Subsidence Due to Liquefaction, *Special Issue of Soils and Foundations*, Vol: 1, pp. 99–108.

Yasuhara, K., Hazarika, H., and Fukutake, K. 2006. Applications of Scrap Tires in Civil Engineering, *Doboku Gijutsu*, Vol. 61, No. 10, pp. 79–86 (in Japanese).

Youwai, S., and Bergado, D.T. (2003). Strength and Deformation Properties of Rigid Rubber Tire- Sand Mixtures, *Canadian Geotechnical Journal*, Vol. 40(2), pp. 254–264.

Scrap Tire Derived Geomaterials – Opportunities and Challenges – Hazarika & Yasuhara (eds)
© 2008 Taylor & Francis Group, London, ISBN 978-0-415-46070-5

Evaluation of effect of tire chips in the cement stabilized soil using X-ray CT

Y. Mitarai
Toa Corporation, Yokohama, Japan

Y. Nakamura & J. Otani
Kumamoto University, Kumamoto, Japan

ABSTRACT: The cement stabilizing is one of the usual recycling methods of soft soil such as dredged clay. It is able to get a high strength and low permeability by cement stabilizing, but the cement treated soil is showed the brittle mechanical property, it is shown the failure at very small deformation and a large crack is observed after failure point in a comparison with natural clay. From results of authors' previous studies, we found that it is possible to change its ductility by mixing tire chips in the volume of 10 to 20%. And it is able to see the deformation situation of the specimens of the unconfined and tri-axial compression test. In this paper, it is shown the X-ray CT images of these specimens during these tests, and from these results, it is able to see and evaluate the effect of mixing tire chips in the cement treated soil.

1 INTRODUCTION AND OBJECTIVE

In Japan, a large amount of dredged soils is discharged constantly from work of winding and deep excavating waterway and anchorage. The amount of dredged soil a year whole of Japan is about 20 millions m^3 to 30 millions m^3. The various disposal and usefully recycling method of dredged soil have been considered and carried out. The cement stabilizing is one of it, and recently the cement treated soil is used as a reclamation or back-filling material. etc, substance for a good quality geomaterial such as mountain-sand and sea-sand.

The cement stabilization is possible to get a large strength and low permeability by the effect of cement hydration, but it has brittle mechanical property such as soft rock. Generally speaking, it will be shown the failure at very small deformation and a large crack will be observed after failure point in a comparison with natural clay. The anthers have been studied on an improvement method of its brittle property. From the results of authors' previous studies, it was found the mixing tire chips into the cement treated soil is able to change its belittle into toughness.

The purpose of this study is to evaluate the failure properties of cement stabilized soil containing tire chips under unconfined and tri-axial compression using X-ray CT scanner. The specimens used in this study were cement stabilized soil both with and without tire chips. A series of compression tests was conducted with these specimens. Failure formation under compression was discussed in evaluating the failure properties of the materials. The effect of tire chips in the cement stabilized soil on the failure properties of the composite material was investigated.

2 MATERIALS AND TEST PROCEDURE

2.1 *Materials and mix proportion of cement stabilized soil with adding tire chips*

Dredged soil, cement and tire chips were used to make the cement stabilized soil. The dredged soil was Tokyo Bay clay (Table 1). The water content of it was controlled to 280% by mixing seawater.

Table 1. Physical properties of dredged Tokyo Bay clay.

Specific gravity of soil particle		g/cm³	2.717
Grading	Gravel sand	%	0
	Silt	%	27.5
	Clay	%	72.5
Consistency	wl	%	100.3
	wp	%	42.2

Figure 1. Tire chips (2 mm Ave.)

Table 2. Mix proportion of cement treated clay.

Target strength qu (kN/m²)	Mix proportion (g/l)		
	Soil particle	Sea water	Cement
250	318	896	57
400	316	890	62

And it was used normal portland cement, and it's particle density was 3.16 g/cm³. The tire chips were cut in pieces from used tires removing textiles and steel wires. The grain size was approximately 2 mm (Figure 1), and its particle density was 1.15 g/cm³. Tire rubber's Poisson's ratio is approximately 0.5, the modulus of elasticity is 4–6 MN/m², and the tensile strength: 15–35 MN/m².

The targets of unconfined compressive strength of the cement stabilized soil at 28 days of curing time were qu = 250 kN/m² and 400 kN/m² under the curing condition of 20° ± 3°C and relative humidity 95% to 100%. The specimen used in this study included different percentage tire chips, such as f = 0%, 9.1%, 16.7% and 23.1% (f = (tire chips)/(whole of composite material): volume %). The size of each specimen was fixed to 50 mm in diameter and 100 mm in height. The mix proportion is shown in Table 2.

2.2 Test procedure

The unconfined compression test was conducted under constant speed of 1% strain per minute. And tri-axial compression test was conducted under condition which back pressure was 100 kPa, confining pressure was 200 kPa (more than 2 hour), and undrained shear was conducted under constant lateral pressure and constant speed of 0.3% strain per minute.

In this study, in order to scan CT-image of specimens during compression test, it was developed the spatial apparatus shown in Figure 2. Then, it is able to scanning at random time under compression test. In this study, CT scanning was conducted at once stopping shear motor and rocking loading shaft.

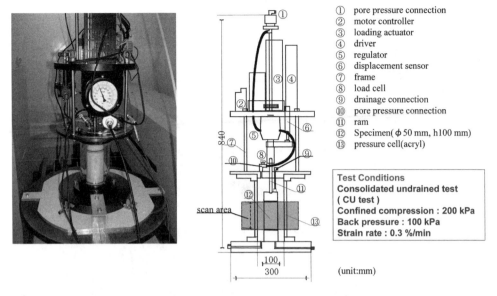

①	pore pressure connection
②	motor controller
③	loading actuator
④	driver
⑤	regulator
⑥	displacement sensor
⑦	frame
⑧	load cell
⑨	drainage connection
⑩	pore pressure connection
⑪	ram
⑫	Specimen(φ 50 mm, h100 mm)
⑬	pressure cell(acryl)

Test Conditions
Consolidated undrained test
(CU test)
Confined compression : 200 kPa
Back pressure : 100 kPa
Strain rate : 0.3 %/min

(unit:mm)

Figure 2.   Triaxial compression test apparatus for X-ray CT.

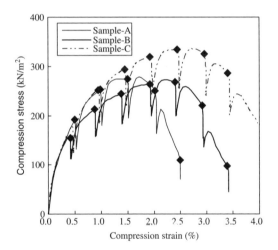

Figure 3.   Stress-strain curve.

3   RESULTS

3.1   *Unconfined compression test*

Figure 3 shows stress-strain relations of unconfined compression tests, in this figure, the stress relaxation part was at the time of CT Scannimg. Figure 4 shows CT-images of the specimen under unconfined compression test and these CT-images were scanned at the center of each specimen. Generally in CT-images, difference of density is shown in its difference tone, and then, light or whity part shows high density, and dark or gray part shows low density.

There cases are differ in mix ratio of tire chips, but each case is same in mix proportion of cement treated clay parts, qu $= 250 \, kN/m^2$. Sample-A is the case of cement treated clay without tire chips ($f = 0.0\%$), Sample-B is the case of $f = 16.7\%$ and Sample-C is the case of $f = 28.6\%$.

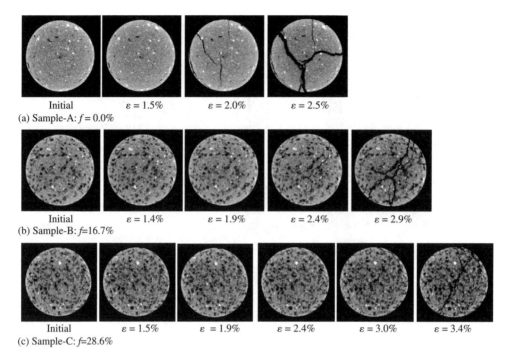

Initial $\varepsilon = 1.5\%$ $\varepsilon = 2.0\%$ $\varepsilon = 2.5\%$
(a) Sample-A: $f = 0.0\%$

Initial $\varepsilon = 1.4\%$ $\varepsilon = 1.9\%$ $\varepsilon = 2.4\%$ $\varepsilon = 2.9\%$
(b) Sample-B: $f=16.7\%$

Initial $\varepsilon = 1.5\%$ $\varepsilon = 1.9\%$ $\varepsilon = 2.4\%$ $\varepsilon = 3.0\%$ $\varepsilon = 3.4\%$
(c) Sample-C: $f=28.6\%$

Figure 4. CT-images of cross section of specimen center (unconfined compression test).

From Figure 3, it is obtained that the larger the percentage of tire chips in the mixture is, the lager the failure strain becomes. In this paper, the failure strain means the stress at the peak of the unconfined stress in the stress and strain relations.

From Figure 4, in the case of $f = 0.0\%$, it is able to see cracks at 2.0% strain. As shown in Figure 3, failure strain is about 1.7%, then, it is consider that crack was observed at the failure point or after failure point. On the other hand, in the case of $f = 16.7\%$ and $f = 28.6\%$, it was not able to see crack at the point of strain $= 2.0\%$. Then, these results is imply that the deformation condition of specimen was changed and improvement ductility of cement treated clay by adding tire chips.

Figure 5 shows 3-dimentional images of specimen after failure point in each case. In the case of $f = 0.0\%$, large and thick cracks were observed, then. it is consider that stress and strain was concentrated in the specimen. On the other hands, in the case of $f = 16.7\%$ and 28.6%, it was not observed large crack such as in the case of $f = 0.0\%$, it is able to see sliding surface.

3.2 Tri-axial compression test

Figure 6 shows stress-strain relations of tri-axial compression tests. In this figure, it is able to see that in the case of f = 0.0%, failure point is at about 2.0% strain, and in the case of $f = 16.7\%$, failure point is at about 4.0% strain. This result shows a tendency similar to that in Figure 3 which is in case of unconfined compression test. However, stress-strain curve s of tri-axial test is not shown large strain softening such as the case of unconfined compression test, such as Figure 3.

Figure 7 and Figure 8 show CT-images of the specimen under tri-axial compression test, figure 7 is the case of $f = 0.0\%$ and figure 8 is the case of $f = 16.7\%$.

From Figure 7, it is able to see cracks at 2.0% strain which is failure point. The lager the strain is, the larger the crack grows. However, from Figure 8, it is not able to see any large and visible crack which is not concerned with growing strain. Then, it is considered that the lateral confining stress is effective to improve ductility of cement treated clay, moreover adding tire chips.

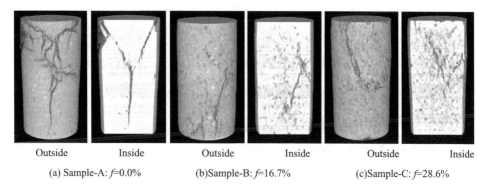

| Outside | Inside | Outside | Inside | Outside | Inside |

(a) Sample-A: f=0.0%   (b)Sample-B: f=16.7%   (c)Sample-C: f=28.6%

Figure 5.   3-dimentional CT-images.

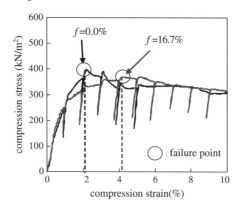

Figure 6.   Stress-strain curve (tri-axial compression test).

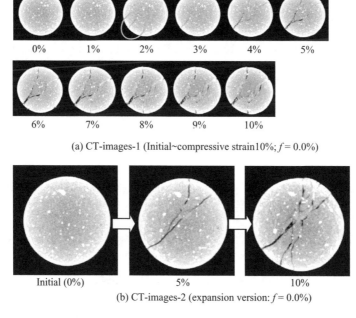

0%   1%   2%   3%   4%   5%

6%   7%   8%   9%   10%

(a) CT-images-1 (Initial~compressive strain10%; f = 0.0%)

Initial (0%)   5%   10%

(b) CT-images-2 (expansion version: f = 0.0%)

Figure 7.   CT-images (tri-axial compression test; f = 0.0%).

227

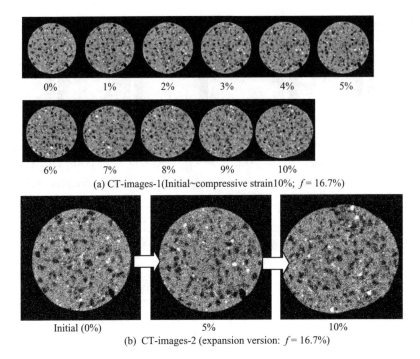

(a) CT-images-1(Initial~compressive strain10%; $f = 16.7\%$)

Initial (0%)   5%   10%

(b) CT-images-2 (expansion version: $f = 16.7\%$)

Figure 8.   CT-images (tri-axial compression test; $f = 16.7\%$).

Consequently, these results imply that tire chips shredded in specimen, prevent stress concentration and locally growth of large crack during compressive deformation, and the effect of that becomes remarkable by lateral confining stress.

4 CONCLUSIONS

In this study, the failure properties of cement treated clay with adding tire chips are discussed based on unconfined and tri-axial compression test with using X-ray CT scanning. Then, failure behavior of cement treated clay is brittle; however, by adding some tire chips, its failure behavior is able to change brittle into toughness. It is also obtained that the effect of that becomes remarkable by lateral confining stress. Therefore, tire chips shredded in specimen has the effect of prevention stress concentration and locally growth of large crack during compressive deformation.

REFERENCES

Kikuchi, Y., Nagatome, T., Mitarai, Y. and Otani, J. (2006): Engineering Property Evaluation of Cement Treated Soil with Tire Chips using X-ray CT Scanner, Proc. of 5th ICEG Environmental Geotechnics Vol. 2, pp. 1423–1430.

Kikuchi, Y. (2006): Investigation of Engineering Properties of Man-made Composite Geomatrials with Microfocus X-ray CT, *International Workshop Advances in X-ray Tomography for Geomatrials*, pp. 53–78.

Otani, J. and Obara, Y. (2004): *International Workshop on X-ray CT for Geomaterials – GeoX2003 –*, A.A. Balkema.

Mitarai, Y., Yasuhara, Y., Kikuchi, Y. and Ashoke K. K. (2006): Application of the Cement Treated Clay with Added Tire Chips to the Sealing Materials of Coastal Waste Disposal Site, *Proceeding of 5th ICEG Environmental Geotechnics* Vol. 1, pp. 757–764.

Seismic response analysis of caisson quay wall reinforced with protective cushion

T. Takatani

Maizuru National College of Technology, Kyoto, Japan

ABSTRACT: A finite element analysis of liquefaction process is carried out to investigate the seismic performance of a caisson structure reinforced with tire chips cushion. An earthquake acceleration wave is used as an input earthquake ground motion to evaluate the effect of tire chips cushion on the seismic response of gravity type quay wall and the improvement of its seismic performance. The effect of tire chips cushion on the displacement behavior of the caisson structure, acceleration response and pore water pressure response in the soil is investigated. The analytical results show that a caisson structure reinforced with tire chips cushion has a significant seismic performance in comparison with an unprotected caisson structure.

1 INTRODUCTION

In the 1995 Hyogo-ken Nanbu earthquake, many caisson quay walls were reported to have suffered damages due to liquefaction in the backfill soil behind the caisson walls and the foundation soil beneath them (JGS/JSCE 1996), and also structure damages were attributed to the seaward ground movement induced by the strong inertia force (Inagaki et al. 1996). It is therefore necessary to improve the seismic performance of a caisson quay wall structure to reduce earthquake damages. Takatani et al. (1996) and Takatani & Maeno (2000) have numerically investigated the seismic performance of a caisson type quay wall with pile foundation during the strong earthquake motion to improve its earthquake resistance.

In recent years, there are many attempts to utilize scrap tire chips as earthquake resistant geo-materials by taking into consideration its major characteristics such as the compressibility, the lightweight and so on, because recycled products derived from scrap tires are utilized as a well-graded coarse grained geomaterials in the civil engineering. These particular characteristics of tire chips can be utilized as a geoinclusion behind the rigid and massive structure, and also are effective in reducing the seismic load acting on the structure during earthquake motion (Hazarika et al. 2003, Hazarika & Okuzono 2004). In order to investigate the seismic performance of a caisson quay wall reinforced by tire chips cushion, a series of model shaking table test (Hazarika et al. 2005) has been conducted on a gravity type caisson quay wall protected with a cushion made from tire chips. It is therefore very important to make an accurate estimation of the seismic performance of a caisson structure reinforced with tire chips cushion by the model shaking table test in the laboratory and the numerical analysis. The purpose of this paper is to numerically investigate the seismic performance of a caisson structure reinforced with tire chips cushion.

A finite element analysis of liquefaction process is carried out to investigate the seismic performance of a caisson structure reinforced with tire chips cushion. This numerical analysis is two-dimensional non-linear finite element method based on an effective stress theory to simulate liquefaction process of saturated soil under undrained condition. In this paper, an earthquake acceleration wave measured at the 1995 Hyogo-ken Nanbu earthquake, Kobe, Japan, is used as an input earthquake ground motion to make an accurate evaluation of the effect of tire chips cushion on the seismic response of gravity type quay wall and the improvement of its seismic performance.

Displacement response of the caisson structure and pore water pressure response in the soil around it were obtained by this numerical analysis. The effect of tire chips cushion on the displacement behavior of the caisson structure, stress-strain response and pore water pressure response in the soil is investigated.

2 ANALYTICAL METHOD

In this finite element analysis, a non-linear relationship between shear stress and shear strain of soil element is accurately expressed by a multi shear spring model (Towhata & Ishihara 1985) and the Masing rule for loading and unloading curves is employed so as to adjust the amplitude of hysteresis damping for the multi shear spring model. Also the cyclic mobility model (Iai et al. 1990), which is of a generalized plasticity-multiple mechanism type, is adapted to simulate excess pore water pressure. Pore fluid is assumed to be non-compressibility, and also the viscous boundary technique is used to create the infinite of seabed ground in this analysis. There are three governing equations of a kinematic equation between soil and pipeline, pore water input/output balance equation and dynamic water pressure wave propagating equation for pore fluid. Pore water pressure can be expressed by an increment of volumetric strain of soil skeleton because of undrained condition, and also dynamic water pressure wave propagating equation for pore fluid can be represented by the technique that the effect of pore fluid can be taken into consideration by using an additional mass to the soil-structure kinematic equation.

Figure 1 shows a finite element mesh for a caisson quay wall with tire chips cushion considering liquefaction phenomenon. The caisson quay wall structure is $3.5\,m$ in width and $7.0\,m$ in height as shown in Figure 1. The analytical domain is $19.25\,m \times 56\,m$ and consists of three layers, that is, backfill sand, foundation soil and crushed stone areas. In this analysis, tire chips cushion is assumed to be elastic and its young's modulus, Poisson's ratio and density are employed $1,6250\,kN/m^2$, 0.333 and $0.611\,t/m^3$, respectively. The joint element is employed in this analysis to represent a slip phenomenon at contact area between structure and foundation soil. The unit tangential stiffness for normal and shear directions, K_n and K_s, for a joint element are used 1.0×10^5 (kN/m) and 1.0×10^4 (kN/m), respectively. The friction angle of joint element is assumed to be 25 degree in this analysis. Before liquefaction analysis for earthquake motion, the self-weighted analysis for caisson structure-soil interaction problem is carried out under the completely drained condition to obtain the initial effective stress of each soil element.

Material properties for three soil layers used in this paper are shown in Table 1. Bulk modulus of pore water K_f, porosity of soil skeleton n, mass density ρ, Poisson's ratio ν, and limiting value of damping factor H_m are assumed to be 2.2×10^6 (kPa), 0.45, 19.1 (kN/m^3), 0.33 and 0.3, respectively.

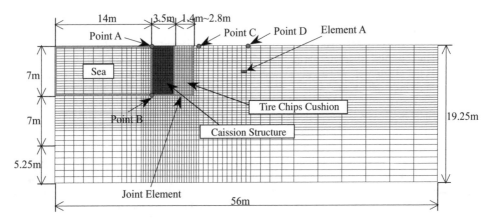

Figure 1. Finite element mesh for caisson quay wall with tire chips cushion.

The liquefaction analysis in this paper uses a strain space plasticity approach for cyclic mobility (Iai et al. 1990) in order to represent the realistic hysteretic damping factor under cyclic loading. In this approach, actual cyclic shear mechanism is decomposed into a set of one dimensional virtual simple shear mechanism. Material properties of dilatancy S_1, w_1, c_1, p_1 and p_2 shown in Table 1 are five parameters to define the cumulative volumetric strain of plastic nature for representing cyclic mobility. In general, when the value of w_1 is larger, the generation of excess pore water pressure is restrained. Excess pore water pressure occurs more rapidly when the value of p_1 is smaller. While, the value of p_2 is smaller when the state of sand layer becomes more dense. Both the pore water pressure accumulation and the shear strain in sand layer trend to increase as the value of p_2 becomes larger. These parameters can be easily obtained by the undrained cyclic shear test results, which are commonly available in the practice of soil dynamics and earthquake engineering, and also define the correlation between the liquefaction front parameter (Iai et al. 1990) and the normalized shear work. The liquefaction front parameter is given by a function of shear work, and Towhata and Ishihara (1985) obtained the correlation between the shear work and the excess pore pressure, and also concluded that the correlation is independent of the shear stress paths with or without the rotation of principal stress axes.

3 NUMERICAL RESULTS

Figure 2 indicates an acceleration wave record measured at the 1995 Hyogo-ken Nanbu earthquake whose maximum value is $810 Gal$. This earthquake acceleration wave record shown in Figure 2 is

Table 1. Material properties for foundation soils.

		Backfill sand	Foundation soil (Liquefaction area)	Foundation (Crushed stone)
Depth (m)		0.0–7.0	7.0–14.0	14.0–19.25
Initial shear modulus, G_{ma} (kPa)		84,900	108,000	152,000
Elastic tangent bulk modulus of soil skeleton, K_{ma} (kPa)		221,000	281,000	396,000
Friction angle, ϕ (degree)		40	41	42
Material parameters for dilatancy	S_1	0.005	0.005	–
	w_1	5.5	9.2	–
	c_1	1.6	1.3	–
	p_1	0.5	0.5	–
	p_2	1.02	0.92	–

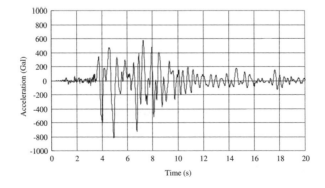

Figure 2. Horizontal acceleration wave record (The 1995 Hyogo-ken Nanbu earthquake).

Table 2. Numerical simulation cases.

	Backfill sand	Case 1	Case 2	Case 3
Type 1	Liquefaction	No tire chips cushion	Tire chips cushion (1.4 m width)	Tire chips cushion (2.8 m width)
Type 2	No Liquefaction	No tire chips cushion	Tire chips cushion (1.4 m width)	Tire chips cushion (2.8 m width)

employed as an input ground motion in order to investigate the effect of tire chips cushion on the seismic performance of caisson structure with tire chips cushion. In addition, liquefaction analyses for some numerical simulation cases shown in Table 2 are carried out to investigate the effect of the width of tire chips cushion. In this paper, the width of tire chips cushion is assumed to be 1.4 m or 2.8 m. Type 1 is the case that the backfill sand behind the caisson quay wall structure shown in Figure 1 is assumed to liquefy, and Type 2 is the case that the backfill sand is assumed not to liquefy.

Figure 3 shows the deformations after earthquake for Cases 1, 2 and 3 in Type 1. It is observed from theses figures that the horizontal deformation of backfill sand area is apparently large in all cases because of its liquefaction phenomenon due to earthquake. Accordingly, the horizontal displacement of caisson structure is larger than the vertical displacement of it due to both the horizontal deformation of backfill sand area and the inertia force. However, both horizontal and vertical displacements at the top of caisson structure have a tendency to decrease with increasing the width of tire chips cushion. In addition, the inclination of caisson structure in Case 3 is different from those in Cases 1 and 2. This is because that the caisson structure behavior in Case 3 due to the deformation of backfill sand area is smaller than those in Cases 1 and 2 because of reducing the seismic load due to the elastic behavior of tire chips cushion. This fact implies that the feasibility of tire chips cushion will decrease the deformation of caisson quay wall during/after earthquake, and that the deformation of caisson quay wall greatly depends on the width of tire chips cushion.

Figure 4 indicate the deformations after earthquake for Cases 1, 2 and 3 in Type 2. The horizontal deformations of backfill sand area in all cases are quite smaller than those in Type 1 because the backfill sand area does not liquefy in Type 2. However, both horizontal and vertical displacements at the top of caisson structure in Type 2 are almost the same in Type 1 and trend to decrease with increasing the width of tire chips cushion. Also, the inclination of caisson structure for each case in Type 2 is almost the same in Type 1 regardless of whether the liquefaction occurs in backfill sand area or not. Moreover, it can be observed from Figures 3 and 4 that the deformation of foundation soil beneath the caisson structure and the backfill sand in Type 1 is almost the same in Type 2. This is because that the liquefaction in foundation soil occurs for both cases, and that only the horizontal acceleration shown in Figure 2 is used and the vertical acceleration wave is not done in this numerical analysis.

Displacement responses at four nodal points in Cases 1, 2 and 3 are shown in Figures 5, 6 and 7, respectively. Four nodal points are Nodal Points A, B, C and D in the analytical mesh shown in Figure 1. Vertical displacement responses at Nodal Points A and B in Cases 2 and 3 are much larger than those in Case 1 because of the existence of tire chips cushion and its width. It can be observed from these figures that the inclination angle of the caisson structure after earthquake in Case 3 is different from Cases 1 and 2. On the other hand, horizontal displacement response at Nodal Point C in Case 3 is smaller then those in Cases 1 and 2, because it is affected by the width of tire chips cushion. The variation of horizontal displacement at Nodal Point D in Case 3 is much larger than those in Cases 1 and 2 because of the width of tire chips cushions, and the vertical displacement response in Case 3 is almost the same in other cases.

Figures 8, 9 and 10 show the stress-strain relationship, the effective stress paths, shear strain variation and pore water pressure response at Element A shown in Figure 2 for Cases 1, 2 and 3

232

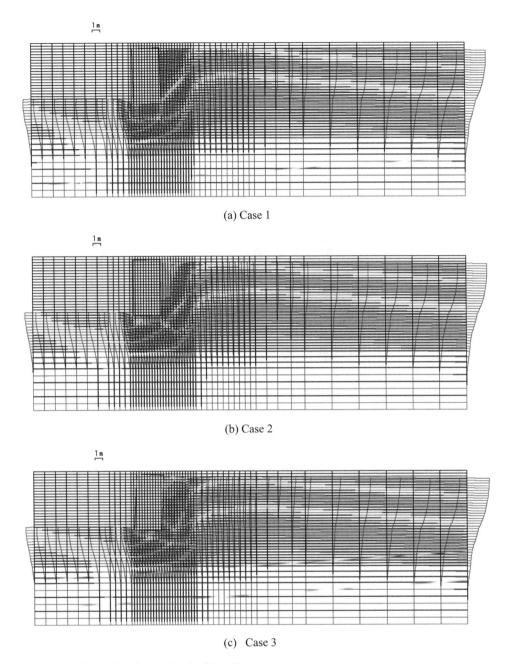

(a) Case 1

(b) Case 2

(c) Case 3

Figure 3.   Deformation after earthquake (Type 1).

in Type 1. Judging from Figure 3, it can be considered that Element A is greatly affected by the liquefaction. As can be observed from stress-strain responses, the range of shear stress is almost the same in three cases, while the range of shear strain responses in Cases 1 and 2 are quite different from that in Case 3. In particular, residual shear strain response in Case 3 is contrary value for those in Cases 1 and 2. This is because that shear strain response is affected by the width of tire chips cushion. On the other hand, pore water pressure responses in all cases are almost the same

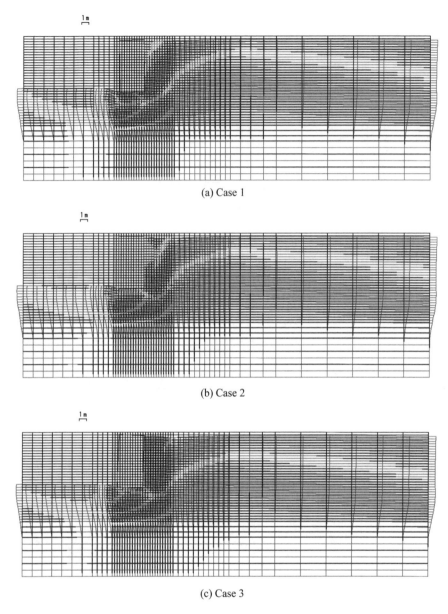

(a) Case 1

(b) Case 2

(c) Case 3

Figure 4. Deformation after earthquake (Type 2).

because the shear stress response is almost the same in three cases and also the distance between tire tips cushion and Element A is large. Consequently, it should be noted that shear strain response in backfill sand area behind the caisson structure is sensitive to the width of tire chips cushion, and that shear strain response gradually changes from minus value to plus one with increasing the width of tire chips cushion. On the other hand, there seems to be needed any countermeasures to reduce pore water pressure accumulation in backfill sand area because the deformation of caisson structure during/after earthquake motion is significantly affected by the liquefaction due to pore water pressure build-up.

234

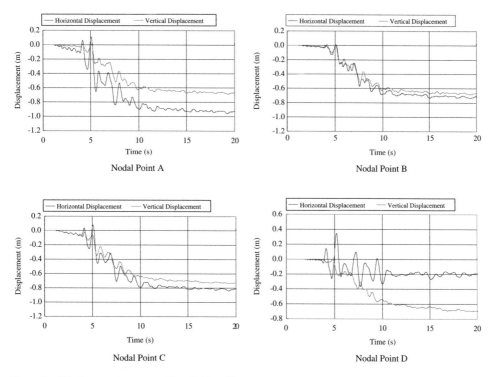

Figure 5. Displacement response (Case 1, Type 1).

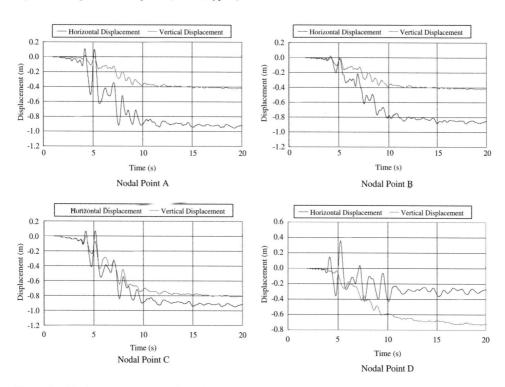

Figure 6. Displacement response (Case 2, Type 1).

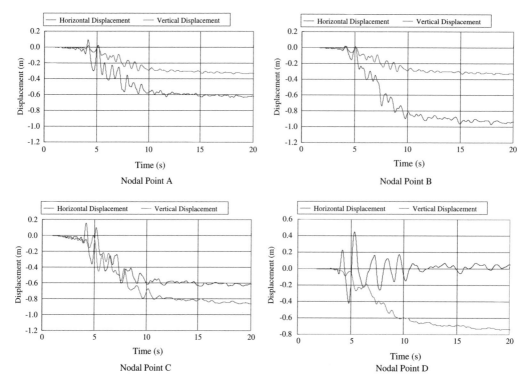

Figure 7. Displacement response (Case 3, Type 1).

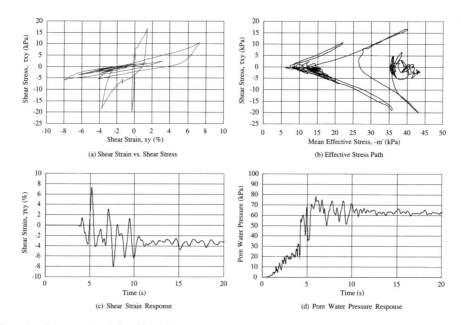

Figure 8. Stress-strain relationship and pore pressure response at Element A (Case 1, Type 1).

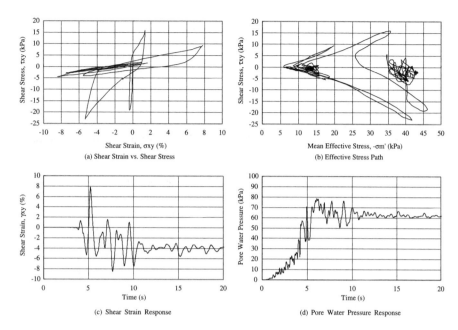

Figure 9.   Stress-strain relationship and pore pressure response at Element A (Case 2, Type 1).

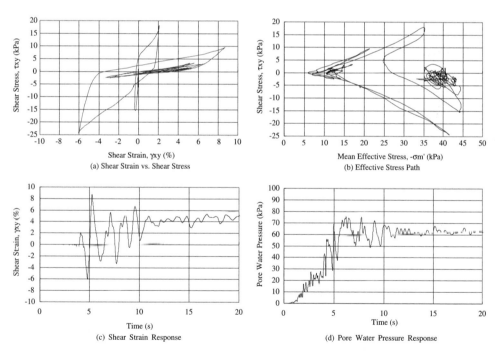

Figure 10.   Stress-strain relationship and pore pressure response at Element A (Case 3, Type 1).

4   CONCLUSIONS

In this paper, a two-dimensional non-linear finite element analysis of liquefaction process, which is based on an effective stress theory, was carried out to investigate the seismic performance of a

237

caisson structure reinforced with tire chips cushion. An earthquake acceleration wave measured at the 1995 Hyogo-ken Nanbu earthquake was used as an input horizontal earthquake ground motion to investigate the seismic response of a gravity type quay wall and the improvement of its seismic performance. The effect of tire chips cushion on the displacement behavior of the caisson structure, stress-strain response and pore water pressure response in the soil was numerically investigated.

In summary, the following conclusions can be made based on the results presented in this paper.

1) The earthquake-induced permanent displacement of a caisson structure protected with tire chips cushion is less than that of an unprotected caisson structure. Also, the larger the width of tire chips cushion behind the caisson structure is, the smaller the earthquake-induced permanent displacement of the caisson structure is.

2) The analytical results show that a caisson structure reinforced with tire chips cushion has a significant seismic performance in comparison with an unprotected caisson structure. This kind of construction technique using tire chips cushion behind a caisson structure will lead to an effective seismic design of caisson quay wall and improve its seismic performance.

3) There seems to be a need for any countermeasures to reduce pore water pressure accumulation in the backfill soil behind the caisson structure because its deformation during/after the strong earthquake motion may be significantly affected by pore water pressure response.

Although an earthquake horizontal acceleration wave was employed in numerical simulations and tire chips cushion material was assumed to be elastic in this paper, more simulations using several acceleration waves and a non-elastic tire chips cushion material may be needed to make some concrete conclusions concerning the feasibility of tire chips cushion to the geomaterials behind the caisson quay wall structure to improve its seismic performance.

ACKNOWLEDGEMENTS

This research is a part of the project by Grants-in-Aid for Scientific Research (No. 18206052) from the Ministry of Education, Culture, Sports, Science and Technology in Japan. The author gratefully acknowledges the financial support. Special thanks are also due to Dr. Y. Kato at Maizuru National College of Technology for his valuable discussion.

REFERENCES

Hazarika, H. & Okuzono, S. 2004. Modeling the behavior of a hybrid interactive system involving soil, structure asn EPS geofoam, *Soils and Foundations*, 44(5), 149–162.

Hazarika, H., Okuzono, S. & Matsuo, Y. 2003. Seismic stability enhancement of rigid non-yielding structures, *Proc. 13th Int. Offshore Polar Eng. Conf.*, Honolulu, 2, 697–702.

Hazarika, H., Sugano, T., Kikuchi, Y., Yasuhara, K., Murakami, S., Takeichi, H., Karmokar, A.K., Kishida, T. & Mitarai, Y. 2006. Model shaking table test on seismic performance of caisson quay wall reinforced with protective cushion, *Proc. 16th Int. Offshore Polar Eng. Conf.*, San Francisco, 309–315.

Iai, S., Matsunaga, Y. & Kameoka, T. 1990. Strain space plasticity model for cyclic mobility, *Report of Port and Harbour Research Institute*, 29(4), 27–56.

Inagaki, H., Iai, S., Sugano, T., Yamazaki, H. & Inatomi, T. 1996. Performance of caisson type quay walls at Kobe port, *Special Issue, Soils and Foundations*, 1, 119–136.

Japanese Geotechnical Society (JGS) and Japan Society of Civil Engineers (JSCE) 1996. Joint report on the Hanshin-Awaji Earthquake Disaster (in Japanese).

Takatani, T., Maeno, Y. & Kodama, H. 1999. Seismic behavior of a caisson type quay wall on a pile foundation, *Proc. 9th Int. Offshore Polar Eng. Conf.*, Brest, 1, 652–659.

Takatani, T. & Maeno, Y. 2000. On liquefaction countermeasure in the soil around a caisson type quay wall on pile foundation, *Proc. 10th Int. Offshore Polar Eng. Conf.*, Seattle, 2, 582–589.

Towhata, I. & Ishihara, K. 1985. Modelling soil behaviour under principal stress axes rotation, *Proc 5th Int. Conf. Num Method Geomech*, Nagoya, 523–530.

Characterization of engineering properties of cement treated clay with tire chips

T. Nagatome & Y. Mitarai
Toa Corporation, Yokohama, Japan

Y. Kikuchi
Port and Airport Research Institute, Yokosuka, Japan

J. Otani & Y. Nakamura
Kumamoto University, Kumamoto, Japan

ABSTRACT: The objective of this study is to investigate failure properties of cement treated clay mixed with tire chips under shear deformation using X-ray CT scanner. The specimens used in this study were cement treated clay both with and without granular materials, which were tire chips, siliceous sand and expanded poly-styrofoam beads, respectively. In this paper, a series of triaxal compression test was performed with these four specimens and failure formation was discussed from triaxial compression test results and X-ray CT data. According to the experimental results, the effect of tire chips on the engineering properties of cement treated clay was then investigated.

1 INTRODUCTIONS

In Japan, research on active way of using dredged clay has been conducted for more than ten years. The main purpose of this kind of research is to convert dredged clay into a high strength mixing cementing material and to reduce the earth pressure on retaining walls and revetments. This material has high strength and favorable characteristics for geo-materials but it is very brittle, so research on improving the ductility of cement treated clay has been conducted. The authors considered that the addition of tire chips to cement treated clay offers an effective means of improving its ductility, enabling use in structures where deformation is anticipated (Mitarai et al. 2004, Kikuchi et al. 2006).

Meanwhile, an X-ray Computed Tomography (CT) scanner, which is known by the name of medical diagnostic methods, has been used even for engineering purposes as a nondestructive testing method (Otani and Obara 2003, Desrues et al. 2006).

The purpose of this study is to investigate failure properties of cement treated clay with tire chips under shear deformation using X-ray CT scanner. The specimens used in this study were cement treated clay both with and without granular materials, which were tire chips, siliceous sand and expanded poly-styrofoam beads, respectively. Here, a series of triaxal compression test was conducted for these four specimens and failure mechanism was discussed from triaxial compression test results and X-ray CT data. According to the experimental results, the effect of tire chips on the engineering properties of cement treated clay was then investigated.

2 SPECIMENS AND TEST PROCEDURE

2.1 *Specimens*

Dredged slurry, cement and tire chips were used to make the cement treated clay. The dredged slurry was Tokyo Bay clay. The physical properties of this clay are shown in Table 1. The water

Table 1. Physical properties of dredged clay.

Soil particle density		g/cm³	2.7
Gradation	Gravel fraction	%	0.0
	Silt fraction	%	27.5
	Clay fraction	%	72.5
Consistency	Liquid limit	%	100.3
	Plastic limit	%	42.2

(a) Tire chips   (b) Gravel   (c) Expanded poly-styrofoam beads

Figure 1. Granular materials.

Table 2. Characteristics of granular materials.

	Tire chip	Gravel	Expanded poly-styrofoam beads
Particle density (g/cm³)	1.15	2.70	0.03
Poisson's ratio	0.5	0.2	0.1–0.2
Modulus of elasticity (MN/m²)	5 level	10 level	more than 1000
Particle form	shured	uneven	sphere
Deformation property	elastic	rigid	plastic

content of it was controlled to 2.8 w_L with adding sea water. Cement was normal portland cement, the particle density of which was 3.16 g/cm³. Cement treated clay with tire chips was adding tire chips to cement treated clay. The tire chips were cut in pieces from used automobile tires to a grain size of approximately 2 mm as shown in Figure 1(a). Comparison materials of tire chips as elastic material were prepared both siliceous sand as rigid material and expanded poly-styrofoam beads as plastic material. The size of siliceous sand and expanded poly-styrofoam beads was approximately 2 mm as shown in Figures 1(b), (c). The material characteristics of these granular materials are shown in Table 2.

The cement treated clay was made by mixing three materials which are soil, seawater and cement. In this study, the targets of unconfined compressive strength of cement treated clay at 28 days curing time were 400 kN/m² and 800 kN/m². These strengths were designed generally application range of cement treated clay. The designated mixing condition was shown in Table 3. And each granular material was mixed 20% to the volume of cement treated clay (16.7% to the volume of specimen). Molds of 5 cm in diameter and 10 cm in height were filled up with the mixture. The molds were placed in a high moisture closed box more than 95% of relative humidity. A series of triaxial compression tests was conducted after 28 days from these specimens were prepared.

Table 3. Designated mixing condition.

| Target of unconfined compressive strength q_{u28} (kN/m^2) | Mixing condition (g/L) | | | Cement treated clay | |
	Soil particle	Seawater	Cement	Water content (%)	Wet density (g/cm^3)
400	316	890	62	282	1.268
800	313	883	8	282	1.284

2.2 Test procedure

A series of triaxial compression tests was performed with these specimens. First, the failure mechanism of the cement treated clay with each granular material was investigated with triaxial test. And the different strength of cement treated clay with each granular material was also inspected. In this case, the specimen was confined 300 kPa compression. Water pressure of 100 kPa was applied to the specimen under consolidated condition. In this time, the specimen was not yield under consolidated condition, because their consolidation yield stress is larger than the effective confining pressure which was 200 kPa. After consolidation condition, undrained shear test was conducted to 10% axial strain. Shearing speed was 0.1 mm/min.

Next, a series of triaxial test was conducted for some specimens, and the failure mechanism was investigated by CT scanning the specimens during the process of confined compression. The specimens were prepared the targets of unconfined compressive strength of cement treated clay were 400 kN/m^2. A series of triaxial compression tests was conducted with these specimens using a triaxial compression apparatus for the X-ray CT scanner. This apparatus was designed in such a way that the triaxial tests could be performed in the X-ray shield box. In this case, the specimen was confined 300 kPa compression. Water pressure of 100 kPa was applied to the specimen under consolidated condition. After consolidation condition, undrained shear test was conducted to 10% axial strain. Shearing speed was 0.3 mm/min. When nondestructive inspection was conducted with the X-ray CT scanner during triaxial compression was stopped momentarily at planning axial strain while scanning was performed.

3 FAILURE PROPERTIES OF CEMENT TREATED CLAY WITH TIRE CHIPS

3.1 Triaxial test with different strength of cement treated clay with tire chips

Figure 2(a) shows the relationship between the stress and axial strain with low strength cement treated clay (CTC) both with and without granular materials. In this case, the target unconfined compressive strength of cement treated clay was 400 kN/m^2. The principal stress difference of each specimen was the same level. And the axial strain required to reach peak strength of cement treated clay mixed with tire chips was the largest among the four specimens. The stress under shear deformation of cement treated clay mixed with granular materials was not suddenly dropped compared with cement treated clay.

Figure 2(b) shows the relationship between the excess pore water pressure and axial strain with these specimens. The excess pore water pressure of cement treated clay with tire chips was slightly low pressure compared with other specimens.

Figure 3(a) shows the relationship between the stress and axial strain with high strength cement treated clay both with and without granular materials. Here, the target unconfined compressive strength of cement treated clay was 800 kN/m^2. The principal stress difference of cement treated clay was the larger than that of it with granular materials. And the axial strain required to reach peak strength of cement treated clay mixed with tire chips was the largest among the four specimens. This tendency was pronounced compared with low strength specimen.

Figure 3(b) shows the relationship between the excess pore water pressure and axial strain with these specimens. The excess pore water pressure of cement treated clay and that of it with gravel

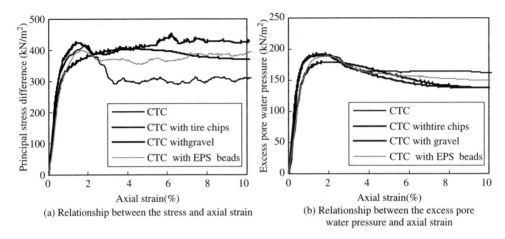

(a) Relationship between the stress and axial strain

(b) Relationship between the excess pore water pressure and axial strain

Figure 2.　Results of the strength of CTC was 400 kN/m².

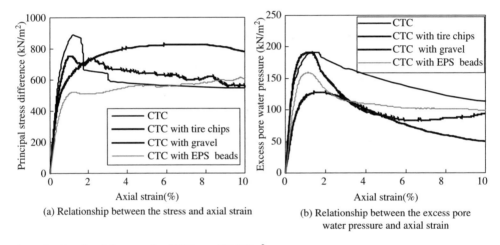

(a) Relationship between the stress and axial strain

(b) Relationship between the excess pore water pressure and axial strain

Figure 3.　Results of the strength of CTC was 800 kN/m².

was high pore water pressure compared with other specimens. And the excess pore water pressure was low with cement treated clay with expanded poly-styrofoam beads, in turn of that of it with tire chips. It was considered that the inclusion in the cement treated clay was prevented occurrence of large excess pore water pressure. And this tendency was prominence at the case of high strength cement treated clay.

In this way, the failure property of cement treated clay was different by inclusion. This property was also changed with the strength of cement treated clay. And it was improved ductility that addition of tire chips like elastic material to cement treated clay.

3.2　Triaxial test with cement treated clay with tire chips using X-ray CT scanner

Figure 4 shows the relationship between the stress and axial strain with cement treated clay both with and without granular materials. The principal stress difference of each specimen was a clear drop in each scanning point. As mentioned above, compression was interrupted at each scanning point. So it is due to the relaxation of the specimen at this time. Figure 4 shows a tendency similar to that in Figure 2(a) and Figure 3(a). That is, the axial strain required to reach peak strength of cement treated clay mixed with tire chips was the largest among the four specimens.

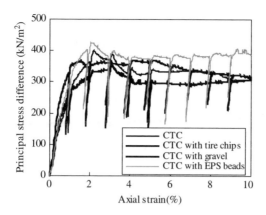

Figure 4. Relationship between the stress and axial strain.

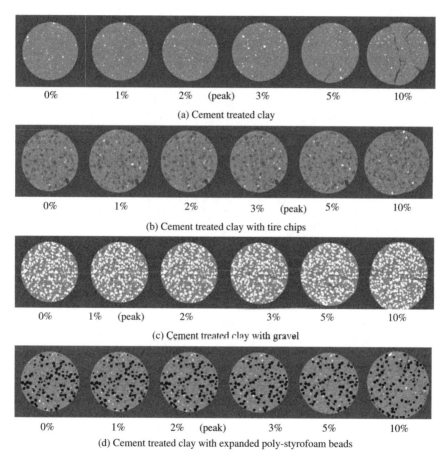

0%   1%   2% (peak)   3%   5%   10%

(a) Cement treated clay

0%   1%   2%   3% (peak)   5%   10%

(b) Cement treated clay with tire chips

0%   1% (peak)   2%   3%   5%   10%

(c) Cement treated clay with gravel

0%   1%   2% (peak)   3%   5%   10%

(d) Cement treated clay with expanded poly-styrofoam beads

Figure 5. Cross sectional images.

Figure 5 shows the cross sectional images of each specimen at initial condition and at planning axial strain. These images were chosen at the development of cracks were the most remarkable section. The CT images were drawn darker for lower densities and lighter for higher densities. The parts shown in black sphere are expanded poly-styrofoam beads. The pure white indicates the

243

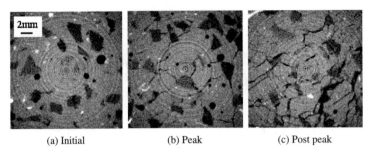

|(a) Initial|(b) Peak|(c) Post peak|

Figure 6.   High magnification images of cross section.

presence of a cement lump, piece of shell and siliceous sand. And rough gray parts approximately 1-3 mm in size are tire chips.

The image of each specimen displays no large change from initial condition to stress peak condition. Then the wide cracks occurred in both the cement treated clay and cement treated clay with siliceous sand after the stress peak condition. The cracks of cement treated clay with expanded poly-styrofoam beads occurred partly in specimen after the stress peak condition. And it was not observed that wide cracks occurred in cement treated clay with tire chips. Figure 6 shows the high magnification images of the cross section during another unconfined compression test result of cement treated clay with tire chips (Kikuchi et al. 2006). In this image, the micro cracks occurred in around tire chips in the specimen. Thus, it was not also observed micro cracks in the images of cement treated clay with tire chips. It was considered that the micro cracks were developed one after another in cement treated clay with tire chips during triaxial compression test. In this way, the failure mechanism of cement treated clay was different with inclusion properties.

4   CONCLUSION

In this study, the failure properties of cement treated clay with tire chips are discussed based on triaxial compression test. At first, failure behavior of cement treated clay both with and without granular materials during shear deformation. Then the internal structure and failure mechanism during the process of compression were investigated visually using X-ray CT scanner. The following conclusions were drawn from this study:

1) The failure properties of cement treated clay are different by mixing inclusion materials.
2) It was not observed wide cracks in cement treated clay with tire chips under triaxial compression. So, it was thought that the stress under shear deformation of cement treated clay with tire chips was not suddenly dropped.
3) The ductility of cement treated clay was improved by addition of tire chips. It was considered that the elastic of tire chips was effective during shear deformation.

REFERENCES

Desurues, J., Viggiani, G. and Besuelle, P. (2006): *International Workshop Advances in X-ray Tomography for Geomatrials*, ISTE.
Kikuchi, Y., Nagatome, T., Mitarai, Y. and Otani, J. (2006): Engineering Property Evaluation of Cement Treated Soil with Tire Chips using X-ray CT Scanner, Proc. of 5th ICEG Environmental Geotechnics Vol. 2, pp. 1423–1430.
Otani, J. and Obara, Y. (2004): *International Workshop on X-ray CT for Geomaterials – GeoX2003 –*, A. A. Balkema.
Mitarai, Y., Yasuhara, Y., Kikuchi, Y. and Ashoke K. K. (2006): Application of the Cement Treated Clay with Added Tire Chips to the Sealing Materials of Coastal Waste Disposal Site, Proc. of 5th ICEG Environmental Geotechnics Vol. 1, pp. 757–764.

Scrap Tire Derived Geomaterials – Opportunities and Challenges – Hazarika & Yasuhara (eds)
© 2008 Taylor & Francis Group, London, ISBN 978-0-415-46070-5

The numerical simulation of earth pressure reduction using tire chips in backfill

K. Kaneda, H. Hazarika & H. Yamazaki
Port & Airport Research Institute, Yokosuka, Japan

ABSTRACT: It is known that the use of compressible material (e.g tire chips) on the backfill can significantly reduce the load against retaining wall. This reduction of earth pressure is achieved through a mechanism of simulating the quasi-active or the intermediate active state. When the soils move to active state, the active earth pressure is reduced. If there exists a highly compressible material between soil and retaining wall, the backfill soil approaches the active state. In this paper, this mechanism has been numerically evaluated. In the numerical simulation, the modified Cam clay model with super-subloading yield surface and rotational hardening concept (SYS Cam clay model) was used. The SYS Cam clay model, which was introduced by Asaoka et al. (2002), is the elasto-plastic model that takes into the account the soil structure, overconsolidation and anisotropy. Using this model, the typical responses of sand with various densities can be described. The earth pressures at rest for both backfill conditions (only soil and tire chips and soil) were calculated and the reduction mechanism made clear.

1 INTRODUCTION

There has recently been much discussion of environmental problems, and this has catalysed efforts to recycle wastes created by the construction industry (Humprey and Manion (1992)). This report describes the experimental use of finely tire chips as a backfill for a retaining wall. Hazarika et al. (2005) are engaged in an energetic series of studies of this material, investigating reductions in earth pressure and its effects on the stability of retaining walls. Hazarika et al. (2005) has shown the mechanism by which highly compacted tire chips function as a buffer to lower earth pressure. The soil in the vicinity of a wall enters an active state, or what has also been called a "quasi-active state," in which it is deformed by this buffer. This report presents a numerical analysis of that mechanism.

The elasto-plastic constitutive model employed by the authors is a super-subloading, yield-surface-modified Cam-Clay model with rotational hardening developed by the soil mechanics laboratory of Nagoya University. The model accounts for soil structure, overconsolidation and anisotropy. "Structure" here means the bulkiness of the soil; given the same void ratio, bulkier soil can bear a greater load, and given the same load, bulkier soil can have a higher void ratio. A benefit of this model is that it explains a wide variety of phenomena, from the differences in shear behavior between loose sand and dense sand to the differences in mechanical behavior between naturally sedimented clay and intermediate soils. The reader can read the references for more detailed discussions (Asaoka et al. (2002)).

In the present study, a finite element analysis was carried out employing finite deformation theory and the above elasto-plastic model (Asaoka et al. (1994)). The numerical calculations are shown and the mechanism of moderation of earth pressure is described with reference to Hazarika's findings in his experiments with tire chips as backfill behind retaining walls.

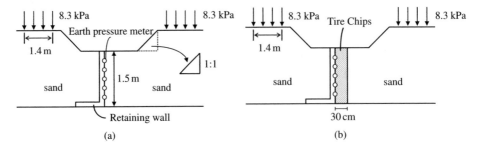

Figure 1. Boundary condition (Experiment).

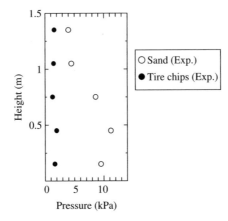

Figure 2. Reduction of earth pressure (Experiments).

2 EXPERIMENTS

Let us begin with a brief description of Hazarika's field experiments (Hazarika et al. (2006)). The reader should see the references for more details, but two experiments are of particular note here. As Fig. 1 shows, in case (a), a 1.5 m high wall was installed and Soma No. 6 sand was packed on each side of the wall to a void ratio of 0.53. A load of 8.3 kPa was then applied to the sand from above and the earth pressure on the wall was measured with earth pressure meters (10 cm in diameter) that had been placed at intervals of 30 cm. In case (b), a similar wall and backfill were installed, except that the backfill on one side of the wall was replaced on one side with tire chips to a thickness of 30 cm. The backfill was again packed to the same void ratio and loaded at 8.4 kPa from above to measure the resulting earth pressure on the wall. Well-graded Masado was chosen as the backfill material. The backfill material has the in place average dry density, $\rho_d = 1.696 \, t/m^3$ an relative density, $D_r = 76 \, \%$ with an angle of internal friction of 40 degrees. Figure 2 provides the experimental results: the unfilled circles represent the earth pressures under sand alone, and the filled circles, those under sand and tire chips. The combination of sand and tire chips resulted in clearly lower earth pressures. In addition, the retaining wall does not move under the experiments.

3 NUMERICAL SIMULATIONS

3.1 Calculation conditions

Table 1 displays the elasto-plastic parameters, the evolution parameters and the initial conditions. The reader should check the references to learn the meanings of these parameters.

Table 1. Material parameters.

[elasto-plastic parameters]	
Compressuin index λ	0.05
Swelling index κ	0.017
Critical state constant M	1.00
Void ratio at $p' = 98\,kPa$	1.51
Poisson's ratio	0.3
Soil density	2.65
[Evolution parameters]	
Degradation parameter of overconsolidated state m	0.05
Degradation of parameter of structure a	1.0
Degradation of parameter of structure b, c	1.0
Evolution parameter of βb_r	3.0
Limit of rotation m_b	0.7
[Initial conditions]	
Initial void ratio e	0.53
Initial value of $1/R^*$ (state of structure)	3.0
Initial coefficient of earth pressure at rest K_0	0.6
Initial slope of axis of rotation	0.231
Coefficient of permeability k (cm/sec)	4.0×10^{-2}

Let us turn to a description of how the soil was simulated. As shown in Fig. 3(a), first, a horizontal base was created. The void ratio was set at 0.53 to match the experimental conditions and the state of the structure and initial anisotropy were set as constant. The confining pressure due to the weight of the sand itself was also calculated and the overconsolidation ratio of the soil was found. The soil was excavated in order to create the inclined volume at the back of the backfill shown in Fig. 3(b). The wall portion of the model was fixed in the horizontal direction (the soil was permitted to shift in the vertical direction). Then, an 8.3 kPa load was imposed in the downward direction and calculations were continued until steady state was reached in order to estimate the earth pressures upon the retaining wall. Next, the 30 cm thick zone of sand behind the wall was instantaneously changed into tire chips (c) (Noda et al. (2005)).

The forces on the nodes of the finite elements representing the sand were calculated, the forces due to the weight of the tire chips replacing the sand were calculated, and the differential forces resulting from these two components were applied to the nodes. Finally, a wedge of the "soil" was removed (d).

Figure 4 presents the initial conditions of the soil. The tire chip volume was treated simply as an elastic body with a high compressibility. The Young's modulus of the volume was varied with the confining pressure. Three elastic constants were used in this calculation at intervals of 5 m, as shown in Table 2. The table indicates the constants in the order of shallow, medium depth and deepest depth considering relativity of confined compression of tire chips. Once the above replacement had been completed, the soil volume was subjected to the load of 8.4 kPa, the simulation was continued until steady state was attained, and the earth pressure on the wall was calculated. Figure 5 shows the mesh used to simulate the volume after excavation and the boundary conditions.

3.2 Calculation results

Figure 6 presents the results from Case 1 in Table 3. The unfilled circles are the soil pressures under sand, and the filled circles are the soil pressures under the combination of sand and tire chips. As seen in Fig. 2, a lower earth pressure is predicted when there were tire chips between the wall and the sand. Figure 7 shows the distribution of shear strain under sand only (a) and in combination with

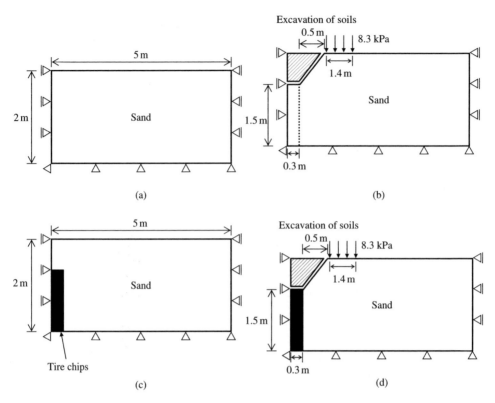

Figure 3.   Stage of initial soil ground.

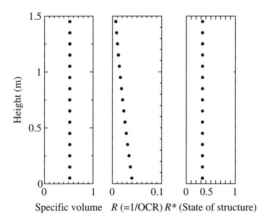

Figure 4.   Initial conditions.

the tire chips (b). The shear strain due to the load is greater in the upper portion of the soil volume in (a), while in configuration (b), the sand body has shifted toward the retaining wall, forming a distribution resembling a shear face. This explains why the earth pressure was reduced, as the soil had deformed in the active state.

248

Table 2. Elastic parameters.

	Case 1	Case 2	Case 3
[Elastic parameters]			
Young's modulus E (kPa)	40,60,80	400,450,500	40,60,80
Poisson's ratio ν	0.1	0.1	0.4
Density (g/cm^3)	1.15	1.15	1.15
Initial coefficient of earth pressure at rest K_0	0.1	0.1	0.6

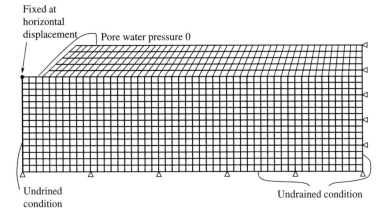

Figure 5. Boundary conditions.

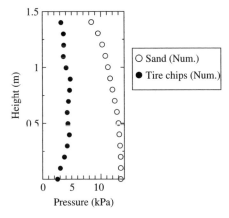

Figure 6. Reduction of earth pressure (Numerical simulation).

Figure 8 presents the results of calculations when the material constants of the tire chips were varied. The volumetric and elastic parameters of this case are shown in Table 2. When Poisson's ratio ν was 0.1 and the elastic parameter was high (Case 2), the earth pressure was reduced in the deeper portion of the volume, but there was little change in the upper portion. When the elastic

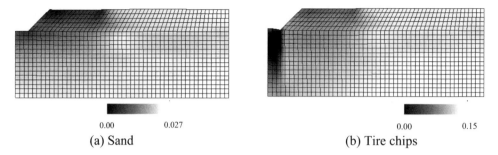

(a) Sand   (b) Tire chips

Figure 7.   Distributions of shear strain.

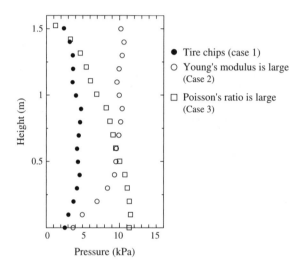

Figure 8.   Case study.

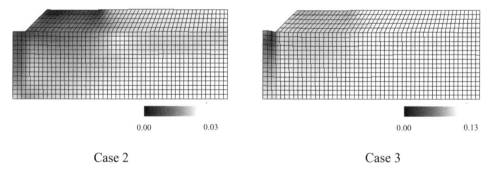

Case 2   Case 3

Figure 9.   Distributions of shear strain.

parameter was the same as in Fig. 6, but ν was raised to 0.4 (Case 3), there was a reduction in earth pressure in the upper portion but no effect in the deep portion.

Figure 9 provides the shear-strain distribution. Only the upper portion shows a high shear strain when ν is high. When the elastic constants are high, the shear strain distribution resembles that in Figure 7(b), but its magnitude is lower, with no sand deformation.

250

REFERENCES

Asaoka, A., Nakano, M. and Noda, T. (1994): Soil-water coupled behavior of saturated clay near/ at critical state, *Soils and Foundations*, 34(1), pp. 91–105.

Asaoka, A., Noda, T., Yamada, T., Kaneda, K. and Nakano, M. (2002): An elasto-plastic description of two distinct volume change mechanisms of soils, *Soils and Foundations*, 42(5), pp. 47–57.

Hazarika, H., Sugano, T., Yasui, K., Mae, Y. and Ejiri, A. (2005): Retaining Structure with Artificianl and Recycled Geomaterials as Sandwiched Cushion, Symposium on Artificial Geomaterial, Fukuoka, pp. 77–82.

Hazarika, H., Kohama, E., Suzuki, H. and Sugano, T. (2006): Enhancement of Earthquake Resistance of Structure using Tire Chips as Compressibel Inclusion, Report of the Port and Airport Research Institute, Vol. 45, No. 1.

Humprey, D.N. and Manion,W.P. (1992): Engineering properties of tire chips for lightweight fill, Ground, Soil Improvement and Geosysnthetics, ASCE, Vol. 2, pp. 1344–1355.

Noda, T., Tashiro M., Takaine T. and Asaoka A. (2005) : Deformation analysis of piled raft foundation in terms of settlement reduction and load sharing ratio, Journal of Geotechnical engineering, No. 799, pp. 37–50 (in Japanese).

Scrap Tire Derived Geomaterials – Opportunities and Challenges – Hazarika & Yasuhara (eds)
© 2008 Taylor & Francis Group, London, ISBN 978-0-415-46070-5

Isotropic pressure loop test and constant effective stress test on used tire rubber chips

J. Yajima & N. Kobayashi
Meisei University, Tokyo, Japan

ABSTRACT: Investigations on the use of granulated scrap tire rubber chips as geomaterials are in progress recently. Assuming that tire rubber chips could be used as geomaterials. However, used tire rubber chips can be thought a material character to be fundamentally different from the geomaterials. Therefore, the following methods have been used in order to evaluate the effectiveness of tire rubber chips of $D_{50} = 2.0$ mm and $D_{50} = 6.0$ mm in size; a constant effective test under constant effective stress but different buck pressures, and an isotropic loop test under various conditions of buck pressure isotropic effective stress. As of the test results, both $D_{50} = 2.0$ mm and $D_{50} = 6.0$ mm sized tire rubber chips proved to be equally applicable like common soil as geomaterials in the constant effective test. In the isotropic loop test, it has been concluded that tire rubber chips have a non-compressive behavior under isotropic pressure conditions, and that they have a depict elastic type behavior.

1 INTRODUCTION

The total amount of use of waste tire in year of 2004 was approximately 103,000,000, and its total weight was 1,040,000 tons. Due to the recent reuse of the waste tire grapples with material recycling, its recycling rate has become up to 90% (Bridgestone (2006)). Elastic pavement and asphalt rubber are used as the effective reuse of waste tires in the field of civil engineering, however, we have just started to develop the technology to make an effect reuse of waste tires as ground material recently in Japan.

This paper presumes that used tire rubber chips smashed from waste tires could be used as geomaterials. Since used tire chips are fundamentally different from geomaterials, it is examined if the Terzaghi's principle of effective stress can be applied, and then an isotropic pressure loop test and constant effective stress test, with used tire rubber chips of $D_{50} = 2.0$ mm and $D_{50} = 6.0$ mm in diameter, were made, in order to examine what would happen under constant effective stress.

2 MATERIAL PROPERTIES AND SPECIMEN CONDITION OF USED TIRE RUBBER CHIPS

2.1 *Material properties of used tire rubber chips*

The photographs 1 and 2 show used tire rubber chips of $D_{50} = 2.0$ mm and $D_{50} = 6.0$ mm, and Table 1 shows physical properties of used tire rubber chips. Grain size accumulation curve is shown in Figure 1. The minimum densities test of used tire rubber chips was carried out in the same man-ner as the minimum and maximum densities test of sand (JISA 1224), and its maximum densities test was done in the same manner as the test method for soil compaction using a rammer. The minimum and maximum densities examination of used tire rubber chips were tested in the same manner as the test method of the minimum densities.

Photograph 1. $D_{50} = 2.0$ mm used tire rubber chips.

Photograph 2. $D_{50} = 6.0$ mm used tire rubber chips.

Table 1. Physical properties of used tire rubber chips.

used tire rubber chips	$D_{50} = 2.0$ mm	$D_{50} = 6.0$ mm
specific gravity: Gs (g/cm^3)	1.15	1.21
minimum density: ρ_{dmin} (g/cm^3)	0.551	0.487
maximum density: ρ_{dmax} (g/cm^3)	0.749	0.679

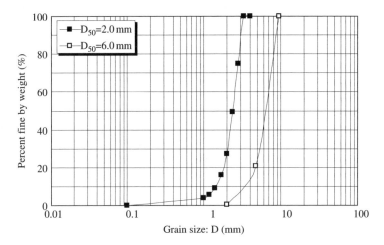

Figure 1. Grain size accumulation curve of used tire rubber chips.

The test results shows that the density of $D_{50} = 2.0$ mm used tire chips particles is 1.15 g/cm^3, the minimum density is 0.55 g/cm^3, and that the maximum density is 0.75 g/cm^3. It was also found that the density of $D_{50} = 6.0$ mm used tire chips particles is 1.21 g/cm^3, minimum density is 0.49 g/cm^3 and maximum density is 0.68 g/cm^3. And it can be concluded that, as for the grain size accumulation curve, both $D_{50} = 2.0$ mm and $D_{50} = 6.0$ mm are curves like sand of the single grain diameter.

2.2 Specimen condition of used tire rubber chips

The size of specimen used for the triaxial compression apparatus is 100 mm in length and 50 mm in diameter. Taking the minimum and maximum densities shown in Table 1 into consideration, the specimen was made by being dropped in the vertical direction with rohto several times to be

Table 2. Physical properties of used tire rubber chips.

used tire rubber chips	$D_{50} = 2.0$ mm	$D_{50} = 6.0$ mm
dry density: ρ_d (kN/m^3)	6.34	6.43
void ratio: e	0.818	0.888

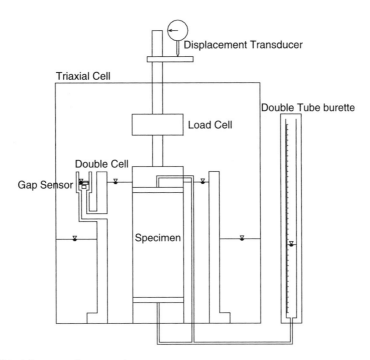

Figure 2. Triaxial compression apparatus.

compacted, so that the relative density of the specimen used in this test could be around 85%. As a result, the initial solid state of the specimen installed with a triaxial compression apparatus was made like Table 2. Then, a level of saturation was made increased by running water through the specimen in upright position with a vacuum method. As a result, the B value became more than 0.95 under the condition of confined pressure 50 kPa or more.

Shown in Figure 2, the triaxial compression apparatus used in this test can measured the outer volume with a gap sensor of double cell.

3 CONSTANT EFFECTIVE STRESS TEST

3.1 *Test method*

Since used tire rubber chips are different as material characteristics from geomaterials, constant effective stress test was carried out to confirm if a principle of effective stress was formed in used tire rubber chips. As the test condition, the effective confining pressure (σ'_c) was kept constant with $\sigma'_c = 100$ kPa, and back pressure (σ_{BP}) was changed to 100 kPa and 200 kPa for used range. A consolidated-drained triaxial compression test (CD-test) and a consolidated-undrained triaxial compression test (CU-test) were performed in this state.

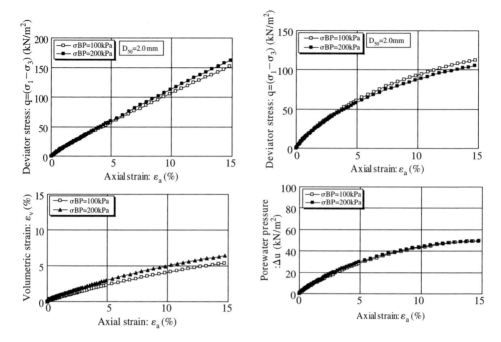

Figure 3. $D_{50} = 2.0$ mm-constant effective stress test. (CD-test)

Figure 4. $D_{50} = 2.0$ mm-constant effective stress test. ($\overline{CU}$-test)

3.2 $D_{50} = 2.0$ mm shear strength behavior

The shear test result of a consolidated-drained condition is shown in Figure 3, and the shear test result of a consolidated-undrained condition in Figure 4. Judging from these results, the relations between deviator stress ($q = (\sigma_1 - \sigma_3)$) and axial strain ($\varepsilon_a$) in the consolidated-drained condition are the same in case effective confining pressure ($\sigma'_c = 100$ kPa) is fixed, even if back pressure (σ_{BP}) is different from 100 kPa and 200 kPa. And the relations between volumetric strain (ε_v) and axial strain (ε_a) are the same in case effective confining pressure ($\sigma'_c = 100$ kPa) is fixed. At the same, the relations between deviator stress ($q = (\sigma_1 - \sigma_3)$) and axial strain ($\varepsilon_a$) in the consolidated-undrained condition are the same in case effective confining pressure ($\sigma'_c = 100$ kPa) is fixed. The relations between pore water pressure (Δu) and axial strain (ε_a) are the same as well.

From these results, all the $q - \varepsilon_a$ relations, the $\varepsilon_v - \varepsilon_a$ relations, and the $\Delta u - \varepsilon_a$ relations are the same in the case of $D_{50} = 2.0$ mm, if σ_{BP} is different but σ'_c is the same, and the principle of effective stress consists in the same way as geomaterials for used tire rubber chips.

3.3 $D_{50} = 6.0$ mm shear strength behavior

Figure 5 shows the shear test result under the consolidated-drained condition, and Figure 6 shows the one under the consolidate-undrained condition. Judging from these results, just like the case of $D_{50} = 2.0$ mm, the relations between deviator stress ($q = (\sigma_1 - \sigma_3)$) and axial strain ($\varepsilon_a$) under the consolidated-drained condition are the same in case effective confining pressure ($\sigma'_c = 100$ kPa) is fixed, even if back pressure (σ_{BP}) is different from 100 kPa and 200 kPa. And the relations between volumetric strain (ε_v) and axial strain (ε_a) are the same if the effective confining pressure ($\sigma'_c = 100$ kPa) is fixed. At the same time, the relations between deviator stress ($q = (\sigma_1 - \sigma_3)$) and axial strain ($\varepsilon_a$) under the consolidated-undrained condition are the same in case the effective confining pressure ($\sigma'_c = 100$ kPa) is fixed, even if back pressure (σ_{BP}) is different from 100 kPa and 200 kPa. And the relations between pore water pressure (Δu) and axial strain (ε_a) are the same

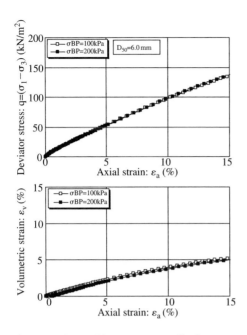

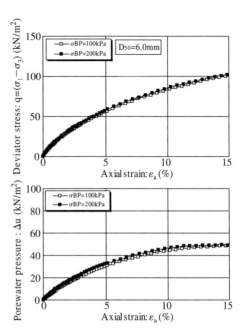

Figure 5. $D_{50} = 6.0$ mm-constant effective stress test. (CD-test)

Figure 6. $D_{50} = 6.0$ mm-constant effective stress test. ($\overline{CU}$-test)

in case the effective confining pressure ($\sigma'_c = 100$ kPa) is fixed. It can be concluded from all those results, in case the D_{50} is 6.0 mm and that the effective confining pressure is the same even if back pressure is different, the q − ε_a relations, the ε_v − ε_a relations, and the Δ u − ε_a relations are the same and the principle of effective stress consists in the same way as geomaterials for used tire rubber chips.

4 ISOTROPIC PRESSURE LOOP TEST

4.1 Test method

An isotropic pressure loop test was carried out under different conciliations of effective pressure (σ'_c) and back pressure (σ_{BP}) so as to examine what kind of volumetric strain would be shown when used tire rubber chips of $D_{50} - 2.0$ mm and $D_{50} - 6.0$ mm get isotropic pressure. The same test was carried out for general saturated clay and rubber for comparison. An isotropic pressure loop test by pressure changing was carried out as shown in Figure 7. Stress path was moved clockwise (A-B-B'-C-D-E-B-A) and anticlockwise (A-B-E-E'-D-C-B-A). The movement of double burette and double cell was measured under each stress condition. First it was measure in the route of B-B'-C and E-E'-D when a sewer valve was closed under the condition of B and E, and it was shifted to that of B' and E'. Then, a sewer valve was opened under the condition of C and D.

4.2 Saturated clay and rubber isotropic pressure behavior

Figure 8 shows the volumetric strain (ε_v) of stress path A-B-B'-C-D-E-A, and Figure 9 shows the volumetric strain (ε_v) of stress path A-B-E-E'-D-C-B-A.

In the case of clay, as effective confining pressure (σ'_c) increases in the route (B-B'-C, E-E'-D), as a matter of course, while sinking may proceed, volumetric strain (ε_v) occurs in both double burette and double cell. As effective confining pressure decreases in the route (D-E, C-B), volume

257

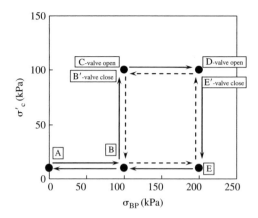

Figure 7. Isotropic pressure loop test.

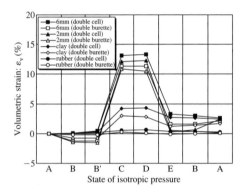

Figure 8. Stress path A-B-B′-C-D-E-B-A.   Figure 9. Stress path A-B-E-E′-D-C-B-A.

expansion occurs and the residual volumetric strain, which is a state of final isotropic pressure (A), occurs. As back pressure (σ_{BP}) increases in the route (A-B, C-D, A-B-E), volumetric strain (ε_v) in the minus direction is formed in double burette because water is supplied to the specimen. As back pressure decreases in the route (E-B-A, D-C, B-A), volumetric strain in the plus direction is formed in double burette because water is discharged from the specimen. The volumetric strain of double cell is zero in the route (B-B′, E-E′), which turns out that the soil particles of cohesive soil are non-compression. In the case of rubber, volumetric strain is less than 1% in all the routes. Judging from this result, a rubber simple substance is a non-compressive material.

4.3 $D_{50} = 2.0\,mm$ isotropic pressure behavior

As effective confining pressure (σ_c') increase in the route (B-B′-C, E-E′-D)s, volumetric strain (ε_v) occurs in both double burette and double cell because, as a matter of course, sinking may proceed. As effective confining pressure decreases in the route (D-E, C-B), volume expansion occurs and volumetric strain (ε_v) is zero under the state of final isotropic pressure(A). Therefore, $D_{50} = 2.0\,mm$ used tire rubber chips shows elastic reaction in a isotropic pressure loop test. As back pressure (σ_{BP}) increases increase in the route (A-B, C-D, A-B-E), volumetric strain (ε_v) in the minus direction is formed in double burette because water is supplied to the specimen. As buck pressure decreases in the route (E-B-A, D-C, B-A), volumetric strain in the plus direction is formed in double burette because water is discharged from the specimen. Then, the volumetric strain of double cell is zero in the route (B-B′, E-E′) under the increased effective confining pressure in the condition that a drain valve was closed. Judging from these, the rubber particles of the $D_{50} = 2.0$ mm used tire rubber chips are non-compressive.

4.4 $D_{50} = 6.0$ mm isotropic pressure behavior

As effective confining pressure (σ'_c) increases in the route (B-B'-C, E-E'-D), volumetric strain (ε_v) occurs in both double burette and double cell, because sinking may proceed. As effective confining pressure decreases in the route (D-E, C-B), volume expansion occurs and the remaining distortion as a state of final isotropic pressure (A) occurs to some extents. This reaction is different from the case of $D_{50} = 2.0$ mm, and the same as that of clay. $D_{50} = 2.0$ mm shows the movement of the elasticity because a deviation between the particles doesn't occur, but residual volumetric strain occurs in $D_{50} = 6.0$ mm because a deviation between the particles occurs. Then, the volumetric strain of double cell is zero in the route (B-B', E-E') under the increased effective confining pressure in the condition that a drain valve was closed. Judging from these, the rubber particles of the $D_{50} = 6.0$ mm used tire rubber chips are non-compressive.

5 CONCLUSION

The following can be concluded through the constant effective stress test and the isotropic pressure loop test toward used tire rubber chips of $D_{50} = 2.0$ mm and $D_{50} = 6.0$ mm.

(1) The deviator stress-axial strain relations, the volumetric strain-axial strain relations and the pore water pressure-axial strain relations remain the same in the constant effective stress test, in case effective confining pressure is the same even in different back pressures from the used tire rubber chips of both $D_{50} = 2.0$ mm and $D_{50} = 6.0$ mm. Therefore, the principle of effective stress consists in used tire rubber chips as well as the soil.
(2) In the isotropic pressure loop test, used tire rubber chips of $D_{50} = 2.0$ mm shows elasticized reaction, but a deviation between the particles occurred. In the case of used tire rubber chips of $D_{50} = 6.0$ mm, residual volumetric strain occurred due to the creep between particles of tire rubber chips.
(3) In the isotropic pressure loop test, used tire rubber chips of $D_{50} = 2.0$ mm and $D_{50} = 6.0$ mm as well as rubber material are non-compressive just like soil.

REFERENCES

Bridgestone (2006). Social & Environmental Report 2006 (in Japanese)
Yajima, J., Ogura. K., Ashoke K., Karmokar and Yasuhara, K. (2006). Mechanical evaluation of used tire rubber chips as geomaterial. *Japanese Geotechnical Journal*. Vol. 1, No. 1 pp. 1–7 (in Japanese)

Scrap Tire Derived Geomaterials – Opportunities and Challenges – Hazarika & Yasuhara (eds)
© 2008 Taylor & Francis Group, London, ISBN 978-0-415-46070-5

An elasto-plastic constitutive model for cement-treated clays with tire-chips

S. Murakami, K. Yasuhara & T. Tanaka
Department of Urban & Civil Engineering, Faculty of Engineering, Ibaraki University, Japan

Y. Mitarai & T. Kishida
Research and Development Center, Toa Corporation, Japan

ABSTRACT: This paper presents an elasto-plastic constitutive model based on the subloading surface concept and damage theory for cement-treated clays with tire-chips. The applicability of the proposed model has been investigated by comparison of the calculated results with experimental results observed in undrained triaxial compression tests on cement-treated clays mixed with tire-chips. The proposed model well explains that the mixture of tire-chips with cement-treated clay improves its toughness.

1 INTRODUCTION

Cement-treated marine clays mixed with tire-chips have some advantageous features as geo-material. Mixing with tire-chips which are made from used tire, cement-treated clays vary from a brittle geo-material to a tough one. This kind of cement-treated marine clay mixed with tire-chips is expected to be as a geo-material with capability for subordinate to unexpected large deformation in application as a barrier geo-material used for coastal waste disposal reclamation because cemented-treated marine clays without tire-chips have brittleness and disturb to reduce the capability against the deformation. However, it is necessary for performing numerical analysis and evaluating the applicability to develop a mechanical model for describing behaviour of the new advantageous geo-material that is the cement-treated marine clay mixed with tire-chips.

The purpose of this study is to establish an elasto-plastic constitutive equation for the above-mentioned cement treated clays mixed with tire-chips. The constitutive model based on the subloading surface concept and the damage theory has been proposed. By introducing the damage theory, the proposed model considers that artificially cementation of a cement-treated clay decreases during shear. In order to represent the smooth stress-strain curve, the subloading surface concept has been introduced. In addition, it is easy to describe the loading criterion by using the concept. Combing the subloading surface concept and the damage theory with an elasto-plastic constitutive equation, the model can represent stress-strain relation with strain-hardening and – softening characteristics which cement-treated clays have in general. The applicability of the proposed model has been investigated by comparison of the calculated results with experimental results observed in undrained triaxial compression tests on cement-treated clays mixed with tire-chips.

2 FUNDAMENTAL CONCEPTS OF ELASTO-PLASTIC CONSTITUTIVE MODELING FOR CEMENT-TREATED CLAY MIXED WITH TIRE-CHIPS

This study is intended to establish an elasto-plastic constitutive equation for cement-treated clays mixed with tire-chips. The authors proposed an elasto-plastic constitutive model based on the

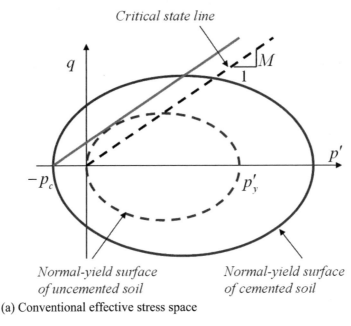

(a) Conventional effective stress space

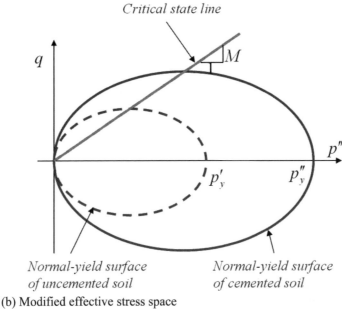

(b) Modified effective stress space

Figure 1.   Diagram of yield surfaces on conventional effective stress space and modified effective stress space.

subloading surface concept and a damage theory in a previous study (Murakami et al., 2006). That model can describe the behaviors of cement-treated clays mixed with and without tire-chips. Nevertheless, the model can not incorporate the effect of confining pressure on the stress–strain relation because Mises's yield function was assumed as the loading function. To consider the dependence of confining pressure, the model has been modified in this study.

Cement-treated soils have higher consolidation yield stress and higher shear strength than the original soils. Figure 1(a) shows the yield surfaces of a cement-treated soil and the original soil

262

on conventional effective mean stress, p', and deviator stress, q, plane. The yield surface of the cement-treated soil expands across the q-axis. Therefore, the cement-treated material is known to be resistant to tensile stress. Considering these effects, let us introduce a new conceptual effective stress for cement-treated soils as

$$\boldsymbol{\sigma}'' = \boldsymbol{\sigma}' + p_c \mathbf{I} \tag{1}$$

where $\boldsymbol{\sigma}'$ is a conventional effective stress tensor and p_c is the bonding stress as internal stress of cemented soils in this study.

Figure 1(b) shows the yield surfaces on the modified effective stress space defined by Eq. (1). The consolidation yield stress of the cemented soil, p''_y, indicates the yield surface size; it must satisfy the following relations.

$$p'_{y,u} + p_c < p''_y \quad when \quad p_c > 0$$
$$p'_{y,u} = p''_y \quad when \quad p_c = 0 \tag{2}$$

In those equations, $p'_{y,u}$ is the consolidation yield stress of uncemented soil. Considering the relations, we assume the following function as p''_y.

$$p''_y = \left(p'_{y,u} + p_c\right) + \alpha_c p_c^{\beta_c} \tag{3}$$

Therein, α_c and β_c are non-negative material constants. Bonding stress decreases with increasing plastic strain. The degradation of bonding stress is irrecoverable without additional admixture with cement. We introduce the following relation for expressing bonding stress degradation.

$$p_c = p_{c,\min} + \left(p_{c0} - p_{c,\min}\right)\exp\left(-m_c \int_0^t \|\dot{\gamma}^p\| dt\right) \tag{4}$$

In that equation, p_{c0} and $p_{c,\min}$ are the initial and residual bonding stress, respectively. m_c is material parameter related with degradation of bonding stress because of plastic shear strain, γ^p. Assuming Eq. (4), we can describe variations of stress-strain curve from brittle behaviour to toughness one by considering degradation of bonding stress with increase in plastic shear strain as shown in Fig. 2. Because bonding stress decreases with increased m_c, the parameter is designated in this study as the brittle index.

In addition to the above relations, the subloading surface concept developed by Hashiguchi (1989) was introduced for describing the smooth stress-strain curve. We can obtain the following loading function on the modified effective stress space by adapting the yield function of the modified Cam-clay model as the shape of the loading function:

$$f(\boldsymbol{\sigma}'') - R \cdot p''_y = 0, \tag{5a}$$

$$f(\boldsymbol{\sigma}'') = p'' \frac{M^2 + \eta''^2}{M^2}, \tag{5b}$$

where M is the slope of critical state line shown in Fig. 1, η'' is the effective stress ratio defined as $\eta'' = q/p''$, and R is the normal-yield ratio defined as the ratio of the size of subloading surface to that of normal-yield surface. Hashiguchi has proposed an evolution rule of R: $\dot{R}$ is assumed as

$$\dot{R} = U(R)\|\mathbf{D}^p\| \quad for \quad \mathbf{D}^P \neq 0, \tag{6}$$

where U is a monotonically decreasing function of R, thereby fulfilling the following condition:

$$U(R) = \begin{cases} \infty & for \quad R = 0, \\ 0 & for \quad R = 1, \end{cases} \tag{7}$$

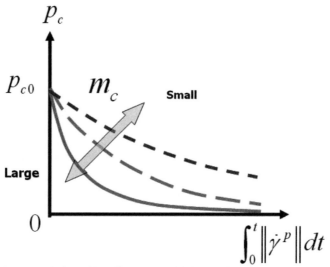

(a) Degradation of bonding stress with increasing plastic shear strain

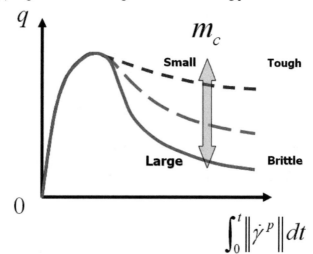

(b) Variations of stress-strain curve depended on brittle parameter

Figure 2.   Diagram of degradation of bonding stress and variations of stress strain curve.

That is,

$$U(R) = -\mu_R \ln R , \tag{8}$$

where μ_R is a material constant. Therefore, we can describe a smooth stress–strain curve.

3   AN ELASTO-PLASTIC CONSTITUTIVE EQUATION AND LOADING CRITERION

Assuming Eq. (1)–Eq. (8) and applying the consistency condition to Eq. (5) and the associated flow rule, we can obtain the relationship between the conventional effective stress rate, $\dot{\sigma}'$, and strain rate $\mathbf{D}$ as

$$\overset{\circ}{\boldsymbol{\sigma}}' = \left(\mathbf{E} - \mathbf{E}_u^p - \mathbf{E}_c^p \right) \mathbf{D}, \tag{9}$$

where $\mathbf{E}$ is the elastic modulus, and $\mathbf{E}_u^p$ and $\mathbf{E}_c^p$ are defined as follows.

$$\mathbf{E}_u^p = \frac{\mathbf{EN} \otimes \mathbf{NE}}{H^P + tr\left(\mathbf{NEN}\right)} \tag{10a}$$

$$\mathbf{E}_c^p = \frac{h_c \mathbf{I} \otimes \mathbf{NE}}{H^P + tr\left(\mathbf{NEN}\right)} \tag{10b}$$

$$\mathbf{N} = \frac{\partial f}{\partial \boldsymbol{\sigma}''} \bigg/ \left\| \frac{\partial f}{\partial \boldsymbol{\sigma}''} \right\| \tag{10c}$$

$$\mathbf{N} = \frac{\partial f}{\partial \boldsymbol{\sigma}''} \bigg/ \left\| \frac{\partial f}{\partial \boldsymbol{\sigma}''} \right\| \tag{10c}$$

$$\dot{p}_c = \left\| \mathbf{D}^p \right\| h_c \tag{10e}$$

$$\dot{p}'_{y,u} = \left\| \mathbf{D}^p \right\| h \tag{10f}$$

The loading criterion in the proposed elasto-plastic constitutive model is expressed as follows.

$$\frac{tr\left(\mathbf{NED}\right)}{H^P + tr\left(\mathbf{NEN}\right)} > 0 \quad \text{for} \quad \mathbf{D}^p \neq \mathbf{0} \tag{11a}$$

$$\frac{tr\left(\mathbf{NED}\right)}{H^P + tr\left(\mathbf{NEN}\right)} \leq 0 \quad \text{for} \quad \mathbf{D}^p = \mathbf{0} \tag{11b}$$

Adopting the subloading surface concept in establishing the constitutive model, the stress state is always on the loading surface during increased plastic strain. In addition, it is easy to describe the loading criterion using the subloading surface concept.

4   APPLICABILITY OF THE PROPOSED MODEL TO CEMENT-TREATED CLAYS MIXED WITH TIRE-CHIPS

For investigating the applicability of the proposed model to cement-treated clays mixed with tire-chips, undrained triaxial compression tests were performed. The cement-treated clays mixed with and without tire-chips were prepared as specimens ($d = 50\,\text{mm}$, $h = 100\,\text{mm}$). The dredged clay was Tokyo Bay clay ($\rho_s = 2.691$ g/cm^3, $I_p = 70.0\%$, percentage content of fine-grained fraction is about 90%). Clay slurry was mixed with seawater to obtain an initial water content of 240%. The cement used was normal Portland cement, whose particle density was 3.16 g/cm^3. The tire chips were made from used tires, and the average grain size was 2 mm by screening: the resultant particle density was 1.15 g/cm^3. The proportions of added tire chips were 0% and 20% of a whole volume of the specimen. All specimens were cured at a high moisture condition of more than 95% and constant temperature of 20°C ± 2°C.

Figures 3 and 4 show experimental results from undrained triaxial compression tests. Comparing cement-treated clay mixed with tire-chips to that mixed without tire-chips, these results show that cement-treated clays vary from a brittle geo-material to a tough one by mixing different amounts of tire-chips. Figures 2 and 3 also show the calculated results using the proposed elasto-plastic constitutive model. The proposed model can represent stress-strain relations of experimental results with differences in consolidation pressure, p_0, and the content of tire-chips. In addition, the proposed model can well explain that the mixture of tire-chips with cement-treated clay improves toughness

265

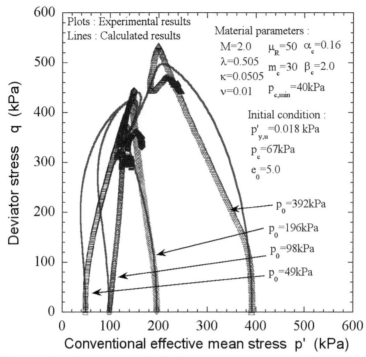

(a) Stress path on the conventional effective stress space

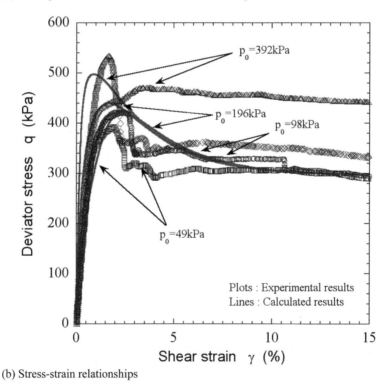

(b) Stress-strain relationships

Figure 3. Comparisons between experimental and calculated results of cement-treated clays without tire-chips.

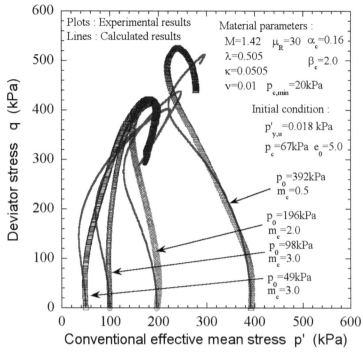

(a) Stress path on the conventional effective stress space

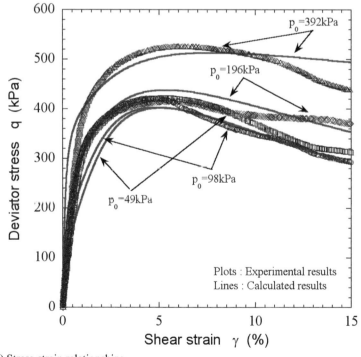

(b) Stress-strain relationships

Figure 4. Comparisons between experimental and calculated results of cement-treated clays mixed with tire-chips.

with mixing with the tire-chips by comparing the brittleness parameter, m_c. By investigating the effects of material parameters determined from experimental results, the parameters included in the proposed model are suitable for evaluating the improvement of toughness of cement-treated clays mixed with tire-chips.

5 CONCLUSION

The proposed elasto-plastic constitutive model based on the subloading surface concept and the damage theory can represent stress-strain relation of the cement-treated clays mixed with tire-chips in undrained triaxial compression tests.In addition, the proposed model can well explain that the mixture of tire-chips with cement-treated clay improves toughness. By investigating the effects of material parameters determined from experimental results, the brittle parameter included in the proposed model is suitable for evaluating improvement of toughness of cement-treated clays mixed with tire-chips.

ACKNOWLEDGEMENTS

This study was partly supported by the Grant-in Aid of the Ministry of Ministry of Land, Infrastructure and Transport, Japan during 2005–2006. We express our sincere appreciation for financial support from that organization.

REFERENCES

Murakami, S., Yasuhara, K., Tanaka, T., Mitarai, Y. and Kishida, T. 2006. Effects of Tire-chips on Cement-treated Clay Behaviour and its Modeling, *Proc. Int. Sym. Geomechanics and Geotechnics of Particulate Media*, 475–481.
Hashiguchi, K. 1989. Subloading Surface Model in Unconventional Plasticity, *Int. J. Solids Structures*, 25, 917–945.

Evaluation of load bearing capacity of shredded scrap tire geomaterials

A.K. Karmokar & H. Takeichi
Central Research, Bridgestone Corporation, Tokyo, Japan

M. Kawaida
Planning Section, Nippon Expressway Research Institute Company Limited, Tokyo, Japan

K. Yasuhara
Faculty of Technology, Ibaraki University, Ibaraki, Japan

ABSTRACT: CBR tests on scrap tire derived rubber chips and shreds have been carried out using a standard test apparatus. CBR values are found to be in the range of 0.3–0.6% which may generally be categorized as poor road aggregates. However, with the use of sand around the tire derived aggregates have shown improving the CBR of the composite mass. Depending on the extent of void-filling by sand around the rubber chips/tire shreds, CBR value of composite mass has found to be increased to a level above 3%. Improvement in the load bearing capacity of tire derived geomaterials to an acceptable level by sand filling method may open up the possibility of the use of shredded scrap tire geomaterials as aggregates in road sub-base/base applications.

1 INTRODUCTION

Number of scrap tires (tires from end-of-life vehicles) in Japan is estimated to about 106 million in 2006 [JATMA (2007)] that constitutes a large quantity of solid waste. The scenario of scrap tire disposition in Japan for 2006 is shown in Fig. 1. While 88% of the annually produced scrap tires are reused and/or recycled, the share in material recycling sector is constituted by only 15%. Therefore, many efforts to recycle scrap tires as geomaterials have been made to increase the share of material recycling of scrap tires.

It is reported that scrap tires have essential engineering properties viz., resilient, water proof, insulation, bondable, durable, and such properties are commonly recognized as beneficial to many civil engineering applications [(Humphrey & Manion (1992)]. In line, various research projects have been initiated to explore the possibility of their use as geomaterials in civil engineering applications [Karmokar et al. (2005), Kawaida et al. (2005), Yasuhara et al. (2006)].

In the present study, an attempt has been made to evaluate the load bearing capacity of shredded scrap tire aggregates with an aim of their use in road structures for sub-base/base applications. A series of CBR (California Bearing Ratio) tests are performed in laboratory for obtaining the CBR% value of scrap tire derived rubber chips and tire shreds materials. As for a measure of improving the load bearing capacity of tire shred geomaterials, specially formed sand-tire shreds mixed specimens viz., replacing voids of tire shreds aggregates by filling of sand are also evaluated in a very similar manner.

The development of potential tire recycling technologies for civil engineering applications is believed to reduce unaccountable tires (about 12% of total scrap tire volume as shown in Fig. 1). Moreover, the increase of material recycle is directly involved with the lowering of CO_2 emission in the environment for which the world community is facing hardship presently.

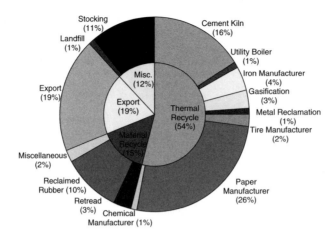

Figure 1.   Japan scrap tire disposition, 2006.

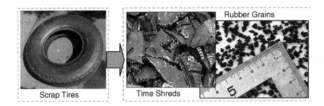

Figure 2.   Scrap tire derived geomaterials.

2   EXPERIMENTAL PROCEDURE

2.1   *Test materials*

Scrap tires could be shredded into tire shreds first and/or then further could be grinded into rubber grains, as shown in Fig. 2. Depending on kind and level of shredding/grinding process, the size of tire shreds and/or rubber grains could be of many kinds. However, both the materials are composed of mainly non-spherical particles. Tire shreds contain steel and textile cords embedded inside the rubber as-it-was in the tire, while rubber grains sample is composed of rubber only materials i.e., steel/textile cords are separated out from rubber. Consequently, tire shreds behave less compressive than rubber grains. The specific gravity of rubber grains and tire shreds are 1.15 and 1.20, respectively.

Two kind of samples, viz., scrap tire derived rubber grains (GRS02) and sand (GSS#3) are included in the experiment for basic study on load bearing capacity. As shown in Fig. 3, the cumulative particle size distribution curves of these two samples are quite identical. Both the samples contain particles of size ranging 1–3 mm, roughly. Photographs, and a few of the physical properties are shown in the inset of Fig. 3.

The scopes of our experiment were also extended to the use of 20–70 mm sized scrap tire shreds (TS2070) for which a series of large scale model load bearing tests were carried out in laboratory. As for a measure of improving the load bearing capacity of tire shred geomaterials, specially formed sand-tire shreds mixed specimens viz., replacing voids of tire shreds aggregates by filling of sand are also evaluated in a very similar manner.

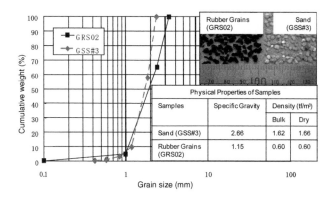

Figure 3. Test samples.

Figure 4. Large scale load-penetration test.

2.2 *Test method*

CBR (California Bearing Ratio) tests were performed in laboratory as a measure of load bearing capacity of scrap tire derived geomaterials. For basic studies, standard CBR test procedure was followed in which a plunger of diameter 50 mm was penetrated at a speed of 1 mm/min onto a compacted specimen of specifications 150 mm diameter and 125 mm high. The specimen was prepared in a cylindrical mold of 150 mm diameter and 175 mm high. Load-penetration curve obtained from such test is used in calculating CBR% for the specimen.

For carrying out tests on TS2070 tire shreds, compacted specimens of size 300 mm in diameter and 250 mm high are prepared in a cylindrical mold of 300 mm diameter and 300 mm high. Figure 4 shows a typical set-up for large scale load-penetration test. A plunger of 100mm diameter was used for penetration at a speed of 1 mm/min onto the specimen in the mold. The apparatus is made capable of producing a longer stroke, if and when necessary. For standard load, load penetration tests on gravel specimen were carried out in laboratory quite similarly. As for the measure of load bearing capacity, apparent CBR% values are calculated by following the standard calculation procedure.

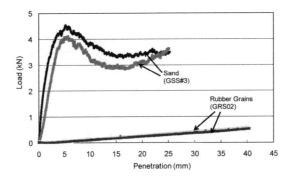

Figure 5.   Typical load-penetration curves.

Figure 6.   Mixing of TC20/GRS02 rubber grains.

3   RESULTS AND DISCUSSION

3.1   *Basic tests*

A few of the typical load-penetration curves recorded during CBR tests on GRS02 rubber grains sample are shown in Fig. 5. For comparison, typical load-penetration curves for GSS#3 sand are also included in the same figure. As may be seen, there exists a vast difference in the levels of penetration resistance for sand and rubber grains, the latter exerts a CBR values ranging 0.3–0.6% only which could generally be categorized as poor road aggregates with respect to the generally specified CBR% of normal road aggregates (CBR value below 3%).

The probable reason may be the high deformability of rubber grains material (typical rubber characteristics) which causes poor compaction of the specimen. Dry and/or wet conditions, as well as the levels of specimen compaction in the mold are found to be only marginally effective in improving CBR% value. Even after a repeated number of compactions of the specimen in the mold, rubber grains material almost rebounds back to its previous condition. Consequently, bulk density of specimen in the mold is only slightly increased (varies only within a narrow range of 0.65–0.75 kN/m^3). In other words, there almost always exists a high void ratio (about 45%) in the compacted mass of rubber grains specimen within the mold.

As shown in Fig. 6, attempts on the use of widely distributed rubber grains in the specimen, prepared upon mixing coarse and fine grains into the bulk, were also found insufficient to increase the compacted density of the composite mass. This leaves still a high void ratio (about 43%) in the specimen. In the separate form of trials, however, the use of fine sand for filling up of voids in the rubber grains specimen had shown reducing the overall void ratio in the composite mass. Depending on the levels of sand filling, void ratio could reach to a level as low as 22% in the composite specimen. The fact of filling-up may be due to higher specific gravity of sand over

Figure 7. Filling voids of TC20 specimen by SS#1 sand.

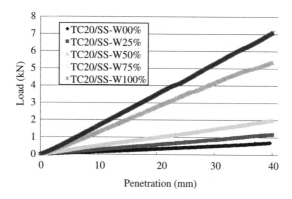

Figure 8. Load-penetration curves of sand filled specimens.

rubber grains, and also due to gap in the levels of particle size distribution of sand with respect to rubber grains used.

3.2 *Improvement of CBR value*

In order to evaluate the levels of improvement of load bearing capacity, a series of CBR tests on rubber grains specimens with various sand filling ratio in the composite mass were carried out. For obtaining efficient and homogeneous filling effect, fine grained sand (SS#1, grain size ranges below 1 mm) was used for filling voids in coarse grained rubber sample (TC20, grain size ranges 4.75 19 mm). The method followed in preparing the specimen is described below.

One-third a volume of the rubber grain specimen was poured in the mold and compacted. Depending on the extent of voids filling, a designated amount of sand was then poured on top of rubber grains layer quite uniformly. Fine and heavy sand could percolate downwards through the voids in the specimen when simply poured on top of the compacted rubber grains layer. A light tamping at the top could enhance this sand percolation process, and consequently, voids at the layer bottom could also be filled up. For obtaining more homogeneous sand filled specimen, rubber grains layer thickness could be adjusted to a lower side. A photograph of sand filled rubber grains specimen in the CBR mold is shown in Fig. 7.

Typical load-penetration traces obtained for different sand filled specimens are shown in Fig. 8. As may be seen, the detected loads for equal level of penetration are increased with the amount of sand added (by weight) for filling up the specimen voids. Further, there exists almost a linear relationship between load and penetration. No peak value was observed though tests were continued for a high level of penetration (about 40 mm).

Three observations were made for each level of sand filled specimens. The values of CBR% calculated for each category are shown in Fig. 9. Depending on the extent of void-filling by

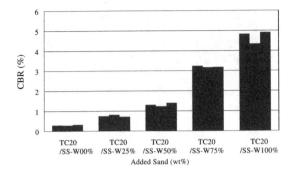

Figure 9.   CBR values of sand filled specimens.

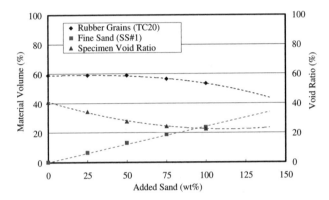

Figure 10.   Relationship between sand added and void ratio.

sand around the rubber grains in the specimen, CBR% value of composite mass has found to be increased manifolds, reaching to an acceptable level generally regarded as normal aggregates for road sub-base/base applications (CBR% 3 or more).

A detailed analysis of the material volumes (rubber grains and sand) and void ratio in the specimen could be seen in Fig. 10. Specimen of rubber grains without sand filling contains about 40% void ratio. However, without much changing the amount of rubber grains in the specimen, sand could be added up gradually for filling the voids present in the mass. Consequently, void ratio in the specimen reaches down to its lowest possible value (about 22%). However, with the increase of sand above this level (about 100% by weight of rubber grains), a portion of rubber grains in the mass has to be replaced for accommodating excess sand which in turn tends to increase the void ratio again in the specimen.

As previously mentioned, the scopes of experiment were also extended to material combinations of TS2070 tire shreds and GSS#3 sand for which a series of large scale load-penetration tests were conducted. Improvements in the load bearing capacity of tire shred materials were also observed quietly in a very similar fashion. However, because of larger size, and/or less flexibility due to embedded steel/textile cords in tire shreds, the range of void ratio manipulated in preparing the sand filled specimens was somewhat different.

Experimentally obtained CBR% values and the corresponding void ratios in the specimen of all tire derived geomaterials are summarized in Fig. 11. Using these data, an average fitting curve was plotted to show the relationship between CBR% and void ratio which could serve as a master curve for obtaining the required CBR% value by controlling void ratio in the specimen. In order to realize higher CBR% value, the specimen has to be compacted for lower void ratio.

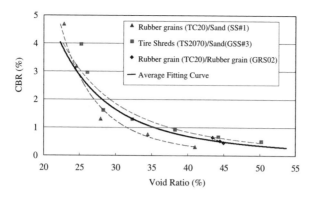

Figure 11. Relationship between void ratio and CBR%.

4 CONCLUSIONS

The theme of this research is to explore the possibility of the use of scrap tire derived geomaterials in road structures for sub-base/base applications. Consequently, an attempt has been made to evaluate the load bearing capacity of shredded scrap tire aggregates.

The CBR values of scrap tire derived geomaterials are found to be in the range of 0.3–0.6% which is generally categorized as poor road aggregates with respect to the specified CBR value of normal road aggregates. The probable reasons may be the high void ratio in the specimen that persists even after severe compaction (poor compaction behavior), and the high deformability of rubber grains/tire shreds (typical rubber characteristics). However, the use of sand around tire derived aggregates (filling up of voids in the specimen) has shown improving the load bearing capacity of the composite mass. Depending on the extent of void-filling by sand around the rubber chips/tire shreds, CBR value of composite mass has found to be increased to a level above 3% which may be categorized as normal aggregates for road applications. Improvement in the load bearing capacity to an acceptable level by sand filling method may open up the possibility of the use of tire shreds as aggregates in road sub-base/base applications.

REFERENCES

JATMA. 2007 – Tyre Recycling Handbook, Tokyo 105-0001, Japan, p. 26.
Humphrey, D.N. & Manion, W.P. 1992 – Properties of tire chips for lightweight fill. *Grouting, Soil Improvement, and Geosynthetics, American Soc. Civil Engineers,* Vol. 2, pp. 1344–1355.
Karmokar, A.K., Hazarika, H., Takeichi, H, & Yasuhara, K. 2005 – Direct shear behavior of tire chips for their use as lightweight geomaterials. *Proc. 40th Geotech. Res. Conf., Japan Geotech. Soc.*, pp. 641–642.
Kawaida, M., Hamazaki, T., Sano, Y., Fujioka, K., Karmokar A.K. & Kawasaki, H. 2005. Compaction and consolidation behavior of tire shreds geomaterials. *Proc. 6th Natl Conf. Env. Geotech., Sapporo,* May, pp. 187–192.
Yasuhara, K., Karmokar, A.K., Kato, Y., Mogi H. & Fukutake, K. 2006. Technology for the application of used tires in soil foundation and soil structures. *J. Found. Engg. & Equipment; General Civil Engg. Res Center,* Vol. 34, No.2, pp. 58–63.

Scrap Tire Derived Geomaterials – Opportunities and Challenges – Hazarika & Yasuhara (eds)
© 2008 Taylor & Francis Group, London, ISBN 978-0-415-46070-5

Undrained and drained shear behavior of sand and tire chips composite material

S. Kawata, M. Hyodo, R.P. Orense & S. Yamada
Dept. of Civil and Environmental Engineering, Yamaguchi University, Ube, Japan

H. Hazarika
Graduate School of System Science and Technology, Akita Prefectural University, Akita, Japan

ABSTRACT: It is estimated that about 88% of 104 million scrap tires generated in Japan in 2005 were either reused or recycled. Because of this, attention has been paid on scrap tires as new ground material, and recent researches have been moving towards this direction. One method of recycling scrap tires is by processing them into tire chips. This paper introduces the results of undrained and drained triaxial compression tests performed to investigate the monotonic shear characteristics of tire chips-sand mixtures of various combinations. Test results showed that if a small quantity of tire chips is mixed with sand, the static strength is influenced greatly.

1 INTRODUCTION

Currently, about 100 million scrap tires are generated in Japan every year as the automobile society develops, and the recycling of scrap tires has become significant. About half of the 88% of the recycled tires is used as fuel because it is cheaper than coal. However the process generates large amount of carbon dioxide and incineration ashes. If the use of waste tires as fuel is continued, it will lead to environmental problem in the near future. On the other hand, scrap tires provide numerous advantages from the viewpoint of civil engineering practices. They have light weight, high elastic compressibility, high vibration-absorption capacity, high hydraulic conductivity, and temperature-isolation potential. Therefore, scrap tire is gaining attention as new ground materials. It is thought that if the above-mentioned environmental problem is to be controlled, one piece can be used a lot of times, and efficiency would be good. Scrap tires can be used in several ways, either as whole, halved or even shredded.

As seen from past investigations, the effective use of waste tires as ground material has advanced greatly in the United States during the first half of the 1990s. For example, waste tires were used as road embankments (Bosscher et al., 1997) and as lightweight backfill in retaining walls (Lee et al., 1999). Furthermore, a standard has been provided on the use of old tires for engineering works through the ASTM Standards (ASTM, 1998). Various researches were also performed in other parts of the world. In recent years, the use of tires in bridge approaches was examined (Bergado et al., 2006), as well as in embankments and retaining walls (Humphrey et al., 2006) and its environmental impact (Tuncer et al., 2006). Although technological advancement in Japan regarding the effective use of old tires as new ground material is still in its early stage, researches on their application as tire chip-mixed solidified soil (Kikuchi et al., 2006), fill improvement (Mitarai et al., 2006) and earthquake-resistant reinforcement (Hazarika et al., 2006) have been actively conducted. In the future, it is necessary to accumulate information on the mechanical characteristics of tire chips for use as construction materials in soil structures, as well as to address stability concerns of tire chip-sand mixtures.

Table 1. Physical properties of soil samples.

Sand fraction	ρ_s (g/cm³)	ρ_{dmin} (g/cm³)	ρ_{dmax} (g/cm³)	e_{max}	e_{min}	D_{50} (mm)	U_c
sf = 1 (Soma sand)	2.645	1.273	1.574	1.077	0.680	0.395	1.65
sf = 0.9	2.576	–	–	–	–	0.399	1.67
sf = 0.8	2.498	–	–	–	–	0.403	1.67
sf = 0.7	2.410	0.939	1.234	1.565	0.953	0.407	1.69
sf = 0.6	2.309	–	–	–	–	0.414	1.72
sf = 0.5	2.192	0.744	0.988	1.948	1.218	0.423	1.75
sf = 0.3	1.892	0.563	0.735	2.361	1.576	0.453	1.91
sf = 0 (Tire chips)	1.150	0.347	0.442	2.318	1.600	0.655	2.72

Considering this background, a series of undrained and drained triaxial compression tests was conducted to understand the monotonic shear characteristics of composite materials containing tire chips and sand mixed at various proportions. Based on the results of the triaxial compression tests, a discussion is presented on the shear characteristics of tire chip-sand mixtures under various confining pressures.

2 MATERIAL USED AND EXPERIMENTAL METHOD

2.1 Physical properties of materials

In this research, the test specimen was made of two types of materials, Souma silica sand No.5 with revised grain size distribution and tire chips. The tire chips were derived from used tires, with metals and fibers removed beforehand, and processed into smaller pieces measuring 2 mm in diameter. The tire chips and Soma silica sand were mixed in various proportions, i.e., the mix ratios of sand to tire chips by volume were set at 10:0, 9:1, 8:2, 7:3, 5:5, 3:7 and 0:10.

Table 1 summarizes the physical properties of the soil mixture, including density of soil particles (ρ_s), minimum and maximum dry densities (ρ_{dmin} and ρ_{dmax}), maximum and minimum void ratios (e_{max} and e_{min}), mean diameter (D_{50}) and coefficient of curvature (U_c), respectively, of the samples used in the experiments. In the table, sf (sand fraction) indicates the proportion by volume occupied by Soma silica sand in the tire chip-sand mixture. Thus, $sf = 1$ indicates samples consisting of sand only, while $sf = 0$ represents sample with tire chips only. The density of the particles of tire chips is 1.15 g/cm³, which is relatively light compared to conventional geomaterials and represents only 2/5 of the particle density of the Soma silica sand. The maximum and minimum dry densities and maximum and minimum void ratios for the composite materials shown in Table 1 are summarized in Figure 1 in terms of their relation with sf. It is evident from the figure that the values of e_{max} and e_{min} show almost the same values within the range of $sf = 0$–0.3, and when $sf = 0.3$–1, the values of e_{max} and e_{min} decrease with increase in sf. On the other hand, both ρ_{min} and ρ_{max} increase in value with increase in sand fraction from $sf = 0$ (tire chip only) to $sf = 1$ (Soma Silica Sand No. 5 only). It is apparent that the lower the dry density of the mixture, the higher is the void ratio because the densities of particles of tire chips and sand differ widely; and this is one of the features of this soil mixture.

Figure 2 illustrates the grain size distribution curve for each sample type. Due to the difference in particle density between Soma silica sand No. 5 and tire chips, the soil mixtures with sand : tire chip mix ratio by volume of 10:0, 9:1, 8:2, 7:3, 5:5, 3:7 and 0:10, have the corresponding sand : tire chip mix ratio by dry unit weight as 100:0, 95:5, 90:0, 84:16, 70:30, 50:50, and 0:100, respectively. Therefore, even if 70% of the entire volume of the sample with $sf = 0.3$ consists of tire chips, the particle size characteristics (D_{50}, U_c) of the soil mixtures as shown in Figure 2 with percent finer by weight in the vertical axis, are much closer to those of pure Soma silica sand than for pure tire chips.

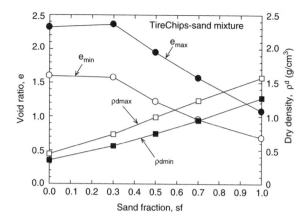

Figure 1. Relations between void ratio, dry density and sand fraction.

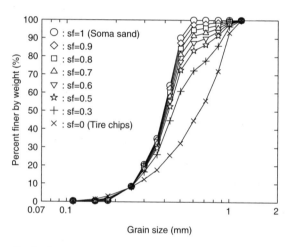

Figure 2. Particle size distribution curves.

2.2 Specimen preparation and experimental method

The tire chip-sand mixture specimen used for undrained and drained triaxial compression tests were prepared by moist tamping method. First of all, the tire chips were washed with a detergent to remove impurities that adhered to their surfaces, and then exposed to warm air to dry for two days. Drying inside an oven with constant temperature of 50°C was attempted, but oil began to ooze out of the tire chips and as a result, warm air was selected to dry them instead. The dried tire chips and Soma silica sand No. 5 were mixed at the prescribed mix ratio by volume. Water was added to the mixture to obtain a sample with initial water content $w = 10\%$ after which the sample was thoroughly mixed again. Membrane was installed in the pedestal of the triaxial apparatus, and the mold 10 cm high and 5 cm in diameter was set up. The test specimen was prepared by placing the soil mixture inside the mold in five layers, with each layer compacted at a prescribed number of times by dropping an iron rammer from a prescribed height to control the compaction energy, Ec, which is given by the following expression.

$$E_C = \frac{W_R \cdot H \cdot N_L \cdot N_B}{V} \qquad (1)$$

279

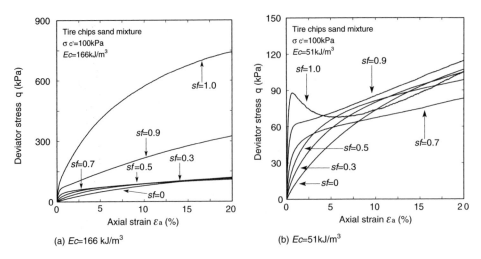

(a) Ec=166 kJ/m^3 (b) Ec=51 kJ/m^3

Figure 3. Undrained triaxial test results showing the relationships between deviator stress and axial strain at confining pressure $\sigma'_c = 100$ kPa.

In the above expression, W_R is the rammer weight ($= 0.00116$ kN), H is the drop height (m), N_L is the number of layers ($= 5$), N_B is the number of drops per layer, and V is the volume of mold (m^3). In the experiments, the test specimens were prepared by adjusting the height of drop H and the number of drops N_B in order to obtain two levels of compaction energy, $Ec = 51$ kJ/m^3 and 166 kJ/m^3. The above compaction energies were chosen such that the relative density of Soma silica sand No. 5 specimen ($sf = 1$) was $Dr = 20\%$ (for $Ec = 51$ kJ/cm^3) and $Dr = 50\%$ (for $Ec = 166$ kJ/cm^3). Because water content has a large effect on the compaction characteristics of soils, a constant initial water content and compaction energy were adopted, and test specimens of tire chip-sand mixtures were prepared with different sf. To saturate the specimens, the voids in the specimens were first filled with CO_2 and de-aired water was allowed to percolate, after which back pressure of 100 kPa was applied for two hours. As a result of this procedure, all test specimens were confirmed to have B-value ≥ 0.95. The saturated specimens prepared as outlined above were then isotropically consolidated at three levels of confining pressure of $\sigma'_c = 50$, 100, 200 kPa, and undrained and drained triaxial compression tests were conducted.

3 BEHAVIOR IN UNDRAINED AND DRAINED CONDITION

3.1 Behavior in undrained condition

Firstly, the results of undrained triaxial compression tests are discussed. Figures 3(a) and 3(b) show the relationships between deviator stress and axial strain for specimens with $Ec = 166$ kJ/m^3 and $Ec = 51$ kJ/m^3, respectively, while the corresponding effective stress paths are shown in Figures 4(a) and 4(b), respectively. Figure 3(a) shows that for compaction energy $Ec = 166$ kJ/m^3, the specimen which includes tire chips shows decreased strength as compared with specimen with $sf = 1.0$ (sand only). The mixtures with $sf = 0$–0.7 indicate nearly the same strengths. For the specimen containing 10% tire chips (i.e., $sf = 0.9$), its shear strength is about half of that of specimen with $sf = 1.0$. It is noted that the shear strength seem to decrease greatly by adding tire chips to sand, even with small quantity. On the other hand, Figure 3(b) showed that for compaction energy $Ec = 51$ kJ/m^3, specimen with $sf = 1.0$ (sand only) indicated strain softening similar to the behavior of loose sand. Comparing the mix ratios, it is seen that the more the value of sf rises, the more the strength increases. When Figures 3(a) and (b) are compared, the shear strengths when the axial strain reaches 20% are almost equal regardless of compaction energy. From Figure 4(a), it is observed that when

280

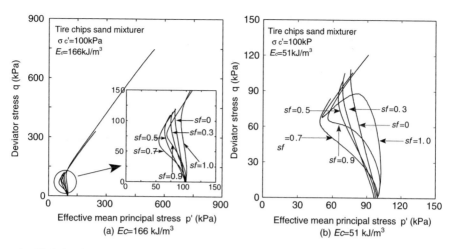

Figure 4. Undrained triaxial test results showing the relationships between deviator stress and effective mean principal stress at confining pressure $\sigma_c' = 100\,\text{kPa}$.

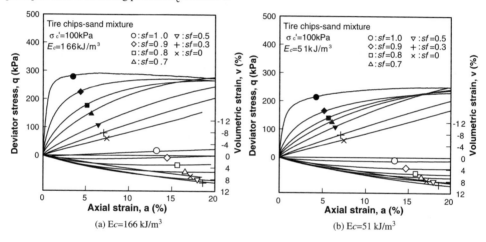

Figure 5. Drained triaxial test results at 100 kPa confining pressure.

$sf = 0.7$, contraction appeared because of negative dilatancy. This is because a part of the link of the grains of sand was cut off by the mixed tire chips, and it is believed that this caused the formation of weak sand structure. However, the contractive tendency deteriorated with the decrease in sand content and it seems that there is no volume shrinkage in the sample of pure tire chips ($sf = 0$).

3.2 Behavior in drained condition

Next, the results of drained monotonic triaxial tests are discussed. Figure 5(a) and (b) show the deviator stress-axial strain and volumetric strain-axial strain relations for specimens with $Ec = 166\,\text{kJ/m}^3$ and $Ec = 51\,\text{kJ/m}^3$, respectively. Figure 5(a) shows that when $sf = 1.0$, the peak deviator stress appears at an early stage of shearing followed by strain softening and the volumetric strain showed dilative tendency. However, when $sf = 0$ (tire chip only), the deviator stress – axial strain relation is virtually linear. Moreover, there is neither peak nor failure, even at 20% axial strain. Similarly, the plots for $sf = 0.3$–0.5 show strain hardening behavior. On the other hand, the inherent behavior of sand slowly appeared as the value of sf rises to 0.7. However, it is obvious from the figure that the behavior of specimen with $sf = 0.9$ is different from that of pure sand. It also noticed that when the axial strain reaches 20%, the shear strength of specimens with $sf = 0.7$–1.0 are almost

281

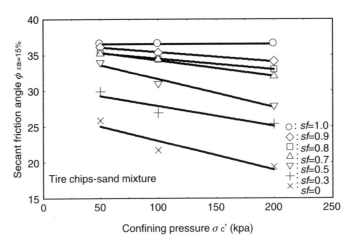

Figure 6. Influence of confining pressure on the secant friction angle.

equal. The volumetric strains of specimens which include tire chips develop a tendency to increase monotonously toward compression and a steady state condition was not reached when $sf \leq 0.7$.

Figure 5(b) shows the relationships between deviator stress and axial strain for specimens with $Ec = 51 \text{ kJ/m}^3$. It is noted that similar results were obtained as those with $Ec = 166 \text{ kJ/m}^3$, indicating that compression energy has no effect on specimens which include tire chips.

Figure 6 illustrates the relation between secant friction angle (at $\varepsilon_a = 15\%$) and confining pressure. From this figure, it can be seen that secant angle increases as the sand fraction increases. When the specimen contained even a small amount of tire chips, there is a decrease in secant angle. Moreover, the secant angle decreases further with the increase in confining pressure when $sf < 0.9$.

4   CONCLUDING REMARKS

In this research, undrained and drained triaxial tests were performed on tire chip-sand mixtures in order to examine their monotonic shear characteristics. The main findings obtained from the test results are as follows.

1. The specimen made of pure tire chips showed linear stress- strain relation and the volumetric strain showed compressive behavior.
2. In the undrained tests, samples containing tire chips showed decrease in strength. Moreover, negative dilatancy was maximum when $sf = 0.7$.
3. The drained test results showed that as sf increased, the rigidity of the specimen also increased. As the volume of tire chips in the sand mixture increased, the secant friction angle showed large decrease with increase in confining pressure.

REFERENCES

ASTM 1998. Standard practice for use of scrap tires in civil engineering applications, *Annual Book of ASTM Standards*, ASTM, 22 (6): 501–520.
Bergado, D.T. & T. Tanchaisawat, P. Voottipruex and T. Kanjananak., Reinforced lightweight tire chips-sad mixture for bridge approach utilization, *Journal of international workshop at IW-TDGM*, 41–55.
Bosscher, P.J., Edil, T.B. and Kuraoka, S. 1997. Design of highway embankments using tire chips, *Journal of Geotechnical and Geoenvironmental Engineering*, 123 (4): 295–304.
D.N. Humphrey ., Tirederived aggregate as lightweight fill for embankments and retaining walls, *Journal of international workshop at IW-TDGM*, 56–79.

Hazarika, H., Sugano, T., Kikuchi, Y., Yasuhara, K., Murakami. S., Takeichi, H., Ashoke, K.K., Kishida T. and Mitarai, Y. 2006. Evaluation of a recycled waste as smart geomaterial for earthquake reinforcement of structure, *Proc., 41st Japan National Conference on Geotechnical Engineering*, 591–592.

Kikuchi, Y., Nagatome, T. and Mitarai, Y. 2006. Failure mechanism during shear of rubber chip-mixed solidified soil, *Report of the Port and Airport Institute*, 45 (2): 87–103.

Lee, J. H., Salgado, R., Bernal, A. and Lovell, C. W. 1999. Shredded tires and rubber-sand as lightweight backfill, *Journal of Geotechnical and Geoenvironmental Engineering*, 125 (2): 132–141.

Mitarai, Y., Kawai, H., Kishida, T., Nagatome, T., Yasuhara, K., Murakami, S., Sugano, T., Hazarika, H., Kikuchi, Y., Tatarazako, N., Takeichi, H. and Ashoke, K.K., Damping capability of impact load by used tire chips, *Proc., 41st Japan National Conference on Geotechnical Engineering*, 595–596.

Tuncer B.Edil., A review of environmental impacts and environmental applications of shredded scrap tires, *Journal of international workshop at IW-TDGM*, 1–16.

Investigation of fracture behavior of cement stabilized soil with tire shreds

Y. Nakamura
Graduate School of Science and Technology, Kumamoto University, Kumamoto, Japan

Y. Mitarai
Toa Corporation Research and Development Center, Yokohama, Japan

J. Otani
Graduate School of Science and Technology, Kumamoto University, Kumamoto, Japan

ABSTRACT: In Japan, total amount of a million tons of scrap tires have been discharged in every year and the rate of recycling this material is about 90%. As far as the technique of recycling this scrap tire is concerned, there are thermal recycle, material recycle, that of reuse and so on. Recently, material recycle has been one of key issue on this problem, because of the effectiveness of recycling capacity and less environmental impact. As a material recycle, there are the methods of using tire itself and shredding tire as a large number of pieces.

The objective of this paper is to develop a new geomaterial using tire shreds with cement stabilized soil. The authors have done the research on the evaluation of deformation property of cement stabilized soil with adding tire chips. From results of authors' previous studies, it was found that it is possible to improve ductility of cement stabilized soil by adding tire chips in the volume of about 10 to 30%. However, the fracture property of such composite material has not been clarified yet. In this paper, the localized failure behavior of this material is investigated using large size rubber balls instead of tire shreds in order to make the behavior clear. Here, the industrial X-ray CT is used for visualizing the fracture behavior of the stabilized soil with adding rubber balls. And in order to evaluate the effect of deformation property with mixing granular-elastic material such as tire shreds and rubber balls, the stabilized soil with adding glass balls as rigid materials are also examined for the comparative study. Consequently, it is clearly shown that the fracture of the cement stabilized soil with adding tire balls is generated by detachment of rubber balls from the cement stabilized soil from the experimental results and those of image analysis of CT images.

1 INTRODUCTION AND OBJECTIVE

In Japan, total amount of a million tons of scrap tires have been discharged in every year and the rate of recycling this material is about 90%. From some yeas ago, the authors have studied on useful recycling method of scrape tire as a civil engineering material or geomaterial. From these results, as the one of the useful recycling method, it was found that it is able to improve the deformation property of cement stabilized soil by adding some tire shreds (Mitarai. et al., 2006).

In this study, it has conducted some unconfined and tri-axial compression test on the cement stabilized soil with adding tire shreds (or tire chips) with X-ray CT scanning. Consequently, it was obtained the ductility of stabilized soil is able to improve by adding tire shreds in the volume of about 10 to 30% as following. (1) The larger the percentage of tire chips in the mixture is, the lager the failure strain becomes. (2) In the case of the only stabilized soil nothing with adding tire shreds,

Table 1. Mix proportion of cement stabilized soil.

	soil particle	sea water	cement	total
Weight (g)	341	875	79	1295
Volume (cm³)	126	850	25	1000

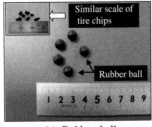

| (a) Rubber ball | (b) Glass ball | (c) Tire chips |

Figure 1. Photograph of the mixture materials.

stress-strain curve of tri-axial compression test shows a sharp strain softening after peak strength point. On the other hand, in the case of adding tire shreds in volume of 10 to 20%, stress-strain curve shows a gradually strain hardening and at after the peak strength point, it shows a gradually strain softening. (3) From the CT-images of specimen during tri-axial test, it was able to see that it is able to prevent strain and localizing and growing a large crack in the specimen during compressive shearing by adding tire shred.

Moreover, in order to evaluate the effect of deformation property with mixing granular-elastic material such as tire shreds, the improved soil with adding silica sands as rigid materials and with adding EPS beads as elasto plasticity material, have been are also examined for the comparative study. From this, it was evaluated that improving deformation property of stabilized soil is generated by elasticity of tire shreds (Mitarai et al., 2006-2 & Nagatome et al., 2007).

In this paper, the localized failure behavior of this material is investigated using large size rubber balls instead of tire shreds in order to make the behavior clear. Here, the industrial X-ray CT is used for visualizing the fracture behavior of the stabilized soil with adding rubber balls. And in order to evaluate the effect of deformation property with mixing granular-elastic material such as tire shreds and rubber balls, the stabilized soil with adding glass balls as rigid materials are also examined for the comparative study.

2 MATERIALS AND TEST PROCEDURE

2.1 Materials and specimen

In this study, the stabilized soil was mixing dredged soil and cement, and its mix proportion was same in each cases. The dredged soil was Tokyo Bay clay ($\rho_s = 2.716\,\mathrm{g/cm^3}$, $w_L = 100\%$, and $Ip = 70$). Its percentage of fine-grained fraction is about 90%. The water content of it was controlled to 280% by mixing seawater. And it was used normal Portland cement. The target strength of cement stabilized soil was $q_u = 400\,\mathrm{kN/m^2}$ (curing 28days), its mix proportion is shown in Table 1.

The composite material was made this stabilized soil and some mixture material. The volume percentage of mixture is same in each case, which is f = 16.7%. Mixture material is tire chips ($\rho = 1.15\,\mathrm{g/cm^3}$, average size is 2 mm), rubber ball ($\rho = 1.26\,\mathrm{g/cm^3}$, uniform size, $\phi = 8\,\mathrm{mm}$) and grass ball ($\rho = 2.531\,\mathrm{g/cm^3}$, uniform size, $\phi = 8\,\mathrm{mm}$)used. The hardness number of tire chips is $H_d = 60$–70 and that of rubber ball is $H_d = 50$–70, then both hardness is about similar. The each photograph is shown in Figure 1. The size of specimen is $\phi = 50\,\mathrm{mm}$ and H = 100 mm.

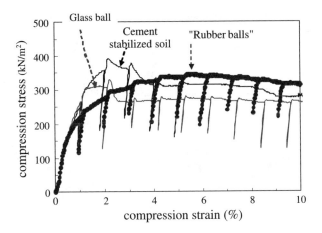

Figure 2. Stress-strain relationships (1).

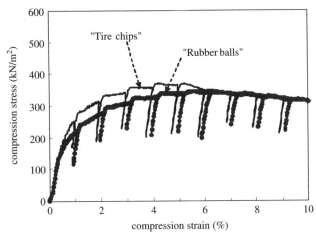

Figure 3. Stress-strain relationships (2).

2.2 *Test procedure*

Tri-axial compression test was conducted under condition which back pressure was 100 kPa, confining pressure was 200 kPa (more than 2 hour), and undrained shear was conducted under constant lateral pressure and constant speed of 0.3% strain per minute.

In this study, in order to scan CT-image of specimens during compression test, it was developed the spatial apparatus (Mitarai et al., 2007). Then, it is able to scanning at random time under compression test. And in this study, CT scanning was conducted at once stopping shear motor and rocking loading shaft.

3 RESULTS

3.1 *Stress and strain relationship*

Figure 2 and Figure 3 are shown the stress and strain relationship of tri-axial compression test. Figure 2 is comparison the case of cement "satirized soil (noting with mixture)" with" rubber balls" and "glass ball". And Figure 3 is comparison "tire chips" with "rubber balls".

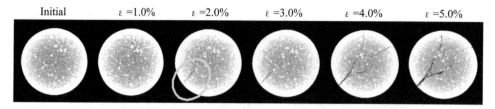

Figure 4(a).   A series of CT-images of "cement stabilized soil" from "Initial" to "5%strain".

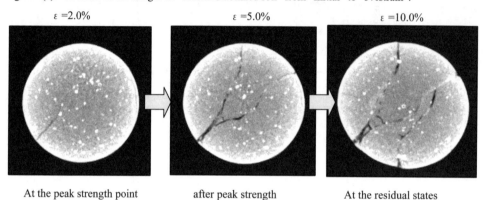

At the peak strength point   after peak strength   At the residual states

Figure 4(b).   Extensional CT-images of "cement stabilized soil".

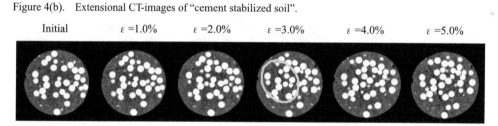

Figure 5(a).   A series of CT-images of "glass ball" from "Initial" to "5%strain".

From Figure 2, in the case of "cement stabilized soil" and "glass balls", the peak strength is shown at about 2%strain and after this point(failure), their stress-strain curve show strain softening. On the other hand, in the case of "rubber balls", the stress-strain curve shows gradual strain hardening and a peak strength strain is lager than other cases and after this point, it shows very gradual strain softening.

From Figure 3, in the case of "tire chips" and "rubber balls", their stress-strain relationship shows almost similar. Then,it is scarcely effective in difference of shape and size of mixtures. It is expected that because, rubber is hydrophobic material, the bonding rubber (tire chips and rubber ball) and stabilized soil is very small.

Therefore, it is implied that at the viewpoint of stress-strain relationship, the ductility of stabilized soil is improved by elasticity of mixtures (tire chips and rubber balls), and it is scarcely effective in difference of these shape and size.

3.2   Comparison of deformation condition by CT-images

Figure 4 to Figure 7 shows the CT-images of each case, which are cross section of specimens' center part. And these are image from 0% strain (=Initial) to 10% strain.

In both case of "stabilized soil" and "glass ball", it is able to see cracks at 2% or 3% strain which is a peak strength point or after it point. After this point, during increasing of strain, the crack is

288

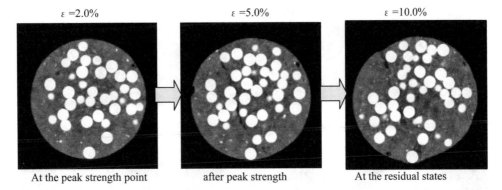

ε =2.0%   ε =5.0%   ε =10.0%

At the peak strength point   after peak strength   At the residual states

Figure 5(b).   Extensional CT-images of "glass ball".

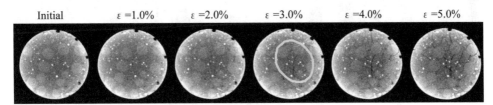

Initial   ε =1.0%   ε =2.0%   ε =3.0%   ε =4.0%   ε =5.0%

Figure 6(a).   A series of CT-images of "Rubber ball" from "Initial" to "5% strain".

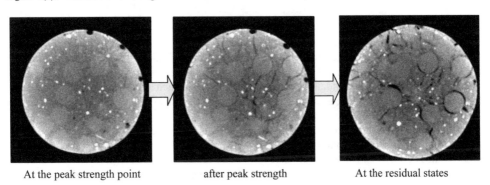

At the peak strength point   after peak strength   At the residual states

Figure 6(b).   Extensional CT-images of "Rubber ball".

growing its length and width. Then, it is expect that the localizing of stress and strain is progressed in the specimen with compressive shearing.

On the other hand, in the case of "tire chips", it is impossible to see visual crack of CT-images during compressive shearing, in spite of over the peak strength point about 5% strain.

In the case of "rubber balls", it is able to see small crack between rubber balls at 3% strain. After this point, other any small cracks are observed in another position with compressive shearing, but localized large crack is not observed as the case of "stabilized soil" and "glass ball". And it is observed that the each crack is thin and stopped at rubber balls. The detachment of rubber balls' surfaces from stabilized soil matrix is not able to see until 5% strain which is the peak strength point. But over this point, the detachment is gradually observed. Then, it is supposed that the strain softening is observed by occurring and growing the detachment between both materials.

In the case of "tire chips", it is not able to see any crack, in spite of same percentage of mixtures. The size of mixtures is small and the distance of each particle is close, then, it is supposed that the crack and detachment is small or thin. Therefore, it is not able to see by CT-images.

Initial ε =1.0% ε =2.0% ε =3.0% ε =4.0% ε =5.0%

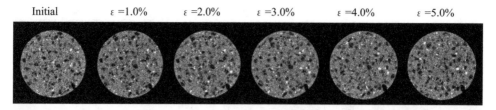

Figure 7(a). A series of CT-images of "Tire chips" from "Initial" to "5% strain".

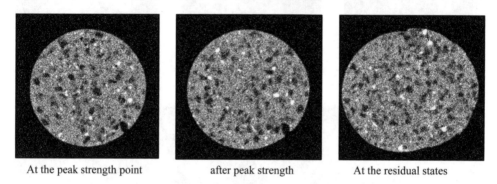

At the peak strength point   after peak strength   At the residual states

Figure 7(b). Extensional CT-images of "Tire chips".

4 CONCLUSIONS

In this study, the failure properties of cement stabilized soil with adding some mixture are discussed based on tri-axial compression test with using X-ray CT scanning. And in order to evaluate the effect of deformation property with adding granular-elastic material such as tire shreds and rubber balls, the stabilized soil with adding glass balls as rigid materials are also examined for the comparative study.

Consequently, it is shown that the improvement the ductility of the cement stabilized soil with adding granular-elastic material is generated by elasticity of mixtures, and it is scarcely effective in difference of shape and size of mixtures.

Moreover, it is clearly shown that the fracture of the cement stabilized soil with adding tire balls is generated by detachment of rubber balls from the cement stabilized soil.

REFERENCES

Desurues, J., Viggiani, G. and Besuelle, P. (2006): *International Workshop Advances in X-ray Tomography for Geomatrials*, ISTE.

Kikuchi, Y., Nagatome, T., Mitarai, Y. and Otani, J. (2006): Engineering Property Evaluation of Cement Treated Soil with Tire Chips using X-ray CT Scanner, *Proc. of 5th ICEG Environmental Geotechnics* Vol. 2, pp. 1423–1430.

Kikuchi, Y. (2006): Investigation of Engineering Properties of Man-made Composite Geo-matrials with Micro-focus X-ray CT, *International Workshop Advances in X-ray Tomography for Geomatrials*, pp. 53–78.

Otani, J. and Obara, Y. (2004): *International Workshop on X-ray CT for Geomaterials –GeoX2003 –*, A.A. Balkema.

Mitarai, Y., Yasuhara, Y., Kikuchi, Y. and Ashoke K. K. (2006): Application of the Cement Treated Clay with Added Tire Chips to the Sealing Materials of Coastal Waste Disposal Site, *Proceeding of 5th ICEG Environmental Geotechnics*, Vol. 1, pp. 757–764.

Scrap Tire Derived Geomaterials – Opportunities and Challenges – Hazarika & Yasuhara (eds)
© 2008 Taylor & Francis Group, London, ISBN 978-0-415-46070-5

Floating structure using waste tires for water environmental remediation

S. Horiuchi, T. Odawara, S. Yonemura, Y. Hayashi, M. Kawaguchi & M. Asada
(Shimizu Corporation)

M. Kato & K. Yasuhara
(Ibaraki University)

ABSTRACT: Geo-materials, such as structural fill and road basement, are expected to be the most suitable target to increase waste utilization, because an enormous quantity could be usable in the earth-works. Waste tire itself might be harmless to the environment, and its stable shape and high durability are the points for the usage. If it could be usable for the environmental cleanup, double merits are available to the society. Sake for the water cleanup, the authors designed a floating geo-structure using waste tire; poly-urethan foam was placed around tire and several grasses were planted inside the void filled with artificial light-weight soil. After four years pre-investigation and a series of safety checks, large size floating structures were placed onto ponds. Observation through the investigations shows a high possibility of this structure for the water cleanup, as follows:

(1) all the grasses planted glow healthy assimilating nutriments from pond without fertilizer,
(2) roots of the grasses spread in the pond, and provide suitable space for fish,
(3) many birds use this structure for their resting and hunting purpose,
(4) biological chain could be provided with this structure, and,
(5) cost of this structure is 1/2 of the conventional products.

1   INTRODUCTION

"If you know your enemies and know yourself, you will win hundred times in hundred battles." This is the famous phrase in " the art of war " written by Sun-tzu, If we apply this phrase for the waste tire utilization as geomaterial, its application corresponds to the war, requirement for the application is the enemy, and waste tire property is yourself. What is the point of the waste tire? Low cost? Stable shape? Chemical stability of rubber? Light-weight? Tensile strength? When waste tire was cut into chips, many merits might be diminished, especially in the cost. This is why the waste tires still being stocked in backyard. Usages in uncut should be investigated.

Photo 1 shows our example for slope coverage; an uncut tire structure was applied in MSW ash landfill site for slope liner coverage, where 23,800 waste tires were used for 7,300 m². As shown in photo 2, a lot of plants are growing even in an early spring. This inspired us an environmental usage; floating structure using waste tires for water environmental remediation. Photo 3 is a prototype unit of the floating structure. Figure 1 shows the cross section of the structure; poly-urethan foam was placed in hexagonal shape around waste tire and several grasses were planted inside its void filled with artificial light-weight soil. Three years experiment of this unit showed an adequate performance, however, environmental impacts of this structure have not been investigated. In this paper, environmental performances of larger size floating structures are reported.

Photo 1.   Slope coverage in MSW ash landfill.

Photo 2.   Plants inside tire coverage.

Photo 3.   Prototype of floating unit.

Figure 1.   Cross section of floating structure

Table 1.   Inner aggregate evaluation; chemical and biological checks using their leachates

Light-weight aggregate	Cr (VI)	B	F	Se	Cu	Zn	Bio-assay
Volcanic stone	<0.005	<0.02	–	–	–	–	–
Cemented coal flyash	<0.005	0.15	0.2	<0.002	<0.01	<0.01	+
Calcinated waste soil	<0.005	<0.02	0.1	<0.002	<0.01	<0.01	+
Rubber chip	<0.005	<0.02	0.1	<0.002	<0.01	0.16	+
Calcinated rock	<0.005	<0.02	0.2	0.005	0.06	0.01	–
Calcinated coal ash	<0.005	<0.02	0.5	<0.002	<0.01	<0.01	–
Environmental standard	0.05	1	0.8	0.01	–	–	–

2   SELECTION OF INNER SOIL

A lot of light-weight aggregates could be usable for the inner soil, however, their negative effect should be severely checked especially on their harmful effluents before the application. Table 1 shows the list of light-weight aggregates and the results of checks using their leachates. All the aggregates pass the environmental regulation. On a biological check using daphnia magna, however, a slight negative effect was detected on three aggregates. As seen in Tab.1, zinc and boron might be the factor of cemented coal flyash and rubber chip. In case of calcinated waste soil, other unknown substances would affected the result.

From the tests on leachates, calcineted rock and calcinated coal ash were selected for the inner aggregates. For the environmental usages, chemical and biological checks are essential before the application, as seen on Table 1.

292

Photo 4.   Planting work on the structure.   Photo 5.   Final view on the game fishing pond.

Photo 6.   Roots of plants after 6 months.

3   FLOATING STRUCTURE ON A GAME FISHING POND

A floating structure of 61 units was introducing on a trout game fishing pond. Photo 4 shows the planting work in the pond. Several grasses, including typha-angustifolia, Japanese iris and watercress, were planted inside of the void space. Final view is shown in Photo 5, where the floating structure covers 2.5% of the pond and it is not adequate for the algae bloom prevention however might be effective on sight remediation, and was friendly to the fish at least.

In this pond, all the game fish gathered around the floating structure. All the game fishers have to through their lures near the structure and a lot of lures were caught by the structure! So, the game fishing in this pond needed higher skills than before.

All the grasses planted glow well assimilating nutriments from the pond. Some birds were coming to this structure for their resting and hunting.

Six months after the introduction, this structure was removed from this pond. During the removing work, shown in photo 6, the effectiveness of this floating structure for the environmental cleanup was investigated. The growth of plants was the first point for the water cleanup. The next is their roots. As seen in photo 6, plants' roots spread widely and could be providing a better water space for small fish and planktons, which ate organic substances from the pond. Between the aggregates in this structure, a lot of pond snails were found. This is the third effect. Because no pond snails were introduced into the structure, this large propagation means that the inner space is very favorable for these creatures. Moreover, there exists a water cleaning system by these creatures.

In this system, some predators are needed for the cleanup. For the plants, human being should cut out from the pond, and birds are effective for fishes and pond snail.

The cost for the floating structure could be saved 50% than the conventional system, because of the waste tire usage.

Photo 7.   Floating structure in biotope.

Photo 8.   Flowers on the structure.

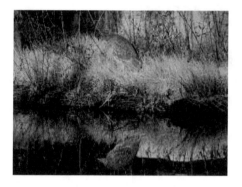

Photo 9.   Night heron on the structure.

Photo 10.   Little-egret on the structure.

Photo 11.   Spotbill ducks beside the structure.

Photo 12.   Spotbill duck in their nest.

4   FLOATING STRUCTURE ON A BIOTOPE

A floating structure of 18 units was installed on a biotope. Photos 7 and 8 show its view. In midsummer, lythrum blooms beautifully on the structure. From winter to spring, a lot of birds came onto the structure for resting, hunting and breeding, as shown in photos 9–12. Birds coming to this structure bring out their foods from the pond. Field observation confirms that insects and fishes also play their role on the environment.

A small floating structure composing only 18 units gives a lot of effect on providing food chain in this biotope.

5 CONCLUSIONS

The observation through two kinds of floating structures using waste tire shows a high possibility for the water cleanup, as follows:

(1) all the grasses planted glow well assimilating nutriments from the pond,
(2) roots of the grasses spread into the pond could provide suitable space for small fish,
(3) inner space of aggregate provides a better space for pond snails,
(4) a lot of birds came to the structures for their resting and hunting,
(5) biological chain could be provided, and,
(6) photogenic view could be presented with this low-cost floating structure.

Scrap Tire Derived Geomaterials – Opportunities and Challenges – Hazarika & Yasuhara (eds)
© 2008 Taylor & Francis Group, London, ISBN 978-0-415-46070-5

An experimental study on two types of foundation block supported on soft soil

Y. Shimomura & N. Sako
Junior College of Nihon University, Chiba, Japan

Y. Ikeda
Taisei Corporation, Tokyo, Japan

M. Kawamura
College of Industrial Technology, Nihon University, Chiba, Japan

S. Ishimaru
College of Science & Technology, Nihon University, Tokyo, Japan

ABSTRACT: This study focuses on the progress of the damping and mitigation performance of foundations that will be built at soft ground sites, and proposes an improved foundation technique which implements backfilling of a damping composite material into trenches dug along a foundation area. The damping material is a mixture of asphalt with crushed stones and scrap tire chips. Forced vibration tests were conducted to confirm the effectiveness of the proposed foundation technique on two types of foundation blocks, that is, CF and IF; the former was constructed by a conventional construction method and the latter applied the above improved foundation technique. Comparing with the response of the both foundations, it is confirmed that the damping material adopted in the IF provides good attenuation and mitigation performance.

1 INTRODUCTION

According to investigative reports of The Hyougoken-Nambu Earthquake in 1995 and The Mid Niigata prefecture Earthquake in 2004, it has been recognized that structures that adopt seismic isolation devices possess remarkable ability concerning the prevention of not only loss of human life but also damage of structures and facilities. However, base-isolated structures can only prevent destruction due to earthquakes when they are constructed on sites having good soil condition. Therefore, foundation improvement work is indispensable for base-isolated structures that will be built on sites having soft ground conditions.

This study focuses on the progress of the damping and mitigation performance of foundations that will be built at soft ground sites, and proposes an improved foundation technique which implements backfilling of a damping composite material into trenches dug along a foundation area. The damping material is a mixture of asphalt with crushed stones and scrap tire chips.

To comprehend the attenuation ability of the improved foundation technique and to verify the effectiveness of this work, we carried out forced vibration tests on two test foundation blocks; one was constructed by a conventional construction technique and the other employs the above mentioned improved foundation technique. Here, the conventional construction technique is a procedure that involves backfilling of dug soil into trenches dug along a foundation. Hereafter, the improved foundation and conventional construction techniques are called IF and CF, respectively. We performed the 3-dimensional simulation analyses adopting a hybrid approach in which two

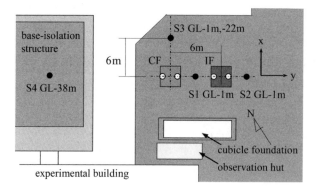

Figure 1.   Schematic view of the test site.

foundation blocks and the adjacent soil regions are modeled by 3-dimensional finite elements. The thin layer element method is applied to the free field ground region that surrounds the above finite element domain.

After the final forced vibration test, we continue seismic observations at the test site. Previously we have obtained many earthquake records, i.e., The Mid Niigata prefecture Earthquake in 2004, The Northwest Chiba prefecture Earthquake in 2005 and so on. In order to confirm the attenuation performance of the damping mixture, analyses of earthquake observation records and their simulation analyses were conducted.

2   EXPERIMENTAL SITE AND FOUNDATION BLOCKS

Figure 1 shows the layout of our experimental site. The experimental site is located at a vacant lot to the north of the experimental building at the Funabashi Campus of Nihon University in Chiba Prefecture, near Tokyo, Japan. The soil profile of the site is illustrated in Figure 2. The ground surface layer 2.8 m and shallower is loamy in the Kanto district, and partially includes backfilling soil 0.1 m deep. Cohesive soil distributes in the range of G.L.-2.8 m to G.L.-3.5 m. Tuffaceous fine sand can be seen in the range of G.L.-3.5 m to G.L.-5.3 m. Silty fine sand and fine sand alternate G.L.-5.3 m and deeper.

The conventional foundation block (CF) had been constructed in August, 2003 (Ishimaru et al., 2004), and the improved foundation block (IF) was built one year after (Ikeda et al., 2005). The IF was constructed 6 m to the east of the CF, both foundations had been supported by four soil cement columns and the existing soil layers. The tips of the improved soil cement columns are located at G.L.-2.8 m, where loam and cohesive soil alternate, and its value for the standard penetration test is five and under.

3   MATERIAL TEST RESULTS

3.1   *Soil cement columns*

The test block and the soil cement columns are shown in Figure 3. The dimensions of both blocks are 2.4 m × 2.4 m × 1.0 m. The diameter of a soil cement column is 600 mm and its length is 2.5 m. Two core rods, 90 mm in diameter and 2.5 m in length, were obtained from a soil cement column which had a material age of 28 days. Figure 4 shows the density and the unconfined compressive strength distributions against depth. The densities distributed uniformly against variations in the depth and were in the range of 1.44 to 1.52 g/cm^3. In comparison with both results of the CF and IF, the unconfined compressive strengths of all the specimens of the latter were slightly smaller than that of the former, and the deformation moduli (E_{50}) and densities of the two test results were

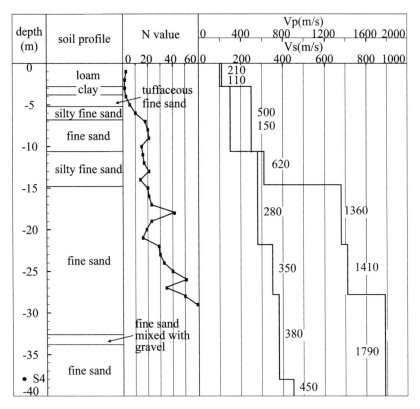

Figure 2. Soil profile of the test site.

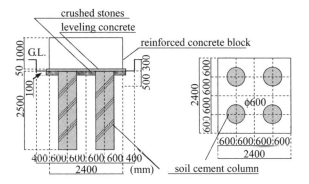

Figure 3. Foundation block and soil cement columns.

almost identical. Properties obtained by the unconfined compressive strength tests are shown in Table 1.

3.2 Scrap tire chips mixed with asphalt and crushed stones

In order to obtain basic material features of the damping material paved into the trenches dug along IF, cyclic triaxial tests have been conducted. Diameters and lengths of the specimens are about 100 mm and 190 mm, respectively. Figure 5 shows examples of hysteresis loops of deviator stresses and axial strains of the damping material for variations of mixed rates of scrap tire chips

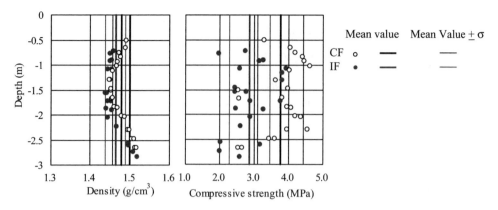

Figure 4.   Density and unconfined compressive strength distributions against depth.

Table 1.   Results of unconfined compressive strength tests of soil cement columns.

Test results	Deformation modulus E_{50} (MPa)	Density ρ (gr/cm^3)	Unconfined compressive strength q_u (MPa)
CF (conducted in 2003)			
Mean value	1462	1.48	3.77
Standard deviation	200	0.02	0.67
IF (conducted in 2004)			
Mean value	1456	1.46	2.86
Standard deviation	130	0.02	0.59

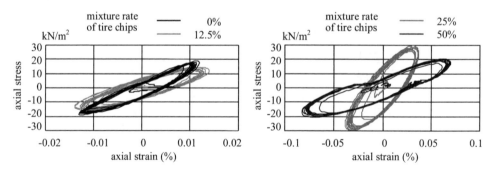

Figure 5.   Hysteresis of axial stresses and strains at various mixture rates of scrap tire chips.

to crushed stones. The former particle sizes are less than 10 mm and the latter are in the range of 5.0 mm to 2.5 mm. The mixed rates of the chips to crushed stones are defined by the ratio of the mass of the tire chips to the total mass (the tire chips and crushed stones). Compared with zero mixed rates of the tire chips, the initial secant moduli of the damping mixture of 12.5% mixed rates were undiminished and its equivalent damping constants increased. Finally, we selected the damping mixture of 12.5% to backfill the trenches dug along the IF. Blending of the damping mixtures is illustrated in Table 2. Figure 6 represents the result of the experiments for various load levels with 50 kPa of confining pressure and 0.1 Hz of frequency. At a small strain level (0.01% and below), it is found that the equivalent Young's moduli and the equivalent damping constants of the damping mixture are about 120 MPa and 20%, respectively.

Table 2. Blending of mixture of asphalt with crushed stones and scrap tire chips (mixed rate: 12.5%).

	Mass ratio rate (%)	Density (gr/cm^3)	Mass (gr)
Crushed stones	58.6	2.7	1983.6
Tire chips	8.4	1.2	284.4
Slow curing	12.0	2.7	406.0
Fine sand	11.0	2.7	372.0
Filler	10.0	2.7	338.0
Total of aggregate	100.0		3383.0
Asphalt	7.5	1.0	274.3
Total (aggregate + asphalt)			3657.3

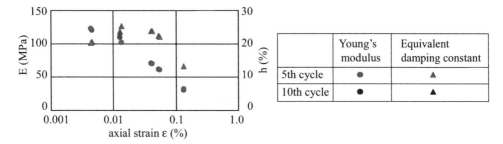

Figure 6. Strain dependence on equivalent Young's moduli and damping constants of specimens.

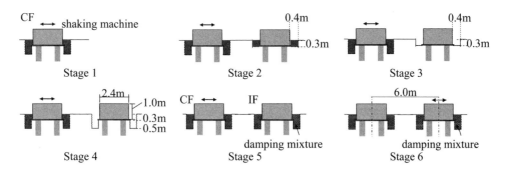

Figure 7. Schedule of forced vibration tests.

4 FORCED VIBRATION TESTS

4.1 Test schedule

To confirm the progress of attenuation performance of the IF, we carried out forced vibration tests under various conditions on the trenches dug along the IF. Figure 7 shows the schedule of the forced vibration tests. The trenches excavated along the CF, whose width and depth are about 0.4 m and 0.8 m, respectively, were backfilled with the dug soil to the ground level. Conditions of the trenches excavated along the IF from Stage 2 to Stage 6 are as follows. First, we backfilled the trenches along the IF with the dug soil to the ground surface level (Stage 2). Then, we excavated the trenches 0.3 m deep (Stage 3). Next, we made the trenches 0.8 m deep (Stage 4). At Stage 5 we filled in the trenches with the proposed damping mixture and carried out a forced vibration test to confirm the mitigation ability of the damping mixture for the oscillation of the IF. The vibration

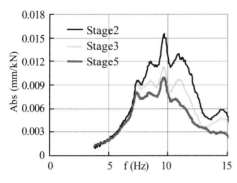

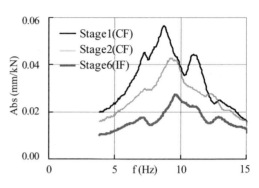

Figure 8.　Resonance curves of IF in stage 2, 3 and 5.

Figure 9.　Resonance curves of CF in stage 1, 2 and IF in Stage 6.

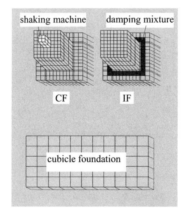

Figure 10.　Models of hybrid approach.

generator was mounted on the CF until Stage 5. We remounted the shaking machine from the CF to the IF and conducted an experiment focusing on the attenuation performance of the damping mixture (Stage 6).

4.2　Tests results

While the CF was oscillated by the shaking machine, forced vibration tests (from Stage 2 to Stage 5) relating to various conditions of the trenches dug along the IF, were carried out. Figure 8 shows amplitude functions of the response displacement curves per unit exciting force for the exciting direction on the upper surface of the IF at Stage 2, 3 and 5. At the peak frequencies, magnitude of amplitudes of the IF at Stage 5 is 15%–30% smaller than that at Stage 2. It was expected that the damping mixture would play an important role in progress of mitigation ability of the structures.

Figure 9 shows displacement resonance curve of the IF at Stage 6, the CF at Stage 1 wherein the CF only existed and at Stage 2. Compared with the results at Stage 1, it can be seen that the resonance frequency at Stage 2 shifts to the high frequency range and the maximum response value is reduced. It is reasoned that the rigidity of the surface soil became high by aging and cross interaction effects. The amplitude at Stage 6 at the peak frequency, 9.6 Hz, is reduced by about 35% compared with the test results of Stage 2. It is herein indicated that the damping mixture backfilled into the trenches dug along the foundation has good attenuation performance.

Table 3. Properties of numerical analysis.

Depth (m)	Vs (m/s)	ρ (t/m^3)	Poisson's ratio	Damping ratio
1.0	90	1.4	0.311	0.03
2.7	90	1.4	0.311	0.03
5.4	150	1.6	0.451	0.03
10.6	280	1.7	0.372	0.02
14.7	280	1.7	0.478	0.02
21.7	350	1.7	0.465	0.02
27.8	380	1.8	0.461	0.02
38.0	450	2.0	0.466	0.02
45.2	420	2.0	0.461	0.02
Material	Vs (m/s)	ρ (t/m^3)	Poisson's ratio	Damping ratio
Foundation blocks	–	2.40	0.20	0.0
Leveling concrete	–	2.40	0.20	0.0
Damping mixture	163	1.68	0.35	0.20

———— CF (test)   ———— IF (test)
———— CF (simulation)   ———— IF (simulation)

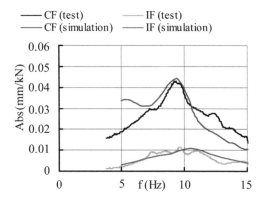

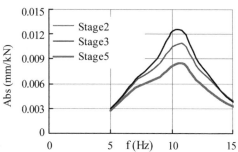

Figure 11. Comparison of response curves of both foundations by test and simulation in stage 2.

Figure 12. Comparison of response curves of IF by simulation in stage 2, 3 and 5.

5 SIMULATION ANALYSIS OF FORCED VIBRATION TESTS

After the experiments, we carried out the 3-dimensional simulation analyses for Stages 2, 3, 5 and 6 using a hybrid approach (Ikeda et al. 2003). In the hybrid approach, both the foundations and soil regions surrounding the foundations, including trenches, are modeled by the 3-dimensional finite element. The thin layer approach is applied to the free field ground regions surrounding the above finite element domain. We also model a cubicle (a power transformation vessel) foundation. Figure 10 shows a bird's-eye view illustration of the hybrid model in Stage 5. Table 3 shows the properties of soil, both foundation blocks and the damping mixture.

Figure 11 illustrates the resonance curves of displacement per unit exciting force, for the exciting direction, on the upper surface of the CF and the IF by the experiments and analysis at Stage 2. The first peak frequency and its amplitude of resonance curves of the CF by the analysis are in good agreement with the experiment results. The amplitude of the resonance curve of the IF by the analyses is reasonably close to the average of the test results, and both frequency dependencies are approximately identical. Figure 12 displays the analysis results of the resonance curves concerning the displacement of the IF at Stages 2, 3 and 5. The tendency expressed by the analysis results is

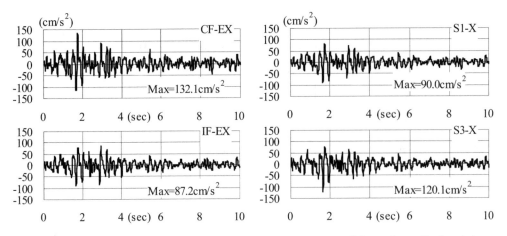

Figure 13.   Examples of acceleration time history records (The Northwest Chiba prefecture Earthquake).

that the amplitude of the response curves becomes small at the peak frequency when the depth of the trench excavated around the foundation block is shallow. The amplitude at the peak frequency at Stage 5 is reduced by about 25% compared with that at Stage 2. This indicates that the attenuation effect of the mixture of damping materials is higher than that of backfilling soil, and the simulation analysis results agree with the test results.

6   ANALYSIS OF EARTHQUAKE OBSERVATION RECORDS

6.1   *Seismograph installation at experimental site*

Figure 2 indicates that the shear wave velocity at G.L.-38 m beneath the base-isolation structure has exceeded 400 m/s. Therefore it is recognized that this point is at the engineering bedrock. Here, we call the engineering bedrock point S4. At the experimental site, we have installed eight seismographs. Two seismographs have been mounted on the eastern and western sides on the surface of each foundation. Three seismographs near both foundations are located at G.L.-1.0 m. Each seismograph, shown in Figure 1, is located on the centerline of the CF and/or the IF; at the midpoint between the CF and IF (S1); 3 m, east from the center of the IF (S2); and 6 m, north from the center of the CF (S3). Another seismograph is installed at G.L.-22 m under the S3.

6.2   *Time histories*

After the final experiment, we have continued earthquake observations at the experimental site. We previously obtained many earthquake records including The Mid Niigata prefecture Earthquake in 2004 and The Northwest Chiba prefecture Earthquake in 2005. The Mid Niigata prefecture Earthquake struck Niigata prefecture on the evening of October 23rd, 2004. It was M_{JMA} 6.8 earthquake. This was the most significant earthquake to inflict heavy damage on Japan after The Hyougoken-Nambu Earthquake in 1995. We obtained records of the main shock wave of the earthquake and its aftershocks. The epicentral distance from the epicenter to the experimental site at Funabashi city in Chiba prefecture is about 230 km. The Northwest Chiba prefecture Earthquake hit the metropolitan area on the evening of July 23rd, 2005, and was an earthquake which exposed the inability of urban sites to withstand earthquake disaster. In addition, it was the earthquake which had the largest acceleration observation recorded in these ten years at our Funabashi campus.

As an example of acceleration time history records, Figure 13 shows the principal motion parts of the main shock records of The Northwest Chiba prefecture Earthquake at the two foundations, and the soil region near the foundations illustrated in Figure 1.

Table 4. Catalog of earthquake records.

No.	Occurrence time (Y/M/D/H/m)	Epicenter	Depth (km)	Magnitude	Epicentral distance (km)	I_{JMA}
1[*A]	2004/10/23/17/56	Chuestu of Niigata pref.	13	6.8	220	III
2[*B]	2004/10/27/10/41	Chuestu of Niigata pref.	12	6.1	200	II
3[*C]	2005/07/23/16/35	Northwest of Chiba pref.	73	6.0	17	IV
4[*D]	2005/07/23/16/42	Northwest of Chiba pref.	69	4.2	13	I

[*A] The main shock of The Mid Niigata prefecture Earthquake.
[*B] An aftershock of The Mid Niigata prefecture Earthquake.
[*C] The main shock of The Northwest Chiba prefecture Earthquake.
[*D] An aftershock of The Northwest Chiba prefecture Earthquake.

Table 5. Peak accelerations and spectrum intensities.

Earthquake

	Component Direction Foundation	Sway X CF	IF	IF/CF	Y CF	IF	IF/CF	Rocking Y CF	IF	IF/CF
No. 1	PA (cm/s²)	28.53	24.61	(0.86)	33.29	27.05	(0.81)	3.78	1.90	(0.50)
	SI (cm/s)	3.73	3.53	(0.95)	3.64	3.47	(0.95)	0.27	0.14	(0.53)
No. 2	PA (cm/s²)	10.99	10.62	(0.97)	7.53	6.93	(0.92)	0.83	0.64	(0.77)
	SI (cm/s)	1.60	1.53	(0.95)	1.56	1.49	(0.96)	0.08	0.04	(0.50)
No. 3	PA (cm/s²)	130.41	88.83	(0.68)	97.57	84.61	(0.87)	11.59	5.55	(0.48)
	SI (cm/s)	1.60	1.53	(0.95)	1.56	1.49	(0.96)	0.08	0.04	(0.50)
No. 4	PA (cm/s²)	14.48	12.71	(0.88)	11.10	10.02	(0.90)	2.07	1.25	(0.60)
	SI (cm/s)	0.73	0.61	(0.84)	0.56	0.50	(0.90)	0.09	0.05	(0.62)

The catalog, the two earthquake records and their aftershocks, is shown in Table 4. Table 5 shows the peak accelerations (PA) and the spectral intensities (SI) (after Housner, 1952), which are estimated by the records observed from both foundations. The sway component represents an average transverse acceleration of two observation points at each foundation and the pseudo rocking component is a rotational acceleration calculated by difference between two vertical records. In terms of the SI that represents an input earthquake energy considering period characteristics of structures, all the SI values of the IF are smaller than those of the CF. In particular, the rocking components of the IF become considerably smaller than the sway components of the IF. This indicates that the proposed damping mixture positively affects the mitigation performance of foundations against earthquake destructivity.

6.3 Transfer functions

Figure 14 displays transfer functions of the CF and the IF to the S4. These transfer functions are calculated by the main shock of The Mid Niigata prefecture Earthquake of 2004. Peak frequencies that can be seen in the transfer functions for x and y directions are the dominant frequencies of the soil region (from the first to the fourth modes), which are above to G.L.-38 m. For x direction, the transfer functions of both foundations are approximately the same. Conversely, for y direction, the transfer function of the IF is smaller than that of the CF in the frequency range of 9.0 Hz and above. This tendency is found in reference to not only the main shock but also aftershocks.

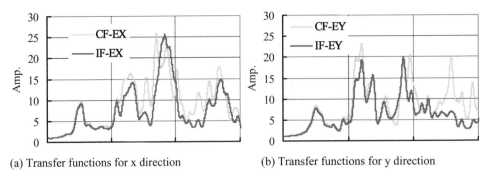

(a) Transfer functions for x direction   (b) Transfer functions for y direction

Figure 14. Transfer functions of CF and IF to S4.

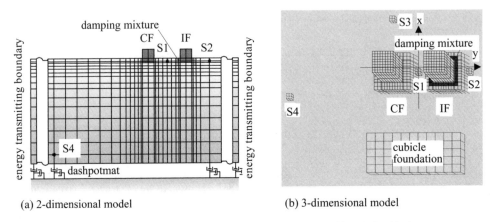

(a) 2-dimensional model   (b) 3-dimensional model

Figure 15. Two models for simulation analyses of earthquake response of foundation blocks.

7 SIMULATION ANALYSIS OF EARTHQUAKE OBSERVATION RECORDS

7.1 *Simulation analysis models*

To grasp the fundamental vibration characteristics of earthquake records of the CF, the IF and the soil region surrounding both foundations, simulation analyses employing the 2-dimensional finite element approach have been carried out. In the 2-dimensional analysis modeling, we have taken into account the y direction that is equal to a parallel direction of the CF and the IF. The S1 and S2 in the soil region near the two foundations and the S4, which locates G.L.-38 m beneath the base-isolation structure, are also modeled. The 2-dimensional model is shown in Figure 15(a). The transfer functions of the two foundations and ground points to the steady state harmonic incident wave defined at the bottom of the dashpot mat underneath the analysis model are evaluated. Table 3 indicates the shear wave velocities of each stratum of the model.

The 3-dimensional analysis using the hybrid model, which has already been utilized for simulation analysis of the forced vibration tests, is also applicable in the simulation of earthquake responses (Shimomura et al., 2005). Figure 15(b) shows the 3-dimensional analysis model. The CF and IF, the S1, S2 and S3 in the soil region near the foundations, the basement of the cubicle, and the S4 are considered the same as the simulation of forced vibration tests. We calculate transfer functions of the foundations or the soil points to the S4 to compare with observation records as well as the 2-dimensional analyses. In the 3-dimensional analysis, to obtain appropriate results, we have taken into account the responses of y direction that are induced not only by the y component but also by the x component of earthquake input motions (Ikeda et al., 2004).

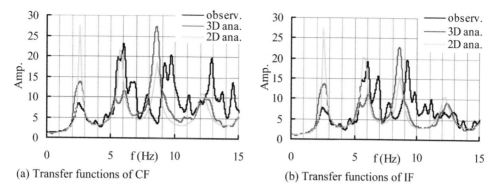

(a) Transfer functions of CF

(b) Transfer functions of IF

Figure 16.   Comparison of transfer functions of CF and IF of observations and analyses for y direction.

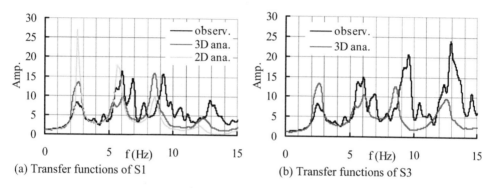

(a) Transfer functions of S1

(b) Transfer functions of S3

Figure 17.   Comparison of transfer functions of S1 and S3 of observations and analyses for y direction.

7.2   *Comparison with observation records and analysis results*

Figure 16 illustrates the transfer functions of both foundations to the S4 calculated by the main shock of The Mid Niigata prefecture Earthquake in 2004. The 2-dimensional analysis and 3-dimensional analyses estimate the peak frequencies of the third mode lower than the observation. According to transfer functions of the observation, amplification of the first mode is smaller than the 2-dimensional analysis, the amplification of the first mode of the 3-dimensional analysis agrees well with the observation. It was found that the third mode's amplification of the IF of the observation is smaller than that of the CF. By results of the 3-dimensional analyses, the same trend can be seen at the first coupled mode of the foundations and the soil region. This might be caused by attenuation performance of the damping mixture. As a result, the transfer functions of both analyses correspond approximately with the observation. In particular, it is confirmed that the 3-dimensional analysis provides more detailed features of the earthquake records than the 2-dimensional analysis except the second dominant frequency range of the ground.

Figure 17 illustrates the transfer functions of the ground points near of the foundation blocks to the S4 estimated by the main shock of The Mid Niigata prefecture Earthquake in 2004. Because there is no significant difference recognized between S1 and S2, only S1 and S3 are shown in the figure. As shown in Figure 17(a), the transfer function of the S1 estimated by the 2-dimensional analysis disagrees with that of the observation. The 2-dimensional and 3-dimensional analyses can simulate approximately the four dominant frequencies of the earthquake. The first mode's amplitude of the 2-dimensional analysis is considerably higher than both the observation and the 3-dimensional analysis. The 3-dimensional analysis simulates the tendency that the fourth mode's amplitude of the S3 is larger than that of the S1.

307

8 CONCLUSIONS

We have proposed an improved foundation technique which is a procedure of backfilling a newly-adopted damping mixture of asphalt with crushed stones and scrap tire chips into the trenches dug along foundations to enhance the attenuation and mitigation performance of structures. According to the results of cyclic triaxial tests, it was found that the equivalent Young's moduli and the equivalent damping constants of the damping material, at the strain level of 0.01 percent and under, became approximately 120 MPa and 20 percent, respectively.

After verification of the high damping ability of the proposed mixture by way of cyclic triaxial tests, forced vibration tests on the CF and the IF were conducted. By the results of the experiment and the simulation analyses, it was confirmed that the damping mixture backfilled into the trenches dug along the IF provided favorable attenuation and mitigation performance. The amplitude of the displacement function on the IF at the peak frequency, 9.6 Hz, was reduced by approximately 35% compared with that of the CF. The evaluation of the spectral intensity of the observation records indicated that the proposed foundation technique would effectively reduced the severe damage of earthquake destructivity. The hybrid approach, used herein, was applied as a viable method to conduct the simulation analyses of not only the forced vibration tests but also the earthquake response.

It is found that the attenuation ability of the proposed damping material during earthquakes is less than at forced vibration tests. This is the reason that the earthquake input motions from the side surfaces of the CF and IF were insufficient due to small foundation block size. So, we are currently executing seismic response analysis of full-scale foundation configurations.

ACKNOWLEDGEMENTS

This research was supported by Grant-in-Aid for Science Research (B), Project No.15360303, 2003–2006, Project No.19360254, 2007, the Ministry of Education, Culture, Sports, Science and Technology, Japan. Further, the research was conducted as a part of the Academically Promoted Frontier Research Program on "Sustainable City Based on Environment Preservation and Disaster Prevention" at Nihon University, College of Science and Technology (Head Investigator: Prof. Ishimaru, S.) under a grant from the Ministry of Education, Culture, Sports, Science and Technology, Japan. We would like to express our sincere appreciation to the members concerned.

REFERENCES

Housner, G. W., 1952. Intensity of ground motion during strong earthquakes, *Caltech Earthquake Engineering Research Laboratory Technical Reports, 5–18.*

Ikeda, Y., Shimomura, Y., Adachi, H., Ogushi Y. and Nakamura M. 2003. An experimental study of mock-up pile foundations Part 2 Dynamic cross interaction of foundations, *Trans. SMiRT17, Paper No. K05-6(CD), August 2003.* Prague, Czech Republic.

Ikeda, Y., Shimomura, Y., Nakamura, M., Haneda, O. and Arai, T. 2004. Dynamic influence of adjacent structures on pile foundation based on forced vibration tests and earthquake observation. *Proc. 13th WCEE. Paper No. 1869, August 2004.* Vancouver, Canada.

Ikeda, Y., Shimomura, Y., Kawamura, M. Ishimaru, S. 2005. A study on damping and mitigation performance of side surfaces of foundations on soft soil – Part1 Forced vibration tests of foundation block with various embedment conditions. *Proc. ISEV2005, 17–23, September 2005.* Okayama, Japan.

Ishimaru, S., Hata, I., Shimomura, Y., Ikeda, Y., Ishigaki, H., Ogushi, Y. 2004. A feasibility study of new type seismic isolation – Composed system of piles covered by pipes and dampers with partial soil improvement -. *Proc. 13th WCEE. Paper No. 2204, August 2004.* Vancouver, Canada.

Shimomura, Y., Ikeda Y., Kawamura M. and Ishimaru S. 2005. A study on damping and mitigation performance of side surfaces of foundations on soft soil – Part 2 Analysis of records of The Mid Niigata prefecture Earthquake in 2004 and its simulations, *Proc. ISEV2005. 25–30, September 2005.* Okayama, Japan.

The experience of long-term performance of reinforced soil structures

A. Zhusupbekov & R. Lukpanov
Eurasian National University, Astana, Kazakhstan

ABSTRACT: The priority task of the development of modern construction is improvement the reliability and longevity of building materials along with economical effectiveness which satisfy mass high volume growth in the term of progressive intensification of constructions. Geosynthetic reliability and durability criterion under the interest of engineers, and reinforced soil model is one of the progressive solution of engineering.

1 INTRODUCTIONS

As it is known reinforcement means to use special elements in soft soil constructions which allow increase the mechanical property of soil. Dealing with soil the reinforced elements redistribute load among construction parts, providing the transmission of stress from the overloading zone to the adjacent underloading zone.

Nowadays there are a lot of different reinforcing materials in a world practice. The most part of them consists of geosynthetic - woven or not woven materials on base of synthetic polymer fiber which are made of polypropylene (PP) or polyester (PET) (F. Tatsuoka etc.). Geosynthetics for soil reinforcement might be as volumetrically geogrid, flat geogrid or geotextil according to their assignments (Figure 1). Although there are many cases when composite materials by combining geogrid with geotextile methods have been used. Geosynthetic material is produced from fabric by the method of needle punching, which provides its high chemical inertness against acid and alkaline, stability against termooxidizely process. The material is fast against ultra-violet rays and it is although green product. Physical and mechanical properties of geosynthetic are shown in Table 1.

Under the highest possible loading, geosynthetic has till 45 percent elongation. It depends on the applicable thickness of material. In this way local damages do not lead to the destruction of materials. Due to the high index of elastic modules, the material can bear considerable load, implementing function of reinforcement at not great deformation.

(a) Geotextile

(b) Volumetrically geogrid

(c) Flat geogrid

Figure 1. Types of geosynthetic reinforced materials.

Table 1. Physical and mechanical properties of geosynthetic.

Characteristics	Geotextile (PET)	Geotextile (PP)	Composite
Surface density, g/м²	250	250	250
Tensile strength, кN\м²	4,2	2,8	8,4
Thickness at the load 2 MPa, мм	3,2	3,2	3,2

Figure 2. Construction of retaining wall within the reinforcement model.

Choice of reinforced material does not depend on its characteristics of strength. The polymer which is produced from reinforced material has substantial degree. By way of illustration geosynthetic made from polypropylene is used in dynamic loading as polypropylene has high index of creep, that is it has ability to long the term extension under the dead load. Therefore the material is used in road building in pavement capacity. Geosynthetic, produced from polyester, with very low index of creep, usually is used in case of static load or exists probability uneven development of settlement in the result of heterogeneousness soil. As example, we can give retaining wall, strengthening of embankment, reinforcment the heterogeneousness soils which have very low index of bearing capacity.

2 REINFORCEMENT MODEL

2.1 Construction of retaining wall within the reinforcement model

Construction of retaining wall within the reinforcement model is usually used to strengthen slope covers of railways and highways in bridge abutments, foundations of different constructions (Figure 2).

As tests have shown the destructing load for these types of constructions exceeds the design load. This is explained as that geotextile possesses high index of tensile strength and follows for deformation of soil of the construction creating general state of stress and increasing construction stability (E.C. Shin etc.). Model of geotextile retaining walls and consisting reinforced elements are given in Figures 3, 4.

Except geosynthetic material to other materials can be used make reinforcement. So during the pullout test chains from non-rusting steel have considerable figures of resistance (Table 2, Figure 5).

The Pullout resistance by chain reinforcement can be defined by the following equation:

$$F_{tc} = F_1 + F_2 + F_3 \qquad (1)$$

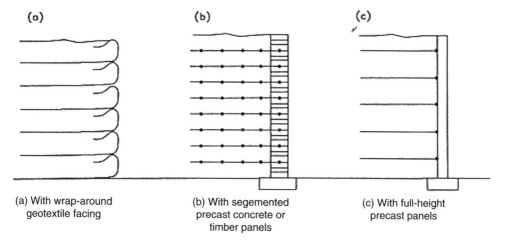

(a) With wrap-around
geotextile facing

(b) With segemented
precast concrete or
timber panels

(c) With full-height
precast panels

Figure 3.   Reinforced retaining wall system using geotextile.

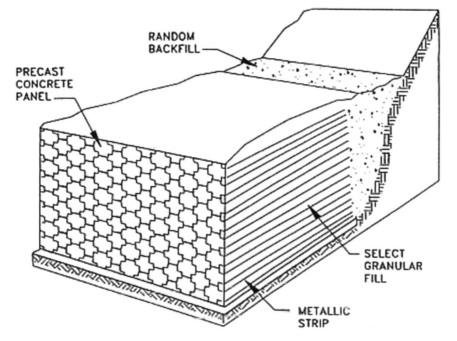

RANDOM
BACKFILL

PRECAST
CONCRETE
PANEL

SELECT
GRANULAR
FILL

METALLIC
STRIP

Figure 4.   Component parts of a reinforced earth wall.

where F_1 = the frictional force between chain and soil skeleton; F_2 = the shearing resistance with including the soil inside the chain; F_3 = the passive resistance in cross sectional area of chain.

The earth pressure resistance of horizontal bar is defined as $F_{ri} \cdot F_{bi}$ is the pullout force with a L type angle.

2.2 *Reinforcement of road building*

In road building reinforcement fulfills the function of layer separation. This permits to increase the index of bearing capacity largely due to of its stress redistribution. By way of illustration – The

311

Table 2. The results of pullout test with different chain lengths.

Length of chain	Vertical pressure (kgf/cm)	Pullout force (kgf)						Sum (kgf)	
		F	F	F	F	F	F	F + F	F + F
2.0 m	0.4	90.93	68.20	81.15	200.27	108.69	360	308.96	560.27
	0.8	101.85	136.39	162.30	400.53	184.45	720	584.98	1120.53
	1.2	152.78	204.58	243.44	600.80	260.21	1080	861.01	1680.80
	0.4	63.49	85.02	101.17	249.68	108.69	360	358.37	609.68
2.5 m	0.8	126.98	170.04	202.34	499.37	184.45	720	683.82	1219.37
	1.2	190.47	255.06	303.51	749.05	260.61	1080	1009.26	1829.05
3.0 m	0.4	76.06	101.85	121.19	299.10	108.69	360	407.79	659.10
	0.8	152.12	203.70	242.39	598.20	184.45	720	782.65	1318.20
	1.2	228.17	305.55	363.58	897.30	260.61	1080	1157.51	1977.30

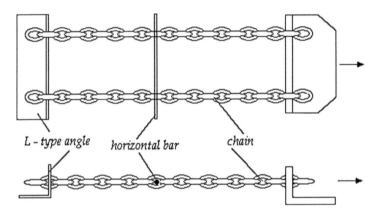

Figure 5. Type of chain reinforcement.

model of reinforcement installation of "The new western road" project (city of Atyrau, Kazakhstan, 2003). In the result of the research, which was held on road building "The new western road" project we want to say that is very difficult to compact natural soil to required coefficient of compaction because the natural soil (loamy soil) has very low index of bearing capacity. To increase of durability and deformation property of road basement the model of reinforcement with the following steps were decided to choose.

1. The natural loamy soil is compacted by road-roller to ultimate level according to required standards. Evening of surface (Filling the pits, pot-holes and another local damage where the water may stay for a long period)
2. Installation of reinforced material (Figure 6)
3. Filling of the soil (Figure 6), with height no more that 200 mm, and its compaction to required coefficient of compaction standard.

The benefit of reinforcement was determined by examine of surface during three years service. The economical efficiency diagram which has been determined by comparing appearance of pits, pot-holes represent on Figure 7.

Initially the reinforced pavement cost more but after a certain period of time the reinforced pavement is a lower total cost.

Figure 6. Installation of reinforced soil of "The new western road" object (Atyrau city, Kazakhstan, 2003).

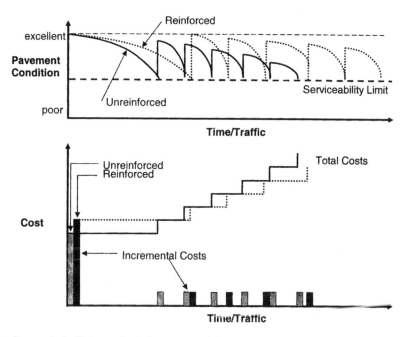

Figure 7. Economical efficiency of reinforcement model.

The next research represents that the effective work of reinforced materials depends on its shape of geosinthetic (Figure 8) besides its type (PET, PP).

Efficiency of geogrid application serviceability with comparing geotextile is represented in Figure 9. For the initial data the appearance of serviceability pits and pot-holes were considered.

However there exists several variations (Figure 10) of choosing reinforced materials for soft soil condition, the final selection is based on technical and economical comparing.

Consequently one of the traditional types of road construction – asphalt pavement has the best characteristics of serviceability but not perfect. Working in various temperature and considerable dynamic load influence lead to the appearance of cracks because of low index of

313

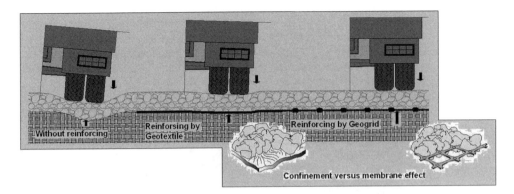

Figure 8. The work of reinforcement model.

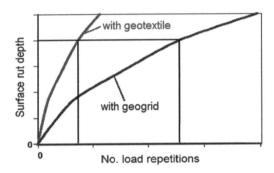

Figure 9. Efficiency of geogrid application serviceability.

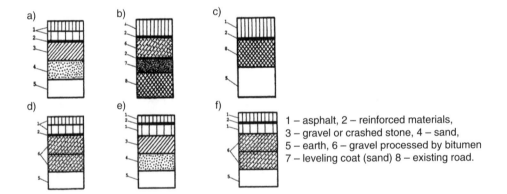

Figure 10. Type of road reinforcement.

asphalt tensile strength. Even the low level of tensile load leads to appearance of crack and decreases serviceability properties and durability of asphalt pavement. Therefore the most progressive solution, based on durability and reliability of construction which excepts such problems, is reinforcement. Influence of reinforced geogrid of asphalt pavement samples are given in Figure 11.

Usually in contrast in non-reinforced asphalt pavement samples where we can see big cracks appear than small distributed cracks will appear in reinforced sample.

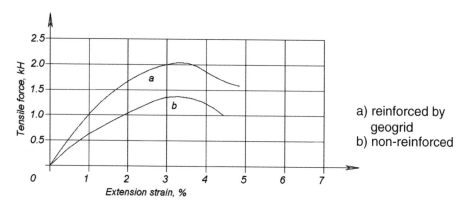

Figure 11.   Dependence extension strain different asphalt pavement from tensile force.

3   CONCLUSION

From the point of economical and technical expediency the reinforcement application is conformed by its wide usage in developed countries of the world and the base of its successful application that will provide to increase its serviceability road period for two times.

As the results of research work show that the application of reinforced construction will be proved from the economical point in case if that height of retained construction are higher than three meters. The price of one meter reinforced wall with reinforcement is cheaper for 2 or three times than the price for one meter reinforced concrete.

ACKNOWLEDGEMENTS

The authors deep thank Professor E.C.Shin (Incheon University, Korea) for his advising and consulting of parts of this research work.

REFERENCES

Shin, E.C. & Young, I. O. 2006. Case Histories of Geotextile Tube Construction Project in Korea. In E.S. Shin & J.G. Kang etc. (ed.), *New Developments in Geoenvironmental and Geotechnical Engineering; Proc. intern. symp., 9–11 November 2006, Incheon*, Korea: IETeC.
Tatsuoka, F., Tatcyama, M. and Koseki, J. 1995. Performance of soil retaining walls for railway embankment. *A Special Issue of Soils and Foundation on Geotechnical Aspects, 1 / January 1995.*

Part 2 Case studies, design methods
and field applications

Using waste tire-soil mixtures for embankment construction

A. Edinçliler
Boğaziçi University, Kandilli Observatory and Earthquake Research Institute,
Department of Earthquake Engineering, Istanbul-Turkey

ABSTRACT: Waste tires are processed to form tire shreds, tire crumbs and tire chips. Another form of waste tire is tire buffings. Tire buffings are the by-product of tire retreading industry. The purpose of using tire wastes is to modify the properties of soil as an additive, rather than using it as fill material. In this study, large-scale direct shear tests were performed to determine the shear strength parameters and deformation behavior of tire buffings-sand mixture having 5%, 10%, 20%, and 30% tire buffings by weight. The laboratory test results show that ten percent tire buffings by weight addition to sand alters the deformation behavior of the mixture by stiffening the material at low strains and softening the mixture at large strains. The contribution of tire buffings addition to sand is mainly observed at low confining stress. Also, a brief literature study on tire shred, tire chips and tire buffings is summarized.

1 INTRODUCTION

Waste tires are an environmental and financial burden in many regions of the world. There are more than 500 million tires stockpiled across the United States, and 270 million more are generated each year. In Canada, this figure is about 28 million (Dickson et al. 2001). About 30% of these waste tires are stocked in landfills (Yoon et al. 2005). Such wastes cannot generally be deposited in landfills since they require large spaces. Recycling and reuse of these waste materials, is increasing worldwide, especially in highway construction. Construction of highways requires large volumes of construction material, so highway agencies are frequent participants in efforts to recycle and reuse waste materials.

Using the tire wastes in highway construction is becoming more popular due to shortages of natural mineral resources and increasing waste disposal costs. Proper utilization of waste and by-product materials in transportation applications requires experience and knowledge regarding the use of these materials. Properties of waste tires such as durability, strength, resiliency, and high frictional resistance are of significant value for the design of highway embankments. Using these wastes in construction requires an awareness of the properties of these materials and the limitations associated with their use.

Two techniques to incorporate waste tires in subgrade/embankment are use of shredded tires as a lightweight fill materials and use of whole tires or their sidewalls for soil reinforcement in embankment construction (Ahmed & Lovell 1992).

When roads are constructed on soft compressible soils, stability and settlement considerations are critical. At locations having poor subgrade support, it is recommended to use alternative lightweight fills such as geofoam, wood chips, tire wastes instead of conventional materials. To reduce the weight of the highway structure at such locations, lightweight materials such as scrap tires, tire chips, tire crumbs and tire shreds are used as a replacement for conventional materials.

Civil engineering applications for scrap tires include lightweight fill, conventional fill, retaining wall and bridge abutment, insulation layer, and drainage applications (Young et al. 2003). In other application waste tires are processed to form tire shreds, tire crumbs and tire chips. These products are being utilized as lightweight aggregates. Another form of waste tire is tire buffings. In contrast to tire chips, crumbs or shreds, tire buffings are the byproduct of tire retread industry. The contact surface of worn tires are stripped off and resurfaced with rubber. The fiber shaped tire buffings are produced during stripping process. Its production requiresminimal energy.

The production of tire shreds or tire chips involves primary and secondary shredding. Shredded waste tires are now being used as subgrade reinforcement for constructing road over soft soils, as aggregate in leach beds for septic systems, as an additive to asphalt, as a substitute for leachate collection stone in landfills, and as sound barriers.

The mixture of tire wastes with soil for embankment construction may not only provide alternative means of reusing tires to address economic and environmental concerns, but also help to solve geotechnical problems associated with low shear strength (Zornberg et al. 2004). It should be noted that in the design of geotechnical projects where tire shreds and tire chips are used either alone or mixing with soil, generally stability and serviceability are of great importance. For service, the compressibility may be more-critical particularly when greater shred and tire contents are used and subjected to low pressures. Edil & Bosscher (1994) have demonstrated that tire shreds and tire shred mixtures are highly compressible at low normal stresses.

The nominal size of the tire shreds depends on the design of the shredding machine (Bosscher & Edil 1995). Tire shreds have various shapes and sizes, typically varying between 50 and 300 mm (ASTM D 6270-98). Processing whole tires through one cycle of shredding produces only large size tire shreds that are on average 150–300 mm long. The tire shreds used in the construction of the embankment were predominantly in the range of 150–300 mm-nominal maximum sizes. The unit weight of different types of compacted tire shreds, as reported in the literature, ranges from 2.4 to 7.0 kN/m^3 (Ahmed & Lovell 1993, Humphrey et al. 1993). These values are approximately 0.1–0.4 times the unit weight of typical soils. Tire shreds have been used either separately or mixed with soil. Mixing of tire shreds with soil reduces both the compressibility and the combustibility of tire shred fills.

Tire chips are pieces of scrap tires that have a basic geometrical shape (i.e., rectangles and squares) and are generally between 12 and 50 mm in size and have most of the wires removed (Federal Highway Administration 1998). According to Humphrey (1999), using tire-chips in civil engineering applications are advantageous because of their low density, high durability, high thermal insulation and in many cases least cost compared to other fill materials.

Previous studies have mainly concentrated on determining engineering properties of pure tire-chips and/or various mixtures of tire-chips with sand as a lightweight fill material (Ahmed 1993, Edil & Bosscher 1994, Humphrey 1995, Masad et al. 1996, Wu et al. 1997, Tatlisoz et al. 1998, Lee et al. 1999, Yang et al. 2002, Youwai & Bergado 2004). They concluded that tire-chips and sand mixtures can be used as a lightweight fill material behind retaining structures and highway embankments over weak or high compressibility soils.

Embankments constructed with soil-tire chip mixtures can potentially have steeper slopes because the backfill has higher shear strength and lower unit weight. Steeper side slopes decrease the volume of material needed. Also, because of using lightweight material, settlement of underlying soil is reduced (Tatlisoz et al. 1998).

Use of tire buffings only for embankment construction will not be feasible due to comparatively lower shear strength of the material. Due to its fiber shape and smaller size, it can be used to modify the properties of soil. The objective of the present study was to evaluate the shear strength properties of the tire buffings and tire buffings-sand mixtures at various compositions and to investigate the potential use of the sand-tire buffings mixtures. A large-scale direct shear testing device was used to determine the shear strength parameters and deformation behavior of tire buffings, tire buffings-sand mixture having 5%, 10%, 20%, and 30% tire buffings by weight (STB5, STB10, STB20 and STB30, respectively). The purpose of using tire buffings was to modify the properties of soil as an additive, rather than using it as a fill material for highway embankment construction.

2 BACKGROUND

2.1 *Shear strength properties of the waste materials*

It is well known that the use of inclusions (or reinforcements) may be used to improve the mechanical properties of earth structures Traditional soil reinforcing techniques involves the use of geosynthetics, metal inclusions oriented in a preferred direction may be used to enhance the stability of the soil. In some cases, randomly distributed inclusions are used to improve the mechanical properties of the composite materials. Gray & Ohashi (1983) performed direct shear test on sand reinforced with both natural and synthetic fibers. Their results showed that fiber reinforcement increases the peak shear strength values.

Shear strength is a fundamental mechanical property that governs fill stability design. The behavior of highway embankments under static loading may be predicted and modeled by using shear strength and deformation parameters obtained from large-scale laboratory tests. In the limit equilibrium analysis the ultimate load leading to failure is determined from the strength parameters. In the elastic analysis the stress strain behavior measured from the laboratory tests are used to predict the response of embankment under static loading.

In the study conducted by Humphrey et al. (1993), tire chips obtained from three different suppliers were used for the large-scale direct shear tests. The tire chip lengths smaller than 72 mm was used in the tests. They have reported friction angles ranging between 19° and 25° and cohesion of 7.7–8.6 kPa. They stated that tire chips are useful in constructing lightweight embankments over soft soils. In addition, tire chips can be used to replace natural aggregate, to improve drainage, end to provide thermal insulation.

Triaxial tests were conducted by Wu et al. (1997) on small tire chips (<40 mm long) to determine the shear strength of five processed scrap tire products having different gradations and particle shapes. They obtained that all five tire chip products have ultimate internal friction angles of 45° to over 60°.

Large–scale triaxial tests by using pure tire shred, pure sand, and tire shred-sand specimens were performed to evaluate the optimum tire shred content and aspect ratio used in tire shred-sand mixtures (Zornberg et al. 2004). They reported that the shear strength increases with increasing tire shred content, reaches a maximum for a tire shred content value in the vicinity of 35%, and then decreased for tire shred contents beyond this value. They concluded that the earth structures under comparatively low confining pressures (e.g. low embankments and retaining walls) can benefit significantly from the addition of tire shreds.

Tatlisoz et al. (1998) conducted the large-scale direct shear tests with tire chips, sand, sandy silt, sand-tire chips and sandy silt-tire chip mixtures. They reported that the shear strength of the sand-tire mixtures increases with increasing tire chip contents of up to 30% by volume. In contrast, the friction angle of the sandy silt-chip mixtures is nearly independent of tire chip mixtures. However the shear strength of the sandy silt-tire chip mixtures increases with tire chip content primarily due to an increase in apparent cohesion.

Ahmed (1993), Humphrey et al. (1993), Edil & Bosscher (1994), and Foose et al. (1996) have reported that sand can be reinforced using tire chips. These reported studies have shown that adding tire chips increases the shear strength of sand, with friction angles as large as 65° being obtained for mixtures of dense sand containing 30% tire chips by volume.

Ahmed (1993) conducted triaxial tests on tire shred-soil mixtures (tire shred size = 25 mm) with various mixing ratios. A tire shred-soil mixture ratio of approximately 40:60 by dry weight (65:35 by volume) was reported to produce maximum shear strength values at low to medium confining stresses.

Foose et al. (1996) investigated the feasibility of using shredded waste tires to reinforce sand. Large scale direct shear tests were conducted on mixtures of dry sand and tire shreds. They have investigated the effect of five factors affecting shear strength such as normal stress, sand matrix unit weight, shred content, shred length, and shred orientation. They found that shred content and sand matrix unit weight were the most significant characteristics affecting the shear strength of the

321

mixture. Foose et al. (1996) obtained an initial angle of friction of 67° for sand reinforced with shreds, whereas the sand alone had a friction angle of 34° at the same sand matrix unit weight. They reported that sand containing shredded tires had higher shear strength than that of sand alone.

Ghazavi & Sakhi (2005) studied the effect of size of waste tire shreds on shear strength parameters of sand reinforced with shredded waste tires. Different shred contents, shred widths, and different aspect ratios have been mixed with the sand at two different sand matrix unit weights and have been tested in large shear box. They have found that regardless of compaction level and shred contents in the mixtures, for a given width, there is only a certain length that gives the greatest value of φ. The absolute greatest value of about $\phi = 67°$ was obtained by using 50% shreds with dimensions of 438 cm at unit weight of 16.8 kN/m³. For a given width of tire rectangular shreds, there is solely a certain length, which gives the greatest initial friction angle for sand-tire shred mixtures. The friction angle of mixtures increases by using optimum shred aspect ratio and by increasing shred contents and mixture compaction. Also, it has been found that shred content, shred width, shred aspect ratio for a given width, compaction, and normal stress are influencing factors on shear strength of the mixtures.

Attom (2006) conducted direct shear tests to study the shear strength behavior of sand-shredded tire mixtures under specific conditions. Shredded tires were passed through US sieve size 4 and mixed with three different types of sands with varying gradations. The three sands were mixed with four different percentages of shredded tires as 10, 20, 30 and 40% by dry weight. Specimens were prepared for the direct shear test at 95% relative compaction and optimum water content. They found that the addition of shredded waste tires increased both the angle of internal friction and the shear strength of the sands.

Edincliler et al. (2004) performed large-scale direct shear tests to determine the effect of tire buffings on shear strength properties. Sand, tire buffings, sand-tire buffing mixture having 10% tire buffings by weight were tested in the dry condition. The sand used in the tests was uniformly graded, medium dense, with a dry unit weight of 15.3 kN/m³. Tire buffings having maximum lengths of 40 mm were used in the tests. The unit weight of the tire buffings was 5.1 kN/m³. The tire buffings did not have any metal pieces in them. The unit weight of the tire buffings-sand mixture was 13 kN/m³. The addition of 10 percent tire buffings by weight to sand increased the internal friction angle from 22° to 33°.

The shear strength of a tire shred–soil mixture is affected mainly by the confining stresses, the tire shred–soil ratio and the density of the mixture. In mixtures of tire shreds and sand, the tire shreds have a reinforcing effect (Edil & Bosscher 1992, Ahmed 1993, Foose 1993, Bernal et al. 1996, Hataf & Rahimi 2005). Humphrey et al. (1993), Foose et al. (1996), Wu et al. (1997) and Tatlisoz et al. (1998), Edincliler et al. (2004), Grazavi & Sakhi (2005), Attom (2006) have reported that sand can be reinforced using tire wastes. These studies have shown that adding tire chips, tire shreds, and tire buffings increases the shear strength of sand with friction angles as large as 54° being obtained for mixtures of dense sand containing 30% tire chips by weight. Comparatively, the corresponding friction angle of sand was only 34°.

From the literature study, it is observed that by-products of tire wastes as tire shreds, tire chips and tire buffings can be used to improve the mechanical properties of the sand (Table 1). Inclusion of these materials into the sand has a reinforcing effect. Also, it has been revealed that tire waste content, aspect ratio, compaction, and normal stress are influencing factors on the shear strength of the mixtures.

3 MATERIAL AND METHODS

3.1 Materials

Sand, tire buffings, and sand-tire buffing mixture, were tested in the dry condition. The sand used in the tests was uniformly graded, medium dense, with a dry unit weight of 15.3 kN/m³.

Tire buffings used in this study are the by-product of the tire retread process and were obtained from tire retread companies in Istanbul (Figure 1). In order to eliminate the uncontrolled effects

Table 1. Waste tire, sand and waste tire-sand mixtures shear strength parameters.

Reference	Material	Unit weight (kN/m^3)	Shear strength parameters
Foose et al. (1996)	%90 sand + %10 tire shreds (15 cm)	16.8	$\tau = 37.9\,\text{kPa}\ (\sigma = 25.5\,\text{kPa})$
	%90 sand + %10 tire shreds (5 cm)	16.8	$\tau = 32.4\,\text{kPa}\ (\sigma = 25.5\,\text{kPa})$
	%70 sand + %30 tire shreds (15 cm)	14.7	$\tau = 11.0\,\text{kPa}\ (\sigma = 25.5\,\text{kPa})$
	%70 sand + %10 tire shreds (5 cm)	14.7	$\tau = 42.1\,\text{kPa}\ (\sigma = 25.5\,\text{kPa})$
Tatlisoz et al. (1998)	%100 tire chips	5.9	$c = 0\,\text{kPa};\ \Phi = 30°$
	%100 sand	16.8	$c = 2\,\text{kPa};\ \Phi = 34°$
	%90s and+%10 tire chips	15.6	$c = 2\,\text{kPa};\ \Phi = 46°$
	%80s and+%20 tire chips	14.5	$c = 2\,\text{kPa};\ \Phi = 50°$
	%70s and+%30 tire chips	13.3	$c = 2\,\text{kPa};\ \Phi = 52°$
Edinçliler et al. (2004)	%100 tire buffings	5.1	$c = 3.1\,\text{kPa};\ \Phi = 22°$
	%100 sand	15.3	$c = 6.9\,\text{kPa};\ \Phi = 33°$
	%90 sand+%10 tire buffings	14.9	$c = 8.7\,\text{kPa};\ \Phi = 29°$

Figure 1. View of tire buffings.

of the particle-size distribution, tire buffings were graded and then mixed according to a desired gradation curve. The tire buffings tested in this study had thicknesses ranging between 1 to 4 mm and lengths ranging from 2 to 40 mm. Due to elongation of tire buffings under stress, the soil-tire composite behavior is compatible. The resulting composite can undergo large deformation without cracking. The unit weight of tire buffings used in this study was 5.1 kN/m^3.

3.2 Large scale direct shear tests

In order to determine the shear strength properties of tire buffings and tire buffings-sand mixture, a large-scale direct shear test equipment (300 mm × 300 mm × 300 mm) was used which was designed and manufactured at Boğazici University (Figure 2). The application of vertical pressure was provided by means of an air-compressor, pressure regulator, and an air bag. The load readings were taken with a moment compensated load cell and axial deflections by means of displacement transducers. In order to provide uniform distribution of normal stress on the sample, the normal stress was applied by an air-bag (Baykal & Doven 2000).

Sand, tire buffings, and sand-tire buffings mixture having 5% (STB5), 10% (STB10), 20% (STB20), and 30% (STB30) tire buffings by weight were tested in the dry condition. The height of

Figure 2. Large-scale direct shear test equipment (Baykal & Doven 2000).

the samples was 30 cm. The sand used in the tests was uniformly graded, medium dense, with a dry unit weight of 15.3 kN/m³. Tire buffings having maximum lengths of 4cm were used in the tests. The unit weight of the tire chips was 5.1 kN/m³. The tire buffings did not have any metal pieces in them. The unit weight of the tire buffings – sand mixtures called STB5, STB10, STB20 and STB30 was 15.19 kN/m³, 14.89 kN/m³, 14.22 kN/m³, and 13.56 kN/m³, respectively (Table 2).

4 RESULTS

Shear stress – displacement curves for sand, tire buffings and sand-tire buffing mixtures at three different vertical stresses (20 kPa, 40 kPa and 80 kPa) are given in Figures 3, 4 and 5. The maximum shear stress values are used to calculate the shear strength parameters.

At low vertical stress (20 kPa), the addition of tire buffings to sand stiffens the mixture at low deformations and shifts the displacement value from 8 mm for sand only, to 34 mm for sand-tire buffings mixtures (STB10). (Figure 3). At shear stresses lower than 16 kPa, the mixture is stiff. The maximum shear strength value were obtained at 10% tire buffings addition to sand (STB10) while minimum value were obtained at STB30.

The shear stress-horizontal displacement curves at medium vertical stress level (40 kPa) are presented in Figure 4. It is seen that the tire buffing addition to sand increases shear stresses at very low horizontal deformations and stiffens the mixture at very low displacements. No clear peak shear stress was observed. The tire buffings addition to sand lowers the ultimate strength. The displacement at failure shifted from 14 mm for sand only to 30 mm for the sand-tire buffings mixture (STB10). At shear stresses lower than 15 kPa, the mixture is very stiff.

At higher vertical stress (80 kPa), STB10 stiffens at very low deformations (Figure 5). At the beginning, sand and STB5 showed the same type of behavior until reaching 12 mm displacement. After that, sand had the higher shear stress value than that of sand-tire buffings mixtures. The ultimate shear strength decreased with the addition of tire buffings. Overall, the tire buffings addition stiffened the initial portion of the shear stress – displacement curve.

As seen in all of the vertical loadings, the addition of tire buffings to sand were effective at very low deformations. As the percentage of tire buffings increase, ultimate strength values drop.

Shear strength data obtained from the experiments are summarized in Table 2. Tire buffings addition to sand increased the internal friction angle from 22° to 29° and cohesion ranged from 3.1 kPa to 15.45 kPa.

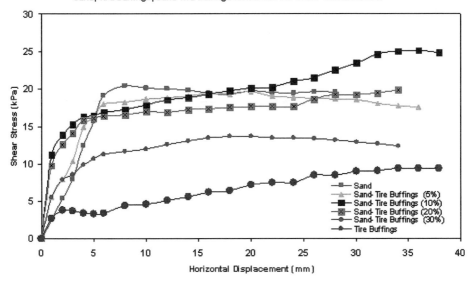

Figure 3.   Shear stress vs. horizontal displacement curves for sand, tire buffings and sand-tire buffings mixtures at vertical stress 20 kPa.

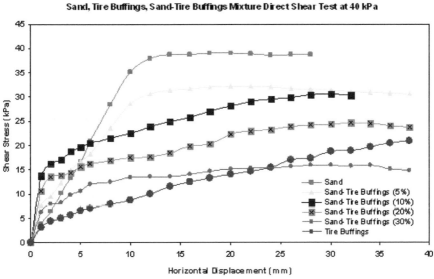

Figure 4.   Shear stress vs. horizontal displacement curves for sand, tire buffings and sand-tire buffings mixtures at vertical stress 40 kPa.

Tire buffings addition to sand increased the initial slope of the shear stress displacement curve and then shear strength is mobilized at lower displacement values. The initial steep shape of the shear stress displacement curve states that there will be more strength available at small displacements when tire buffings are added to sand.

It is found that tire buffings addition to the sand changed the deformation behavior of the mixture by stiffening the material at low strains and softening the mixture at large strains. Even at low weight percentage, such as 10 percent by weight, the deformation behavior of the mixture is altered significantly.

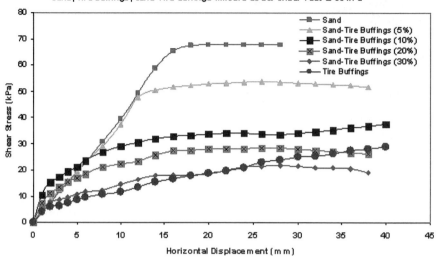

Figure 5. Shear stress vs. horizontal displacement curves for sand, tire buffings and sand-tire buffing mixtures at vertical stress 80 kPa.

Table 2. Summary of shear strength data.

Testing method/ sample size	Material	Unit weight (kN/m³)	Shear strength parameters	
			c (kPa)	Φ (°)
Large Scale	Tire buffings	5.1	3.1	22
Direct Shear	sand	15.3	6.9	33
(300 mm x 300 mm x	5% Tire buffings +	15.19	10.4	28
300 mm)/tire	95% sand			
buffings < 4 cm.	10% Tire buffings +	14.89	8.7	29
	90% sand			
	20% Tire buffings +	14.22	15.5	5
	80% sand			
	30% Tire buffings +	13.56	10.7	8
	70% sand			

5 CONCLUSIONS

This paper investigates the potential use of sand-tire buffings mixture for highway embankment construction. In order to predict the behavior of highway embankments under static loadings, shear strength parameters were obtained from the large scale direct shear test results for the tire buffings.

Tire buffings added to the medium dense sand at various compositions were tested. Based on the results presented, the following conclusions are made:

1. Adding tire buffings to sand changed the deformation behavior and shear strength values.
2. Addition of tire buffings to sand causes a strength increase at low confinement stresses.
3. The laboratory test results show that, ten percent by weight tire buffings addition to sand alters the deformation behavior of the mixture by stiffening the material at low strains and softening the mixture at large strains.
4. Good handling and workability of the material at the laboratory tests make it a good candidate for the actual road application.

326

5. Use of tire buffings additive at the top one-two meters of the highway embankments will improve the performance of the embankments under the traffic load. The traffic load imposes low strains under pavement structure overlaying the embankment. Fiber-shaped tire buffings stiffens the composites at low strains while softening it at large strains. The support under the pavement structure is improved. The same concept can be also used for railroad embankments.

The observed increase in shear strength of tire buffings – sand mixtures showed that tire buffings can be used as fiber reinforcement elements. For the given tire buffings and test conditions, even at low inclusions of tire buffings such as 10 percent, the deformation of the mixture is altered considerably.

REFERENCES

Ahmet, I. & Lovell, C.W. 1993. Rubber soils as lightweight geomaterials. *Transportation Research Record*, 1422. TRB, National Research Council, Washington DC: 61–70.

Ahmet I. & Lovell, C.W. 1992. Use of waste products in highway construction. *Environmental Geotechnology:* 409–418. Rotterdam: Balkema.

American Society for Testing and Materials, 1998. ASTM D 6270-98, Standard practice for use of scrap tires in civil engineering application.

Attom, M.F. 2006. The use of shredded waste tires to improve the geotechnical engineering properties of sands, *Environmental Geology* 49: 497–503.

Baykal, G. & Doven, GA, 2000. Utilization of fly ash by pelletization process: theory, application areas and research results. *Resources, Conservation, and Recycling* 30: 59–77.

Dickson, T.H., Dwyer, D.F., & Humphrey, D.N. 2001. Prototypes tire-shred embankment construction. *Transportation Research Record* 1755, TRB, National Research Council, Washington DC: 160–167.

Edil, T, & Bosscher, P. 1994. Engineering properties of tire chips and soil mixtures. *Geotechnical Testing Journal* 14(4): 453–464.

Edil, T.B. & Bosscher, P. 1992. Development of engineering criteria for shredded waste tires in highway applications. *Final report, Research report No. WI 14-92. Wisconsin Department of Transportation, Division of Highways.*

Edinçliler, A., Baykal, G. & Dengili, K. 2004. Determination of Static and Dynamic Behaviour of Waste Materials. *Resources, Conservation and Recycling* 42(3): 223–237.

Foose, G.J. 1993. Shear strength of sand reinforced with shredded waste tires. M.S.Thesis, Department of Civil and Environmental Engineering, University of Wisconsin. Madion.

Foose, G.J, Benson, C.H. & Bosscher P.J. 1996. Sand reinforced with shredded waste tires. *Journal of Geotechnical Engineering* 122 (9): 760–767.

Gray, D.H. & Oyashi, H. 1983. Mechanics of fiber reinforcement in sand. *Journal of Geotechnical Engineering* 109: 335–353.

Ghazavi, M. & Sakhi, M.A. 2005. Influence of optimized tire shreds on shear strength parameters of sand. *International Journal of Geomechanics* 5 (1): 58–65

Hatal, N. & Rahimi, M.M. 2006. Experimental investigation of bearing capacity of sand reinforced with randomly distributed tire shreds. *Construction and Building Materials* 20: 910–916.

Humphrey, D.N. 1995. Cvil Engineering Applications of chipped tires. North Platte, Nebraska.

Humphrey, D., Sandford, T., Cribbs, M., & Manison W. 1993. Shear strength and compressibility of tire chips for use as retaining wall backfill. *Transportation Research Record, No. 1422, Transportation Research Board, Washington DC:* 29–35.

Humphrey, D.N. 1999. Civil engineering applications of tire shreds. *Proccedings of the tire industry conference, Clemson University.*

Ghazavi, M. & Sakhi, M.A. 2005. Influence of optimized tire shreds on shear strength parameters of sand. *International Journal of Geotechnics* 5 (1): 58–65.

Lee, J.H., Salgado, R., Bernal, A. & Lovell, C.W. 1992. Shredded tires and rubber – sand as lightweight backfill. *Journal of Geotechnical and Geoenvironmental Engineering* 125: 132–141.

Masad, E., Taha, R., Ho, C. & Papagiannakis, T. 1996. Engineering properties of tire/soil mixtures as a lightweight fill, *Geotechnical Testing Journal* 19: 297–304.

Tatlisoz, N., Edil, T.B. & Benson, C, 1998. Interaction between reinforcing geosynthetics and soil-tire chip mixtures. *Journal of Geotechnical and Geoenvironmental Engineering* 124 (11): 1109–1119.

Yang, S., Lohnes, R.A. Kjartanson, B.H, 2002. Mechanical properties of shredded tires. *Geotechnical testing Journal* 25: 44–52.

Yoon, S., Prezzi, M., Siddiki, N.Z. & Kim, B. 2006. Contruction of a test embankment using a sand-tire shred mixture as fill material. *Waste Management* 26: 1033–1044.

Young, H.M., Sellasie, K., Zeroka, D. & Sabris, G. 2003. Physical and chemical properties of recycled tire shreds for use in construction. *Journal of Environmental Engineering* 129 (10): 921–929.

Youwai, S. & Bergado, D.T. 2004. Numerical analysis of reinforced wall using rubber tire chips-sand mixtures as backfill material. *Computers and Geotechnics* 31: 103–114.

Wu, W.Y., Benda, C.C. & Cauley, R.F. 1997. Triaxial determination of shear strength of tire chips. *Journal of Geotechnical and Geoenvironmental Engineering* 123 (5): 479–482.

Zornberg, J.G., Alexandre, R.C. & Viratjandr, C. 2004. *Canadian Geotechnical Journal* 41 (2): 227–241.

Tire buffings – soil liners against hydrocarbon contamination

Gökhan Baykal

Boğaziçi University, Department of Civil Engineering, Istanbul, Turkey

ABSTRACT: Underground petroleum storage tank leaks pose great threat to the environment. Generally the contaminant release is very slow and detection is very hard. Stored gasoline, volatile organic compounds from dry cleaners etc contaminate the groundwater making it very hard and expensive to clean. A retarding composite layer made of tire buffings mixed with clay or sand is developed for this purpose to encapsulate the released material and give enough time for clean up in the case of a leak. The rubber absorbs the organic contaminants physically and expands imposing confining stress on the soil, causing a further decrease in hydraulic conductivity in addition to its retardation capacity. Consolidation permeameter tests are conducted on tire buffings-soil mixtures and hydraulic conductivity values are measured. A Finite Element Model of the composite is made and the effect of swelling of rubber upon contact with gasoline is studied. The design guidelines for site implementation are given.

1 INTRODUCTION

Underground storage tanks are widely used in residential, commercial and industrial areas to store chemicals, gasoline, diesel fuel, fuel oil etc. Although the construction of these storage tanks are regulated, many leak incidents have been reported annually It has been estimated that nearly ten per cent of the three million underground petroleum storage tanks present in the USA are leaking. The early tanks were made of steel so the major cause of leaking was corrosion. Recently fiber wall tanks are used which minimize leakage. In both cases the leakage may occur in very long time and until it is detected the contaminant may reach the groundwater and the clean up process becomes costly. In earthquake prone regions sudden loads emanating from earthquakes may cause tank failure or failure from connections or piping causing sudden release of hazardous chemicals or gasoline. This sudden release of contaminants makes it impossible to clean before the groundwater is contaminated. A flexible layer capable of absorbing the contaminants will allow clean up process to begin in a reasonably long time after the occurrence of the earthquake. This way the risk of contamination will be reduced. For this purpose the excavated area for the placement of underground storage tank is lined by rubber fiber mixed soil. Another critical issue concerning hydrocarbon leaks is its effect on soil hydraulic conductivity. The low dielectric constant of hydrocarbons cause shrinking of diffused double layer of clays causing an increase in hydraulic conductivity. The retardation mechanism of rubber buffings plays a key role in controlling the hydraulic conductivity increase. In this study the effect of low dielectric effluents on soil hydraulic conductivity, the working principle of the proposed liner and recommendations for construction are presented. A finite element model of the soil-tire composite is made and the effect of swelling of rubber buffings upon contact with gasoline is studied.

2 BACKGROUND

2.1 *Tank types and regulated substances*

Underground storage tanks are made of single wall or double wall, steel or fiber material. The older tanks are mostly steel and corrosion is one of the major problems related. Once leaks from these tanks are detected, the old tank is removed, site clean up process is undertaken and a long contamination monitoring program is commenced. The cost of clean up far exceeds the cost of land and because of the high risks involved, the insurance rates increased to unaffordable values. Especially in rural USA, the people have to travel long distances to reach gasoline stations. Most of the small size gasoline stations have been closed because of large budgets needed to take care of liability issues.

For underground storage some substances are regulated and some are exempt from regulations. Gasoline is regulated. It is highly volatile and can contaminate an entire aquifer. Therefore gasoline stations pose the greatest risk due to their gasoline tanks. Other materials typically found in gas stations under regulation are diesel fuel and fuel oil which are less volatile than gasoline and will not migrate as far as gasoline. Heating oil (No. 4–No. 6) is also regulated but is does not migrate. LPG and propane tanks are exempt from regulations but they are regulated for fire safety. Septic tanks are also located underground and they are of concern with regard to public safety. They are exempt form Underground Storage Tank (UST) regulations.

2.2 *Tank failure mechanisms*

The leaks that occur in the underground tanks are grouped in three categories; piping leaks, tank leaks and overfills. The piping leaks are due to brittle steel piping, rigid pipe joints and connections, and shallow embedment of the pipes. Tank leaks mainly occur due to corroding steel. In fiber wall tanks structural defects like cracks are of concern. Under static loads though leaks are not likely in fiberglass tanks provided that the tank is not damaged during installation and the material is not deficient. The third major cause of contamination is due to overfilling. The contamination can take place in the form of free product or volatile organic compounds (VOC) carcinogenic or harmful to human health.

The gasoline seeps into the surrounding soil in downward direction to the groundwater. The distance of travel for the contaminating gasoline will be a function of the size of the spill and the soil type. Once the groundwater is reached "free product" will float: individual components will dissolve into the groundwater. The contaminant will travel horizontally over the groundwater table.

Earthquake regions pose special risk areas for underground petroleum storage tanks. The earthquake loads impose large horizontal stresses on the tank, the pipes and the pipe connections. The approach to decrease this risk should involve retarding the contaminants for a reasonable amount of time until remediation measures are taken. Due to large number of underground petroleum storage tanks found in one area, after an earthquake several remediation jobs should be done which will be impossible due to limited resources. If all tanks are lined with a retarding liner, in a reasonable length of time the remediation process can be completed.

2.3 *Concept of the retarding liner system*

Tire buffings obtained as a by-product of tire retread companies is mixed with sand, clay or waste material (fly ash etc.) and compacted under and around the UST. Rubber expands and holds spilled hydrocarbons; while retarding the contaminant, with the stresses exerted by the expanding rubber a considerable decrease in hydraulic conductivity is obtained. Typically ten percent by weight rubber fiber addition (tire buffings) is adequate to provide low hydraulic conductivity and high retarding capacity. Due to lower unit weight of tire fibers, 10 percent by weight corresponds to 20 percent by volume. One cubic meter of tire fiber-soil liner can retard approximately 200 liters of contaminant.

Tire buffings are obtained as a byproduct of tire retreading industry. The rubber fibers are 10–20 mm long with diameter in the range of a couple of mm. Typical tire buffing samples is shown in Figures 1 a,b,c. The figures show digital images created by a digital 3D plotter using 0.1 mm scanning array. Each scan took approximately 6 hours to complete.

The rough surfaces of the fibers have several advantages. With the rough surface as demonstrated the rubber fibers are anchored to plastic soil easily and its tensile strength is mobilized making the tire buffings soil composite ductile. Another advantage of the rough surface is to increase the

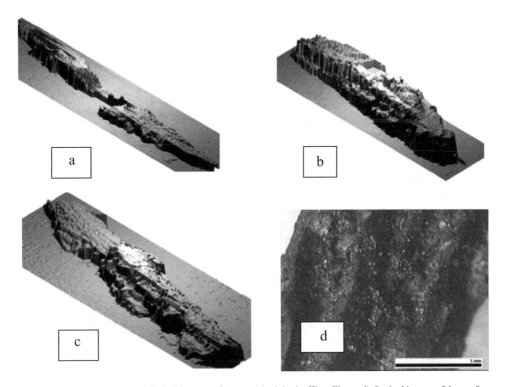

Figure 1. (a,b,c) 3D Scanned digital images of three typical tire buffing fibers, d) Optical image of the surface of a tire buffing fiber at 75 × magnification.

surface area of the fiber causing larger interaction surface with soil. The larger surface area also helps in adsorption of the leaking hydrocarbon. Figure 1d shows optical image at a magnification of 75 times of the tire buffing. The rough surface is easily observed.

The swelling pressure of tire buffing- sand mixture upon contact with gasoline is one of the major design components. Swelling pressure tests must be conducted and the rubber percentage should be decided in a way that in the case of gasoline leakage, the stresses developed due to swelling of rubber will be around 40 kPa. Researchers reported that at confining stress of 38 to 50 kPa there was a major decrease in hydraulic conductivity. There is another controlling factor on the tire buffings percentage to be used with soil; the swelling pressures due to leaking UST's should be less than the weight of the tank and the stored material to ensure there is no uplift. Considering the fact that one cubic meter of composite liner can hold 200 liters of leaking gasoline, the liner dimensions and rubber content can be optimized. The retarding liner can be used around all piping and at pipe to tank connections. The advantage of having a thick composite retarding liner is its capability of holding large amounts of contaminants likely to be released due to sudden impact loads from earthquakes.

2.4 *Previous studies*

The concentration at which organic chemicals significantly alter the hydraulic conductivity of clays has been investigated for neutral organics such as methanol, acetic acid, heptane and TCE. No alteration was observed in the conductivity for liquid mixtures with a dielectric constant greater than 40. Substantial increases in the conductivity were measured for liquid mixtures with a dielectric constant less than 35. For methanol this corresponds to 80% concentration and for heptane and TCE this value corresponds to concentrations above their solubility limit (Bowders & Daniel 1987).

The typical dielectric constant of gasoline is approximately 3. Since the gasoline in storage tanks is in pure form it is expected that, in the case of a leakage, the high concentration will result in major increases in the hydraulic conductivity.

The permeation of pure ethanol at zero vertical stress causes a 100-fold increase in the hydraulic conductivity of a medium plastic clay. This increase can be prevented with the application of 70 kPa effective stress prior to and during permeation of the sample (Fernandez and Quigley 1991). Kaolinite samples permeated with methanol and heptane have a similar break point when vertical compressive stresses of 34–69 kPa are applied (Broderick and Daniel 1990).

At higher effective stresses of 160 kPa, the measured conductivity values of clay permeated with ethanol (3.9×10^{-9} cm/sec) are lower than the original reference water values tested at zero effective stress (6.25×10^{-9} cm/sec). This phenomenon is due to the 'chemically induced' consolidation (and lateral yield) that closes macropores and shrinkage cracks produced by double layer contraction (Fernandez and Quigley 1991).

Since 1990 potential use of rubber fibers in liners against petroleum contamination has been investigated at Bogazici University (Baykal et al.). Tire buffings, obtained as a waste material from the tire retread industry, are mixed with soil and compacted at various weight percentages. The hydraulic conductivity tests conducted on compacted tire buffings-soil mixtures yield lower conductivity values as compared to those of only soil when gasoline is permeated. The controlling factors in these tests are the swelling behavior of rubber and its retardation capacity.

It is known that when rubber fibers are immersed in gasoline they swell to 2.5 times their original volume, with the great majority of this swelling occurring within the first 30 minutes. The behavior of rubber added kaolinite clay permeated with gasoline has also been investigated. Swelling tests on kaolinite-tire buffings composite samples immersed in gasoline have shown an 18% swell under 5 kPa vertical pressure. Swelling pressure tests performed in consolidation devices on kaolinite and kaolinite-tire buffings mixtures have shown that kaolinite +10% rubber samples inundated with water and gasoline have swelling pressures of 54 and 142 kPa respectively (Alpatlı1992).

3 HYDRAULIC CONDUCTIVITY TEST RESULTS

3.1 *Hydraulic conductivity*

Falling head hydraulic conductivity tests on kaolinite and kaolinite added rubber samples in proctor molds and SSRI have shown that an addition of 10% by weight of tire buffings will increase the water conductivity of the liner by less than 2 folds (Baykal and Alpatlı1995 and Baykal et al. 1992). These results indicate that the addition of tire buffings does not cause the development of large pores or cracks in the mixture and that there is a good interaction between kaolinite and rubber (Baykal et al. 1992). Rigid wall hydraulic conductivity tests using specially constructed consolidation cell permeameters have been used with applied vertical stresses between 50 and 450 kPa to illustrate the effects of confinement on the conductivity of rubber-soil samples permeated with both water and gasoline.

These experiments show that as the applied vertical stress is increased, the conductivity to both water and gasoline decreases (Alpatlı 1992).

The effect of vertical stress on the hydraulic conductivity is presented in Figures 3a, 3b and 4. Previous findings of tests conducted in Sealed Single Ring Infiltrometer (SSRI), proctor and small size consolidation cells have also been plotted in these figures (Baykal and Alpatli 1995 and Baykal et al 1992).

Figure 3a shows only kaolinite samples permeated with water and gasoline. At stresses lower than 50 kPa kaolinite samples permeated with gasoline (K in G) have conductivities nearly 2 orders of magnitude higher than those of the samples permeated with water (K in W). However, at stresses higher than 50 kPa the reverse is true. Gasoline causes large increases in conductivity only under conditions of low vertical stress. The results of tests conducted in compaction molds (Baykal et al 1992) and those in smaller sized consolidation cells (Alpatlı1992) are also plotted in Figure 3a. Figure 3a also shows that when the stress imposed on a kaolinite sample prior to permeation with

Figure 2. Specially designed large size pneumatic consolidation permeameter for testing gasoline permeation through tire buffings – soil composite (CBR mold size).

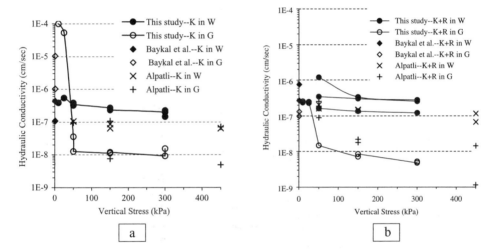

Figure 3. Change of hydraulic conductivity with vertical stress; a) Kaolinite b) Kaolinite-Rubber.

gasoline is increased from 25 to 50 kPa, the conductivity decreases by three orders of magnitude. Thus, vertical stresses of 25–50 kPa is a breakpoint at which the load applied appears to strengthen the soil mass to a point where it is able to resist changes associated with chemical permeation.

Figure 3b shows hydraulic conductivity versus stress values for tire buffings added kaolinite samples (K + R) permeated with water and gasoline. At low vertical stresses the conductivities of both water and gasoline permeated samples are nearly equal and are in the order of 3×10^{-7} cm/sec. The performance of the composite liner is satisfactory at low confining pressures for both contaminated and uncontaminated conditions. confining pressures the gasoline permeated samples have conductivity values approximately one order of magnitude smaller than the same samples permeated with water. This can be explained by the confinement produced when rubber expands upon contact with gasoline. The swelling pressure of kaolinite samples with 10% rubber additive immersed in water and gasoline are 54 and 142 kPa respectively (Baykal et al). Since larger confining pressures are obtained when these samples are permeated with gasoline than with water, there is a larger restriction to flow and smaller permeabilities are obtained. However, it is very important that the vertical applied stress is larger than the swelling pressure of the K + R composite otherwise

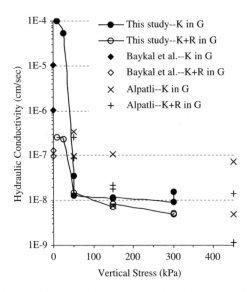

Figure 4. Comparison of hydraulic conductivity values for Kaolinite and Kaolinite+Rubber at increasing vertical stresses.

the sample will swell. In one experiment the applied stresses (10 and 25 kPa) were much lower than the swelling pressure of the composite (142 kPa) and the rubber particles steadily absorbed gasoline and caused swelling of the sample. Although the void ratio increased by 13.3%, a decrease in conductivity of 10% was recorded.

For water permeated samples shown in Figure 3a and Figure 3b there is a comparatively small difference in the conductivities of samples tested at low stresses and those tested at high stresses provided the samples are well compacted. Samples obtained from fields where the conductivity is greater than 1×10^{-7} cm/sec are more sensitive to stress because they contain larger pores which collapse and lead to higher decreases in hydraulic conductivity (Trast and Benson 1995). A similar behavior was seen in one sample with an initially high void ratio and conductivity. This sample experienced a decrease in conductivity nearly 3 times larger than other similar samples.

It may thus be stated that under uncontaminated conditions the composite samples do not have significantly larger conductivity values than those of the kaolinite samples. It would also be beneficial to see how the performance of these two samples compares with each other when contamination occurs. The composite samples perform better than the kaolinite samples at all stress levels (Figure 4).

In these experiments two factors, stress and rubber inclusions, work to prevent excessive increases in conductivity due to gasoline. The application of vertical stresses lower than 50 kPa, prior to pollution does not have a significant beneficial influence on hydraulic conductivity. Rubber, on the other hand, stabilizes the liner such that no increase in conductivity is observed even at stresses as low as 10 kPa.

4   IMAGE ANALYSIS OF TIRE BUFFINGS – KAOLINITE COMPOSITE

Compacted clay liners have brittle behavior and under large deformations, formation of cracks cause detrimental increase in hydraulic conductivity values. Addition of tire buffings makes the clay liner ductile. Shear strength and deformation properties of compacted tire buffings clay liners have been investigated in addition to the investigation of conduction behavior. Global Lab Image processing software was used to observe and demonstrate the ductility of the composite liner

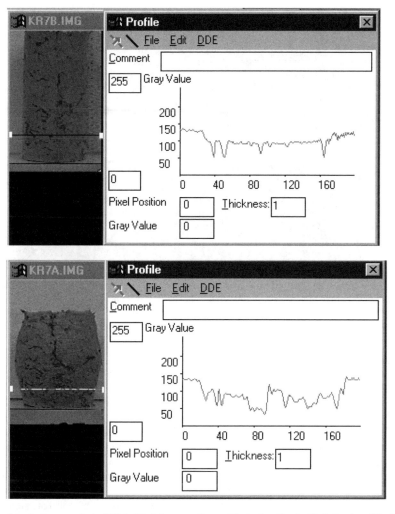

Figure 5.   A typical screen shot of Global Lab Image software illustrating the ductile behavior of tire buffings – soil composite; the lower gray values indicate tire buffings or cracks.

(Figure 5). Poisson ratio was calculated using these digital measurements. Addition of tire buffings increased the undrained shear strength of kaolinite and decreased the amount of cracks (Sarica, 2001).

5   FINITE ELEMENT MODEL STUDY

From the experimental results presented above, it is observed that a breakthrough confining stress of 35–50 kPa is adequate to compensate for the increasing hydraulic conductivity of clay soil due to shrinkage of diffused double layer caused by low dielectric constant of the gasoline. For a successful liner design with tire buffings, it is important to come up with an additive weight percentage of tire buffings which will cause a minimum of 35–50 kPa swelling pressure to confine the surrounding soil in the liner. In addition to retard the leaking gasoline by physical absorption, the swelling pressure exerted by tire buffings upon contact with leaking gasoline on surrounding liner soil

Figure 6. A typical screenshot of FEM analysis results of tire buffings – soil composite; lower stress levels increase from dark to light blue, green, yellow, orange and highest level is shown in red colour.

further causes a decrease in hydraulic conductivity. To be able to demonstrate this phenomenon a finite element model was established and the stresses on the liner soil due to swelling of rubber tire buffings was studied. The amount of tire buffings that can cause a minimum of 35–50 kPa stress in the soil matrix is set to be minimum design percentage for the tire buffings.

ANSYS finite element analysis program version 5.2 was used to model the compacted tire buffings- kaolinite soil composite placed in a consolidation oedometer. Six different finite element models with different rubber area fractions were prepared involving horizontal, vertical and randomly oriented fiber inclusions. The swelling behavior of the composite is a physical phenomenon following the absorption of gasoline by rubber and this mechanism is simulated using a thermal strain analogy with thermal expansion coefficient assigned.

The results which are illustrated by stress contour plots (Figure 6) reveal that, although weak zones exist under 50 kPa vertical pressure, their total area is reduced by the increased overburden pressure and rubber area fraction. In the actual case highly stressed zones act as confinement and shrink the flow paths causing a decrease in hydraulic conductivity.

6 ECONOMIC FEASIBILITY

Underground petroleum tanks are manufactured at different dimensions. A typical size will be considered for comparison. In Figure 7 a typical double lined FRP tank is presented. The tank is overlayed by a 2 m thick soil. For a 3 m diameter and 10 m long tank (70000 litres), with the assumption of 1 m tire buffing soil liner at the bottom, sides and in the front and back parts of the tank a total of 169 cubic meter of liner material is needed. Considering a 20 percent by volume tire buffings inclusion (approximately ten per cent by weight) 34 cubic meters of tire buffings are needed. 17 tons of tire buffings will cost 1700 euros plus transportation costs (market value in Turkey is 100 euros per ton). For a typical tank installation using tire buffings in the composite liner will cost in the range of couple of thousand euros. With the benefit of decreasing risks of groundwater contamination the liability of the gasoline station will decrease causing a potential decrease in insurance premiums. In case of leakage the surrounding liner will be capable of retarding

336

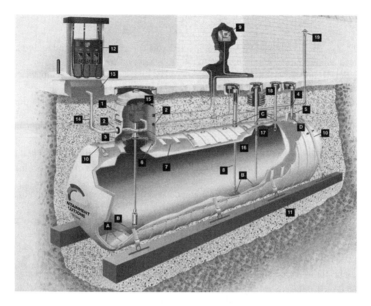

Figure 7.   A typical underground petroleum storage tank (after, google.com).

more than 17000 litres of leaking gasoline. This will provide enough time for clean up process in the case of a leakage.

Another potential use of tire buffings is as a base layer under the pavement of a gas station. A 30 cm concrete pavement will exert a pressure of 7.5 kPa on the base layer. A 15 cm thick tire buffings – sand composite base layer will provide a retarding liner in the case of a surface spill. This layer will be able to hold 7500 litres of gasoline per 100 square meter of pavement area. The cost of tire buffings for this application will be in the range of 750 euros per 100 square meters. Another advantage of the composite base layer will be; in the case of a leakage under the pavement, the swelling pressure developed in the rubber will cause expansion of the pavement and will be immediately noticed for clean up. Although this may bring some extra cost for concrete pavement renewal, the benefit of immediately learning any leakage in time is more critical to prevent potential contamination of the groundwater.

7   CONCLUSIONS

With its retarding capability and swelling behavior, tire buffings may be used as an additive to soil to capture the leaking hydrocarbons from an underground storage tank. The proposed liner system will allow a reasonable amount of time after leakage before clean up process begins. The liner system will be efficient both for very slow leakage rates as well as sudden releases from earthquake loads. Although the cost of tire buffings is not cheap, the total cost of using tire buffings in a gas station project will be in the order of a couple of thousand euros. The tire waste is turned into a superior geomaterial when it is used to retard hydrocarbon contamination.

ACKNOWLEDGEMENTS

All of the studies presented here are conducted by my MSc and PhD students. Their persistence and hard work is appreciated. Bogazici University Research Fund has supported the experimental studies.

REFERENCES

Alpatlı, M. 1992. Hydraulic conductivity of compacted waste rubber-clay mixtures as earthen liners for petroleum based contaminants. *MSc. Thesis*, Istanbul: Boğaziçi University.

Baykal, G., Yeşiller N. and Köprülü K. 1992. Rubber-clay liners against petroleum based contaminants. *Environmental Geotechnology*, pp. 477–481 Rotterdam: Balkema.

Baykal, G., Kavak, A. and Alpatlı, M. 1995. Rubber-kaolinite and rubber-bentonite liners. *Waste Disposal by Landfill, Green 93*, pp. 399–404. Rotterdam: Balkema.

Baykal, G. and Alpatlı, M. 2000. Permeability of rubber-soil liners under confinement. *Geoenvironment 2000, ASCE STP*, Vol. 1, pp. 718–731.

Baykal, G. and Ozkul, Z. 2001. The effect of stress on the hydraulic conductivity of rubber soil liners permeated with gasoline. *Proceedings of the Fifteenth Conference on Soil Mechanics and Geotechnical Engineering, 27-31 August 2001,* Vol. 3, pp. 1955–1958.

Bowders, J.J., Jr. & Daniel, D.E. 1987. Hydraulic conductivity of compacted clay to dilute organic chemicals. *ASCE, Journal of Geotechnical Engineering*, Vol. 113, No. 12: 1432–1448.

Broderick, G.P. & Daniel, D.E. 1990. Stabilizing compacted clay against chemical attack. *Journal of Geotechnical Engineering*, Vol. 116, No. 10: 1549–1567.

Dündar, Ş. 2003. Utilization of scrap rubber tires in civil engineering applications. *M.Sc Thesis*, Istanbul:Boğaziçi University.

Erişen, S. 1997. The finite element analysis of swelling behavior of the soil-rubber composite. *MSc Thesis*, Istanbul:Boğaziçi University.

Fernandez, F. & Quigley, R.M. 1991. Controlling the destructive effects of clay-organic liquid interactions by application of effective stresses. *Canadian Geotechnical Journal*, Vol. 28: 388–398.

Köprülü, K. 1991. Hydraulic conductivity and strength of rubber reinforced fly ash used as liner material. *M.Sc Thesis*, Istanbul: Boğaziçi University.

Mehmetoğlu, D. 1994. Compacted fly ash mixtures as highway safety barriers. *M.Sc Thesis*, Istanbul:Boğaziçi University.

Özkul, Z.H. 1998. The effect of stress on the hydraulic conductivity of rubber soil liners permeated with gasoline. *M.Sc Thesis,* Istanbul:Boğaziçi University.

Sarıca, R. 2001. "Shear strength and deformation behavior of rubber fiber kaolinite mixture", *MSc Thesis*, Istanbul:Boğaziçi University.

Trast, J.M. & Benson, C.H. 1995. Estimating field hydraulic conductivity of compacted clay. *ASCE, Journal of Geotechnical Engineering*, Oct.: 736–739.

Türe, E.Z. 1995. Effect of freeze and thaw cycles on the performance of clay liners, *MSc Thesis*. Istanbul:Boğaziçi University.

Yeşiller, N. 1991. Hydraulic conductivity of compacted waste rubber-clay mixture as earthen liner for petroleum based contaminants. *M.Sc. Thesis*, Istanbul:Boğaziçi University.

EPS-filled used tires as a light weight construction fill material

Isao Ishibashi
Department of Civil and Environmental Engineering, Old Dominion University, Norfolk, VA, USA

Sethapong Sethabouppha
Faculty of Architecture, Chiang Mai University, Chiang Mai, Thailand

ABSTRACT: An innovative EPS-filled used tire as a lightweight construction fill material is introduced. Hollow inner spaces of recycled used tires are entirely filled with EPS (Expanded Polystyrene) to make rather rigid, against vertical compression, but extremely lightweight composite materials. The units could be utilized by stacking those in several layers; to make light weight embankment system on soft ground; to construct a temporal access road system, which could be removed easily after the end of the project and reused again at other projects; to make lightweight backfill in retaining wall system against which the lateral earth pressure would be minimized; and many other applications when both rigidity and lightness are required. In this paper, the focus was placed on lightweight embankment construction on soft ground. Laboratory experiments were conducted to study the response of EPS-filled tires under compressive stresses as a single unit as well as stacks of several layers. By using numerical methods, parameters for designing of pavement structures on EPS-filled tire embankments were determined, and an example of pavement designed on EPS-filled tire embankment was investigated. It was concluded that construction of embankments using EPS-filled tire system is technically feasible though the cost of EPS-filled tire may still be unfavorable due to its manufacturing cost at this moment.

1 INTRODUCTION

Since the disposal of used automobile tires has been an important issue of the environmental concern, several research efforts have been made to explore successful ways to use the discarded tires, especially in geotechnical engineering application. Shredded or whole tires have been used as lightweight fill materials in many experimental roadway projects around the world (e.g. Edil and Bosscher 1994, Hoppe 1994, PIARC 1997, Tweedie et al. 1998). However, those applications had yielded a limited success due to the still relatively high density of shredded tires-soil mixture, and the very high compressibility and potential for progressive deformation of pure shredded tires. Therefore, the uses of discarded automobile tires are still limited. Meanwhile, EPS (expanded polystyrene) has been widely used around the world as an embankment fill material to solve the settlement and stability problems of soft ground, as well as a frost protection layer in cold region (e.g. Esch 1995, Beinbrech and Hillman 1997). However, EPS still possesses a relatively low compressive strength and a very large deformation beyond its yield point.

2 EXPERIMENTAL RESEARCH

2.1 Used automobile tires

A number of used tire samples were randomly picked up from a local tire store in Norfolk, VA, USA for a preliminary study. During the compression tests on used tire samples, it was observed that the tread wall of the tires bends inward around at the mid-height when axially compressed.

Figure 1.   EPS-filled tire.

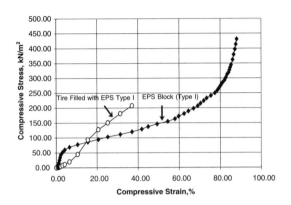

Figure 2.   Stress-strain relationships of EPS-filled tire and EPS cube specimen.

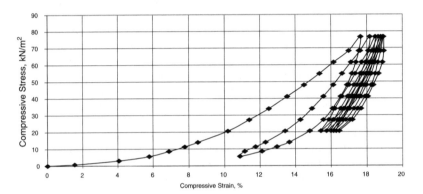

Figure 3.   Stress-strain relationship of EPS-filled tire under a series of repetitive compression.

Under compressive stress, this inward movement should provide confining stress to any material contained inside the hollow space of the tires.

2.2   Trial batch of EPS-filled tires

A number of trial EPS-filled tires were made by injecting the spaces inside the tires with EPS (expanded polystyrene) type I. The process employed the same technique as used for molding regular EPS blocks, which employed placing tires in a large model and injecting EPS in it. Figure 1 shows an EPS-filled tire made by an EPS molder in Radford, VA.

The first trial batch of EPS-filled tires underwent preliminary compression tests. A stress-strain relationship of an EPS-filled tire is shown in Figure 2, where a data for a regular $0.1 \times 0.1 \times 0.1$ m EPS cube sample is also shown. The figure shows an improved yield point of EPS contained inside a used tire in comparison with that of the regular EPS specimen, though there is some softness at the initial portion of the EPS-filled tire curve. This softness might caused by some air pockets which remain unfilled with EPS inside the tire walls. This can be felt by simply pushing the tire by a thumb.

In Figure 3, a stress-strain relationship of an EPS-filled tire under repetitive compression test is shown. The EPS-filled tire underwent a relatively large deformation at the first loading, and partially recovered when unloaded. However, the progressive permanent deformation under the

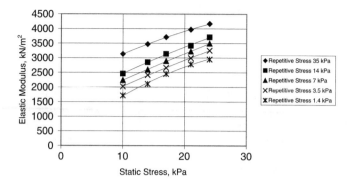

Figure 4.   Elastic modulus of EPS-filled tires as a function of static stress and repetitive stress.

later load repetition had improved. If this material is used for highway embankment, the initial deformation could be overcome by the weight of the overburden pavement structure.

2.3   Prototype batch of EPS-filled tires

The promising trial tests on EPS-filled tires led to the manufacturing of the prototype batch. However, this time EPS type II was used because the anticipated stresses due to pavement system and traffic are within its yield point. To solve the problem of air pockets, sealant foam made of polyurethane, which is readily available from hardware stores, was injected through several holes drilled around the tire walls. Though this foam was not as strong as EPS, it is anticipated to improve the performance of the EPS-filled tires.

2.4   Compression tests of single EPS-filled tire units and stacks of EPS-filled tires

A goal of this study was to obtain the pavement design parameters by utilizing numerical methods, which is to simulate plate loading tests on embankment of EPS-filled tires covered by a layer of sub-grade soil. The design guideline by AASHTO (1998) recommends that the plate loading test could be either the non-repetitive or the repetitive type. For EPS-filled tire system, the repetitive type, which requires to record the deflection of the loading plate at the tenth loading, seems to be more suitable since the EPS-filled tires tend to stabilize after a few rounds of load repetition as observed in the earlier tests.

   One of important parameters needed in the numerical simulation is the elastic modulus. Therefore, repetitive static compression tests of single tire units were carried out and the elastic modulus was calculated by the chord modulus method using the data at the tenth loading. The testing procedure, which includes the time to maintain the deformation rate at the end of each loading and the recovering time at the end of each unloading, was similar to that of the repetitive static plate load test (ASTM D 1195-93).

   Since the elastic modulus of EPS-filled tires could be affected by the initial static stress and the magnitude of the repetitive stress, a series of compressive stresses with various static stresses and repetitive stresses were applied. The elastic modulus of the EPS-filled tire prototypes as a function of static stress and repetitive stress level can be seen in Figure 4.

2.5   Compression tests of EPS-filled tire stacks

A distinct element method (DEM) is used to simulate the plate loading tests. However, the DEM code available for this study is a two-dimensional version. Therefore, the three-dimension stack compression tests (Figure 5) are replaced by two-dimension stack tests (Figure 6). The replacement

Figure 5.   Three-dimensional compression test.

Figure 6.   Two-dimensional compression test.

is made by the principle that at the same layer the applied contact stress on tires are the same in both two- and three-dimension tests, so that the contact areas between the EPS-filled tires at the same layers are kept equal.

The principle of equivalent stresses was proved by the tests of the two- and three-dimensional compression tests of several set of stacks. Under the same magnitude of compressive stresses applied on the top, the deformations of both dimensional tests were similar.

3   NUMERICAL SIMULATIONS

The numerical simulations of the compression tests were performed by using the distinct element method (DEM) code, UDEC. The DEM method was first developed in early 1970's by Cundall (Cundall 1971) for the main purpose of rock mechanics analysis. A review of the use of the DEM on several systems of EPS blocks by Takahara and Miura (1998) demonstrated that the DEM was appropriate for analyses of EPS-filled tire systems because of the discontinuity of EPS-filled tire units, which can be categorized as a discrete system. A discrete element system is comprised of discrete blocks with existence of contacts or interfaces between them. Two springs (shear spring and normal spring) and a slider represent the contact between those blocks. Important material properties of all the EPS-filled tire blocks, and other blocks for the DEM are mass density, shear modulus (G), and bulk modulus (K). G and K are calculated from the elastic modulus (E) and Poisson's ratio (v) of each block by using the following equations.

$$G = E / [2 (1+v)] \tag{1}$$

$$K = E / [3(1-2v)] \tag{2}$$

Normal stiffness (k_n) and shear stiffness (k_s) of the contacts were approximated from G, K, and the zone size of the contacting blocks by using the following criterion.

$$k_n \text{ and } k_s \leq 10 \times [\max \{ (K + 4G/3)/ \min z \}] \tag{3}$$

where min z is the smallest width of the zone adjacent to the contact. Detail of the zone and its width could be seen in the UDEC software manual (Itasca Consulting Group, Inc., 1996). 1.0×10^{10} Pa/m was used for k_n and the variation of k_s had little effect on the simulation result.

The program used for the distinct element analyses in this study was a two-dimensional program. This makes all the blocks rectangular in the analyses, which are different from circular EPS-filled tires in the reality. This problem was solved by adjusting the overlapping length of the blocks in the analyses so that the ratio of the contact area to the full area of the block was equivalent. In the analyses by DEM, the stacks were modeled by half with the boundary fixed in x-direction, and freed in y-direction at the axis of symmetry as shown in Figure 7.

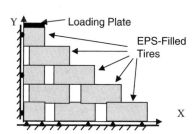

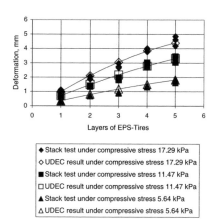

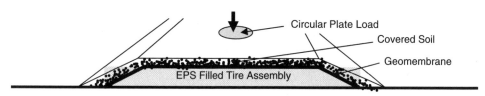

Figure 7. Compression test of EPS-filled tire stack in DEM.

Figure 8. Comparisons of laboratory compression tests and DEM simulations.

Figure 9. System of EPS-filled tire embankment with covering geomembrane and soil.

4 RESULTS OF LABORATORY TESTS AND NUMERICAL SIMULATION

The results from laboratory compression tests and the numerical simulations by DEM are compared in Figure 8. The plot demonstrates excellent agreements between the results from the laboratory tests and the simulations. Therefore, it is concluded that DEM can effectively be used for the simulations of plate loading tests on EPS-filled tire embankments.

5 PAVEMENT DESIGN PARAMETERS BY NUMERICAL SIMULATION OF PLATE LOADING TESTS

The proposed embankment for a highway consists of layers of stacked EPS-filled tires covered by a geomembrane sheet, and a layer of 0.3-m thick covering soil on the top. The purposes of use of the geomembrane sheet are to prevent the soil from falling into the spaces between the EPS-filled tires, and to protect EPS material from the possible gasoline leaked from automobiles and seeped through the cracks in the pavement layer. Trial DEM analyses showed that 10–15 degree reductions in friction angle between the geomembrane and the covering soil or EPS-filled tires do not significantly affect the result of this problem. Thus, the use of geomembrane with rough surface is justified.

In order to obtain the modulus of sub-grade reaction, k_g value, numerical plate loading tests were performed on the top of the covering soil as shown in Figure 9.

The pavement design manual by AASHTO (1998) suggests that the modulus of sub-grade reaction can be determined from either one of two types of plate bearing tests. One is repetitive static plate loading test (AASHTO T221, ASTM D1195-93), another one is non-repetitive plate loading test (AASHTO T222, ASTM D1196-93). In the repetitive test, the k_g value is determined from the

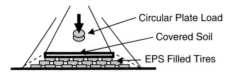

Figure 10. Model A: Circular plate loading on discrete system of EPS-filled tires.

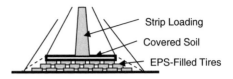

Figure 11. Model B: Strip loading on discrete system of EPS-filled tires.

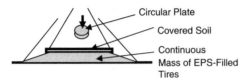

Figure 12. Model C: Circular plate loading on continuous system of EPS-filled tire mass.

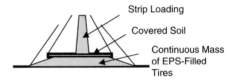

Figure 13. Model D: Strip loading on continuous system of EPS-filled tire mass.

ratio of the load to its corresponding recoverable elastic deformation at the tenth cycle (Rodriguez et al., 1988). On the other hand, the load-deformation ratio at the deformation of 1.25 mm (0.05 in.) is used to determine the modulus of sub-grade reaction in the non-repetitive test. Since EPS-filled tires tend to stabilize after a series of load repetition, the repetitive test method was adapted and thus the k_g value can be determined by the following equation.

$$k_g = q / \delta_v \tag{4}$$

where q = average compressive stress applied on rigid circular plate; and δ_v = vertical deflection at the center of rigid circular plate due to the applied load.

Since UDEC DEM code is two-dimensional code, it is necessary to use an indirect approach to solve three-dimensional circular plate loading test (Model A) as shown in Figure 10. Model A is the system to obtain the design sub-grade reaction modulus k_g. Models B, C, and D are further introduced. Model B (Figure 11) is the same system as in Model A but with a strip loading in stead of a circular loading. This is solvable by UDEC EDM code. Model C (Figure 12) is the same as Model A, but EPS-filled tire system is replaced by an equivalent continuous mass. Three-dimensional Model C is solved by using a finite element code (FEM), KENLAYER, which is capable to analyze deformation, stress, and strain in layered pavement structures (Huang, 1993). Model D (Figure 13) is the same as Model C, but with a strip load in stead of a circular load. This is two-dimensional model and thus UDEC DEM is used.

Based on the solutions for Models B, C, and D, an assumption was made that the ratio of vertical deflection due to circular load in a discrete system (Model A) to the vertical deflection due to strip load in a discrete system (Model B) should be the same as the ratio of those deflections in a continuous system (Model C and D) and thus;

$$\delta_{vA} / \delta_{vB} = \delta_{vC} / \delta_{vD} \tag{5}$$

where δ_{vA}, δ_{vB}, δ_{vC}, and δ_{vD} are the vertical deflections of the plate in Model A, in Model B, in Model C, and in Model D, respectively. δ_{vB} and δ_{vD} were obtained by DEM, while δ_{vC} was solved by using FEM, and thus δ_{vA} can be computed from Equation 5. The value of δ_{vA} was then substituted into Equation 4 to obtain the modulus of sub-grade reaction, k_g.

FEM simulation requires an elastic modulus of the EPS-filled tire systems. The modulus was obtained by modeling compression tests of multiple layers of EPS-filled tires unit using the DEM as seen in Figure 14. The equivalent elastic modulus of EPS-filled tire mass was dependent of the applied stress in a similar manner as that of the single EPS-filled tire unit as shown in Figure 15.

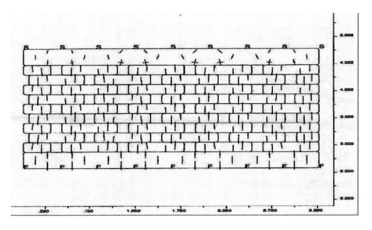

Figure 14.   Compression test of EPS-filled tire mass by DEM (vectors indicate principal stresses).

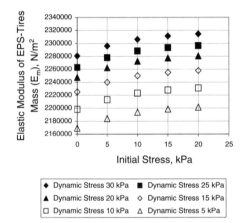

Figure 15.   Equivalent elastic modulus of
EPS-filled tire mass.

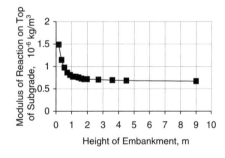

Figure 16.   Modulus of reaction on top of
sub-grade soil covering EPS-filled tires.

The computed moduli of reaction k_g on top of the 0.3-m-thick sub-grade soil covering EPS-filled tires (Model A) were plotted as a function of the height of the embankment in Figure 16. The k_g values appear to be too low to design a pavement structure directly on it. Therefore, a gravel sub-base layer on top of the covering sub-grade soil is necessary before placing any pavement on the top. Figures 17 and 18 present the modulus of reaction k_b on top of various thickness of non-stabilized and stabilized gravel sub-base layer, respectively. These figures were obtained from the graph of k_b values on top of sub-base as function of k_g value on top of sub-grade and sub-base thickness proposed by the 1966 Portland Cement Association (PCA) design manual of pavement structures as complied by Rodriguez et al. (1988).

In Figures 17 and 18, with increasing sub-base thickness, the k_b values increases, and a rigid pavement structure over the sub-base layer can be designed for light traffic. Based on the 1966 PCA method, Figures 17 and 18 can be used for the design in corporation with the axle loads. In conclusion, a typical cross-section of highway designed with EPS-filled tire embankment can be drafted as seen in Figure 19.

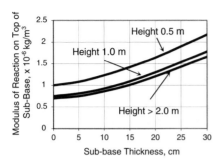

Figure 17. Calculated k_b values on non-stabilized sub-base.

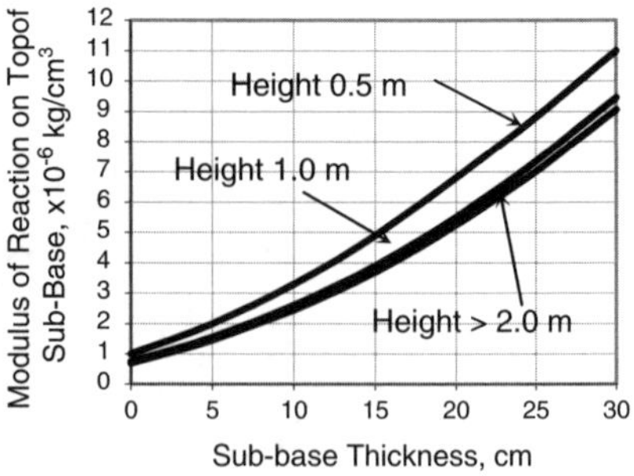

Figure 18. Calculated k_b values on stabilized sub-base.

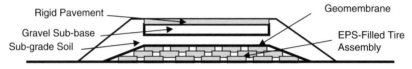

Figure 19. Typical cross-section of highway designed on EPS-filled tire embankment.

6 NUMERICAL EVALUATION OF HIGHWAY DESIGNED WITH EPS-FILLED TIRE EMBANKMENT SYSTEM

The numerical evaluations of designed embankments with EPS-filled tire system were carried out by using the DEM. It is assumed that possible heavy traffic of 100 kN axle loads are running on a highway. In the worst case, the wheel loads may be continuously intense that the tracks of wheel loads can be replaced by strip loads of 68.3 kPa. Figure 20 illustrates a configuration for the investigation, which yielded results leading to following considerations.

6.1 Stresses in embankment components

Under the described traffic, stresses in all EPS-filled tires are within the EPS' creep limit as tested by Duskov (1998). However, the analysis of this system during construction time, when only the sub-grade soil is placed, revealed that tensile stresses are created at lower section of the sub-grade soil and thus it was necessary to design the geomembrane to carry tension of at least 10 kN per one meter length of the highway.

6.2 Vertical stress on the soft foundation soil ground

Total vertical stress on soft foundation clay, which includes the static stress from the weight of embankment and the traffic-induced stress, is critical for the design because this system is intended for solving the problem of soft ground. Figure 21 shows the maximum traffic-induced vertical stresses calculated on soft foundation clay as function of the height of embankment for both EPS-filled tire system and pure soil sub-grade embankment. EPS-filled tire embankment system spread stress over a wider area under the applied load and thus shows slightly smaller maximum stresses than the cases of pure soil embankment. Figure 22 demonstrates the possible maximum total vertical stress (static and traffic-induced) on the soft ground surface under EPS-filled tire embankment system in comparison with stress under regular soil embankment. The system with EPS-filled tires reduces the total vertical stress to about one third to that of ordinary embankment, which is a significant advantage for soft ground conditions. The total vertical stress curve of the EPS-filled tire in this figure can be used as a guideline for the consideration of bearing capacity requirement for the soft foundation ground.

346

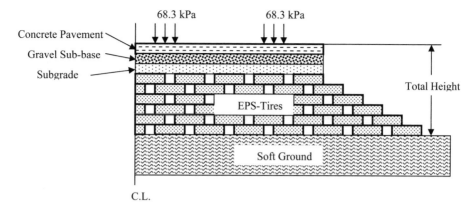

Figure 20. DEM pavement model designed with EPS-filled tire embankment with traffic load.

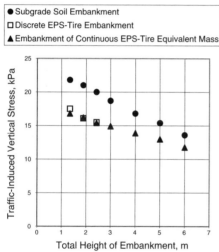

Figure 21. Maximum traffic-induced vertical stresses on soft foundation clay.

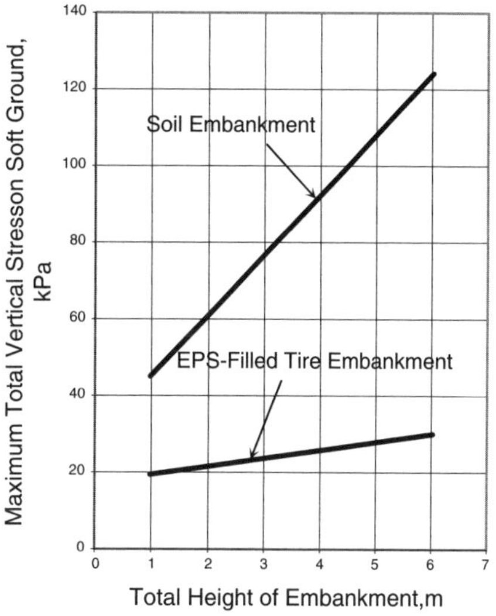

Figure 22. Approximated maximum vertical stress on soft ground surface.

6.3 *EPS-filled tire embankment with vertical cut*

In some cases, when the right of way is limited, side slope has to be avoided. EPS-filled tire embankment with vertical cut at side slope can be constructed with masonry wall. Analysis by the DEM as shown in Figure 23 demonstrates very small lateral and shear stresses on the wall boundary. It was also revealed from a separate analysis that the vertical cut of EPS-filled tire system can stand by itself without any retaining wall under a worst traffic scenario.

7 CONCLUSION

The elastic modulus of EPS-filled tire prototypes in this study was lower than it was initially expected. This is mainly due to the problem of air pockets remaining unfilled with EPS. However, even with the relatively low elastic modulus, EPS-filled tires can still be used as an embankment fill material as evidenced by the numerical analyses. It has been proved also that the total vertical

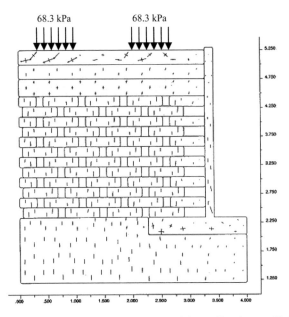

Figure 23. Analysis of EPS-filled tire system with upright retaining wall under a traffic load (vectors indicate principal stresses).

stresses on the foundation ground are reduced to about one third to that of ordinal embankment and thus it is highly beneficial to construct embankment with EPS-filled tires on soft foundation ground.

The next step of this research could be investigation of an experimental full-scaled EPS-filled tire embankment to study more aspect so that the design of highways with the EPS-filled tire system can be improved. At this moment, this system is more suitable for temporary roads because EPS-filled tires may be used repeatedly. The tire walls make the composite durable because they act as a shield protecting the contained EPS from any possible damages such as scratching or punching.

As could be anticipated, the cost of a composite is always higher than a pure material. By analyzing the manufacturing expenses, the cost of this newly invented material may be unfavorable because it includes expensive EPS and injection cost in addition to the cost of collecting and handling used tires. However, the system of EPS-filled tire embankment may be more realistic if the elastic modulus and the strength of the EPS-filled tire composite are improved. This can happen by improving the manufacturing technique so that the EPS foam beads can flow thoroughly inside the hollow space of the tires. One possible way to do so is using a tire as a mold itself to produce each single EPS-filled tire unit individually instead of molding several units in a large mold at a time. For environment concern, this system is worth study because hundreds of thousand of used tires can be kept inside a highway embankment of only 100 meters long.

REFERENCES

AASHTO, 1998, "*Supplement to the AASHTO Guide for Design of Pavement Structures, part II*", American Association of State Highway and Transportation Officials, Washington, DC, 74 p.

Beinbrech, G., and Hillmann, R., 1997, "EPS in Road Construction-Current Situation in Germany", *Geotextiles and Geomembranes*, Vol. 15, pp. 39–57.

Cundall, P.A., 1971, "A Computer Model for Simulation Progressive, Large-Scale Movements in Blocky Rock Systems", *Rock Fracture: Proceedings of the International Symposium on Rock Mechanics* (section II-8), Nancy, France, October 4–6, 1971.

Duscov, M., 1998, "EPS as a Light-Weight Sub-Base Material in Pavement Structures" (2nd edition), Vlaardingen, Netherlands, 251 p.

Edil, T.B., and Bosscher, P.J., 1994, "Engineering Properties of Tire Chips and Soil Mixtures", *Geotechnical Testing Journal*, GTJODJ, Vol. 17, No. 4, pp. 453–464.

Esch, D.C., 1995, "Long-Term Evaluations of Insulated Roads and Airfield in Alaska", *Transportation Research Record No. 1481*, Transportation Research Board, Washington, DC, pp. 56–62.

Hoppe, E.J., 1994, *"Field Study of a Shredded Tire Embankment"*, Interim Report No. FHWA/VA-94-IR1, Virginia Transportation Research Council, Charlottesville, VA.

Huang, Y.H., 1993, *"Pavement Analysis and Design"*, Prentice Hall, Englewood Cliffs, NJ, 805 p.

Itasca Consulting Group, Inc., 1996, *"UDEC: Universal Distinct Element Code, Version 3.0"*, Volume I: User's Manual, Minneapolis, MN.

PIARC, 1997, *"Lightweight Filling Materials"*, Technical Committee on Earthworks, Drainage, Sub-grade (C12), World Road Association, Paris, France, pp. 248–287.

Rodriguez, A.R., Castillo, H. and del, Sowers, G.F., 1988, *"Soil Mechanics in Highway Engineering"*, Trans Tech Publications, Clausthal-Zellerfeld, Germany, 843 p.

Takahara, T., and Miura, K., 1998, "Mechanical Characteristics of EPS Block Fill and It's Simulation by DEM and FEM", *Soils and Foundations*, Vol. 38, No. 1, pp. 97–110.

Tweedie, J.J., Humphrey, P.E., and Sandford, T.C., 1998, "Tire Shreds as Lightweight Retaining Wall Backfill: Active Conditions", *Journal of Geotechnical and Geoenvironmental Engineering*, Vol. 124, No. 11, pp. 1061–1070.

Investigating the strength characteristics of tyre chips – sand mixtures for geo-cellular structure engineering

Ph. Gotteland, S. Lambert & Ch. Salot
3S-R Laboratory, CNRS-UJF-INP, Grenoble, France

ABSTRACT: Waste tyres are more and more widely used in geotechnical applications as backfill material in substitution or in combination with natural soils. Beyond the economical and environmental concerns, these materials can help solving problems with low shear strength soils. In a first part, the study aims at investigating the mechanical behaviour of tyre chip – sand mixtures thanks to triaxial tests. Two factors were studied (i) the tyre chip content, from 0 to 100% by mass and (ii) the orientation of the pieces of tyre, with four varying orientation conditions. In a second part, the presentation will describe the feasibility of using tyre chip-sand mixture as a fill material in cellular structures (gabion technique) for potential civil engineering application. This investigation is part of a large study engaged in an innovative development aiming at constructing the front face, the core and possibly the back face of reinforced structure using gabion technique. Analysis of the mechanical behaviour of the gabion cells filled with geocomposite material, i.e. tyre-sand mixtures, will be presented. Particular interest is placed on the deformation of the cell in comparison with traditional stone filled gabion cells. It is expected that the proposed tyre-sand mixture will have the potential to act as an absorber of energy rather than an energy dissipater.

1 INTRODUCTION

Tyre disposal is a huge challenge faced by waste management engineers, particularly in more economically developed countries where there exist stockpiles of tyres in alarming volumes. Their disposal proves to be a large problem as tyres do not decompose. Waste tyres pose a threat to public health and to the environment under current methods of disposal due to the following three reasons; (i) they occupy large volumes in already overcrowded landfills, (ii) waste tyre storage can caught fire, (iii) waste tyre dumps provide a breeding ground for vermin, including rats and mosquitoes. In this context, the European Union is progressively banishing the disposal of tyres in landfills, in any form, through the directive 1991/31/EC that applies to all countries within the European Union. This directive has prompted interest in new ways to recycle end of life tyres, as for instance in civil engineering applications.

Waste tyres have many properties which result in their being of value from a civil/geotechnical engineering perspective: low density, high strength, hydrophobic nature, low thermal conductivity, durability, resiliency and high frictional strength. It is due to these properties that the use of tyres has been specifically focused towards civil engineering applications such as lightweight material for backfill of retaining structures, drainage layer, thermal insulation layer or reinforcement layer.

Otherwise, due to the increasing expansion of the cities, urban zones, or tourist resorts in mountain sites, it is often requested to build in "vulnerable" sites requiring to protect the construction zones against rockfall hazards. The traditional solutions of works, i. e. reinforced or not reinforced protection dams, dissipate the energy of the boulder mainly thanks to their large mass. Such works have the disadvantage of a significant base to satisfy static stability, which represents important costs. New techniques with reduced base prove very interesting, especially if one associates criteria of reparability, which is facilitated by the modularity of the structure. One innovative development

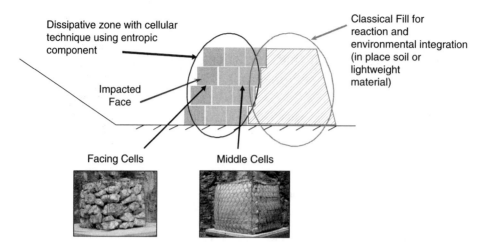

Figure 1.   Concept of the dissipative cellular structure.

for cellular structures is to build the front face, possibly the core and the back face of the structure with cell assembly (Figure 1). A cell is a wire netting envelope filled with the site material (e.g. gravel, stones, sand or composite material entropic such as adapted mixture of scrap tyres, including an appropriate geosynthetic bag). Relative displacements between cells in contact and large deformations within them may take place, making it possible to increase the energy dissipation by friction (Bertrand 2006, Gotteland et al. 2005). Using this technique allows easy repair after rockfall events because of the cellular constitution of the structure and the small area of damage. Cells are commonly called gabions and this technique is often used for the construction of retaining walls and noise protection structures, for protection against hydraulic erosion, and rockfall protection.

This study is part of a program investigating the possible use of pieces of tyre as filling material for structures subjected to impact. The most important mechanical property in the investigation of the suitability of the use of soil – tyre pieces mixtures in geotechnical applications is shear strength. Presented results focus on the characterisation of the shear strength and deformation behaviour of mixtures of sand and tyre ships and aims at determining the influence of different parameters such as tyre content.

2   BACKGROUND

Among all the studies aiming at characterizing tyre pieces and mixtures with soils (Reddy et al. 2001, Edil & Bosscher 1994, Foose et al. 1996) some refers specially to triaxial tests (Zornberg & Cabral 2004, Youwai & Bergado 2003, Wu et al. 1997, Yang et al. 2002). These differ in aims, forms and shapes of the pieces of tyre and test method but provide a basis for confirmation of results observed throughout this testing campaign while allowing for differences.

The internal angle of friction of waste tyre shreds with soil is $33.7° \pm 15°$, as reported in a statistical analysis of 13 sets of investigations by different authors (Reddy et al. 2001). Yang et al. 2002 correlated the initial stress-strain modulus, with the confining pressure. Zornberg et al. (2004) concluded that an optimum tyre shred percentage, where maximum shear strength was attained, occurred at approximately 35% by mass of tyres. The volume variations observed were similar to the results presented by Youwai & Bergado 2003. The factors with the greatest influence on the shear strength property of the material appeared to be normal stress, confining pressure, tyre content and density of mixture. These factors were investigated in the presented study, so as to confirm previously reported results and to further investigate the mechanical behaviour of this geocomposite material.

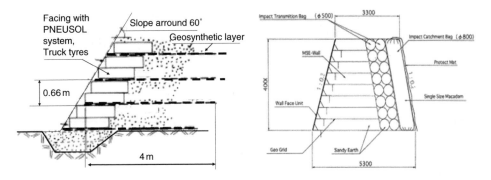

Figure 2. a) French PNEUTEX system for structure protection: entire tyres at the facing and geosynthetic reinforcement. b) Dissipative facing system; GEOROCK wall – (Yoshida, 1999).

Rockfall protection dykes are constructed by ground elevation, from bottom to top. Original roll filled (heavy structure with very high covered area) were progressively replaced by reinforced dykes with stiffened upstream face and geocomposite reinforcement. Classical reinforcements are performed using bi-dimensional materials such as geotextiles, geogrids or meshes. In some techniques, commonly used in France, entire tyres are placed on the upstream face to endure the impact and dump the shock energy (PNEUTEX system, Figure 2a). The design of dykes considers energy levels to which the structure is projected to be submitted. It is conducted without any consideration to dynamics: a static load is deduced from the energy. The designer checks the internal stability of the structure to its dead load and to the equivalent static load. The global stability of the structure, depending on the characteristics of the supporting soil, is also verified. In these structures, the impact energy dissipation is due to the displacement of the structure materials. These structures proved to be efficient but require an important soil volume, as fill materials, and a large covered area. Recently, a new and interesting concept appeared: the dissipative facing (Yoshida, 1999). The principle of this technique is to separate the impacted zone from the downstream part of the structure thanks to a dissipative layer made of wrapped geomaterials (Figure 2b).

The proposed evolution takes into account the dynamic response of the structure for their design. One innovative development for cellular structures is to build the front face, possibly the core and the back face of the structure with cell assembly to localize and diffuse the energy dissipation (Figure 1).

3   MATERIALS

3.1   Tyres ships

The tyre ships (as defined by CEN Workshop Agreement – CWA-14243) used in this investigation were produced using a punching method, where metal cylinders are forced to cut through tyres. The resulting material is composed of rounded pieces (30% by mass) and pieces of no definable shape or dimension, with or without sticking out steel wires (Figure 3). The number of wire layers within the structure of the pieces varied from zero to six.

To be able to characterize tyre chips with laboratory tests it was decided to focus on specimens composed of circular pieces of tyres with sticking out steel wires cut. Indeed, the raw material is rather hard to define, it presents a risk to the equipment, and mainly to the triaxial membrane, and its variability may lead to high tests results variability. The average diameter and thickness values for circular chips were measured to be respectively of 28.1 mm and 10.4 mm. The thickness varies significantly and depends largely on the number of steel belt layers. Tyres containing no steel belt layers generally have a smaller thickness. The unit weight of the rounded pieces ranged from 11

Figure 3. a) Sample of raw material , b) selected circular chips of tyres for TRX tests, c) tyre-sand mixture (M) containing 30% by mass of tyres.

to 15.4 kN/m³, also depending on the presence of steel belt layers, with an average value of about 13.3 kN/m³. These values are consistent with reported data.

3.2 Sand

Tests were performed with a River Sand from the Seine River, classified in Class D in French Standards. As this is a fluvial soil its most important characterisation is that it contains well-rounded grains. It is an unconventional sand of a shell basis, with unit weight 17 kN/m³. All sand used in tests was sieved using a 5 mm grain size sieve. The sand was tested dry.

3.3 Tyre-sand mixtures

Four different types of arrangement were investigated; (i) alternate horizontally and vertically placed tyre pieces (H&V specimens), (ii) horizontal only (H specimens), (iii) vertical only (V specimens) and (iv) no orientation (NO specimens). NO specimens are representative of on-site use whereas H&V lay out aimed at producing well-defined specimens. Except for NO specimens each piece of tyre was placed manually. Pieces of tyre placed on the perimeter of the specimen were selected without steel layers in their structure, in order to reduce the risk of puncture of the membrane. H&V specimens were prepared placing alternatively pieces of tyre horizontally and vertically. For equivalent tyre content, a same mass of tyres was placed in the cell together with a same mass of sand then compacted in order to reach similar densities. The relative unit weight of the sand within a series varied from 16.3 kN/m³ to 17 kN/m³, depending on the tyre content and orientation, except in the case of 50% by mass tyre specimens where it was of 10.5 kN/m³ as it was almost impossible to compact the mixture without any segregation.

4 METHOD

For Triaxial compression tests, nine series of consolidated drained, CD tests were carried out. The tests were conducted in general accordance with French standard (NF P 94-074) under a strain rate of 2 mm/min, measuring the confining pressure, the axial load and the specimen volume variation. Adaptations to the preparation procedure were imposed by the characteristics of the material tested (tyre content and tyre orientation). Tests were interrupted either after the shear strength peak or before a strain of about 20%, due to the limited section of the triaxial cell. The size of the pieces of tyre requires the use of a large-scale triaxial cell to test large specimens: 150 mm in diameter and 300 mm in height, giving an average specimen volume of $5 * 10^6$ mm³. Specimen volume changes during the test were determined measuring changes of water level in a burette connected to the cell pressure line and taking account of the loading piston displacement.

For each series a minimum of three confining pressures was investigated, 50, 75 and 100 kPa so that the strength parameters could be determined. Low confining pressures were applied because materials tested are to be used in up to 4 meters in height civil engineering projects.

Each specimen within a series was prepared with the same pieces orientation and the same tyre content, hence giving a similar unit weight (see Table 1) as described in the previous section. The deviatoric stress was not corrected according to the cross-sectional area change as required by the

Table 1. Scope of the testing program and shear strength parameters, na = Not applicable, H = Horizontal, V = Vertical, NO = No Orientation.

Specimen description			Shear strength parameters			
Series	Tyre chips content (% by mass)	Tyre chips orientation	Unit weight (kN/m^3)	ϕ (°)	C (kPa)	ϕ_{eq} (°)
A	0	na	16.7	40.9	0	40.9
B	15	H&V	15.5	41.1	10	44.5
C	14	H	15.9	42.6	15	44.5
D	14	V	15.9	41.7	7.5	43.5
E	14	NO	15.5	39	13.8	43
F	22	H&V	15.3	36.1	50	45
G	50	NO	11.4	41.5	7.5	43.5
H	100	H	6.8	31	28	38.5
I	100	NO	6.1	19	16.3	25

standard. Indeed, this formula is intended for use with uniform materials e.g. sand only, and using it assumes that a tyre chip-sand mixture or a tyre ships only specimen will behave similarly. Failure of the specimen was deemed to have taken place once a maximum deviatoric stress was attained. However for series I no peak was obtained whereas for series H a peak was observed from 14% strain depending on the confining pressure. For comparison purpose it was decided to consider for series I and H the deviatoric stress at 14% axial strain where other authors consider 15 (Zornberg et al 2004), 10 or 20% strain (Yang at al. 2002). Angle of friction (ϕ) and cohesion (C) were calculated after plotting Morh's circles. An equivalent angle of friction (ϕ_{eq}) was also calculated considering the cohesion as being zero. This data can be used in order to simplify the comparison between the series (Zornberg et al 2004).

5 NUMERICAL MODELLING WITH DEM METHOD

The Discret Element Method (DEM) is a numerical method developed by Cundall (Cundall & Strack 1979) to model granular media. This method is used nowadays in many geomechanical fields such as: granular flows, concrete modelling, wave propagation in particle assemblies, mining, and geotechnical problems. The DEM models the interaction between particles. The basic elements are spherical particles, which are considered as rigid bodies. All the particles can interact with one another through contact or remote interaction. The contact behavior is characterized using a soft contact approach (Cundall ct al. 1992). Contact interactions are governed by the linear elastic contact model in the normal direction and with the linear elastic perfectly plastic model of the tangential direction of contacts. When there is contact, the force between two elements is proportional to the overlap, remaining small compared to the size of the particles. Force and overlap are related by the normal stiffness, k_n, and shear stiffness, k_s. The maximum shear force allowable is dependent on the friction between the two elements in contact and is described using the usual Coulomb law. The friction coefficient is defined as $\mu = \tan \Phi$, where Φ is the friction angle between the elements in contact. The materials as well as sand and tyre pieces are modeled with these contact models.

In order to model the real behaviour of materials, it is not necessary to take into account their real shape. Thus, non spherical elements, called clumps, are created bonding together 2 to 5 spherical particles (Figure 4a); Even if the real grain or piece of tyre shape is not strictly reproduced and the real microscopic behaviour is not simulated, the aim is to find the type of clumps that allows reproducing the experimentally observed macroscopic behaviour with low computing time. The sample modeling principle and the numerical test methods are described in Salot 2007. Tyre pieces

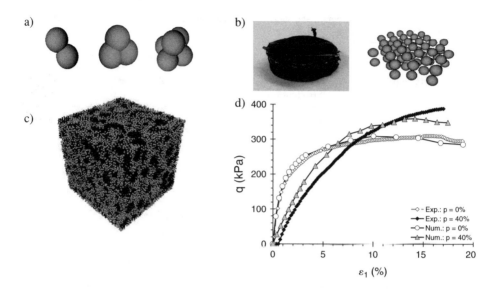

Figure 4. a) Clusters for sand modelling and b) Clusters for piece of tyre inclusions. c) Tri-numerical sample for 40% piece of tyres content d) DEM modelling compared to experimental results.

(a) Series B,   (b) Series F,   (c) SeriesC,

Figure 5. Specimens at the end of the tests.

are modeled using cylindrical clumps illustrated on figure 4b. Spheres are not joined in order to increase interlocking between elements, so to increase the macro-friction. Micromechanical and geometrical parameters are calibrated separately. Ratios of tyres in mixtures with soil are the same as in experiments. Numerical sample is presented (Figure 4c) and results of the modelling are compared with experimental ones for 0 and 40% tyres content (Figure 4d). Evolutions of deviator stress show a good agreement between experimental results and numerical predictions. In the end, this model will allow developing a homogenized constitutive law of these composite materials.

6 RESULTS AND DISCUSSION

Tyre chips-sand specimens tend to have many rupture planes, that follow closely the lay out of the chips within the specimen, particularly where there are vertically placed chips around the perimeter of the specimen (a and b) (see Figure 5).

Shear strength parameters are given in the lasts columns of Table 1. Angles of friction vary from 19 to 42.6°. Cohesions vary from 0 to 50 kPa; It is not a real cohesion but an intercept cohesion resulting from the simplistic Mohr-Coulomb linear model. The cohesion is not negligible

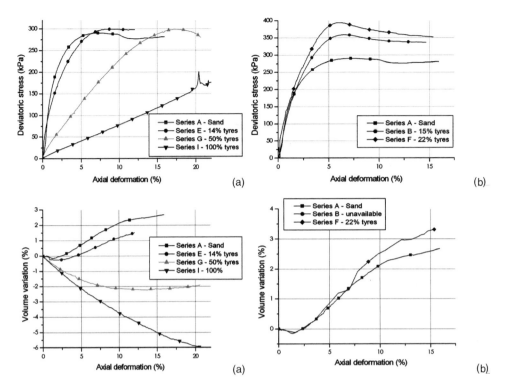

Figure 6. Influence of the tyre content on the behaviour a) NO specimens, b) H&V specimens (75 kPa confining pressure tests).

for series E, F, H, and I mainly, affecting ϕ_{eq} values. Thus, considering ϕ_{eq} values rather than ϕ and C values to characterise the shear strength of the various specimens may lead to different conclusions. Whatever, it appears that the shear strength of a tyre chips only specimen is much less than that of a sand or a sand-tyre chip mixture up to 50% mass. These shear and stress results are in good agreement with results reported (Zornberg et al. 2004) regardless of size and shape of pieces of tyre. Investigating the influence of the tyre content and the chips orientation allows going deeper in the comparison of the different series.

6.1 Influence of the tire contents

Figure 6a shows the influence of the tyre content on the behaviour of NO specimens during 75 kPa confining tests giving results on series E, G, and I together with results on series A. The tyre content greatly affects the deviatoric stress-strain and volumetric-strain behaviour.

The sand shows contraction then dilation, characteristic of a medium to dense sand. The peak on stress-strain graph is well defined at approximately 6% axial strain, although it is unexpectedly high compared to peak obtained at 50 and 100 kPa. The difference in behaviour of 100% tyre specimens with sand is obvious. The 100% tyre NO specimen stress-strain curve shows approximately linear behaviour with an initial modulus of 790 kPa. This value is smaller than what is predicted by Yang's correlation (Yang et al. 2002) giving 882 kPa in these conditions. The volumetric strain is contractive following an approximately linear relationship. This linear behaviour of tyre only specimens is consistent with previously undertaken research (Zornberg & Cabral 2004). With a mix of 14% by mass of tyres the behaviour is similar to that of a medium to dense sand although the tyres provide greater shear strength to the mixture, allowing higher peak strength to be achieved at

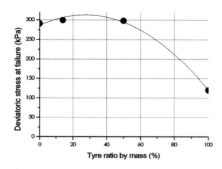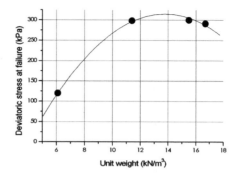

Figure 7. Influence of (a) tyre content and (b) unit weight on the shear strength at failure for NO specimens (75 kPa confining pressure tests).

approximately 8% axial strain. The 50% tyre specimen corresponds more closely to the behaviour of the 100% tyre series, but the stress-strain curve exhibit a peak after an axial strain of 15%, much later than for sand only. Although the behaviour between 14% and 50% specimens vary considerably it should be noted that the peak shear strength achieved is similar, hence it is proposed that between these two percentages a greater peak strength can be achieved.

The same conclusions can be drawn from results obtained with H&V specimens, as shown (Figure 6b) on giving the same presentation for series B and F, together with series A. It gives complementary information with results on a 22% tyre specimen. Increasing the tyre content from 15% to 22% increases the peak shear strength from 359 to 394 kPa. The volume variation for the 22% tyre specimen also follows closely as for a medium to dense sand with a dilatant behaviour following a short contractive period.

Hence it is thought that an optimum percentage mass lies under 50% of tyres which would give a maximum peak shear strength for a particular confining pressure (Figure 7). The composite shear strength envelopes were defined by a second order polynomial with $R^2 > 0.98$, hence a good correlation. The same observations were made at the other confining pressure with an optimum tyre content changing a little; 28 and 36% respectively for 50 and 100 kPa. Taking an average of the three, an overall optimum percentage mass of tyres, which gives maximum peak shear strength is 34%. An average optimum density is 13.5 kN/m^3. These results correspond closely to published optimum percentage mass of 35% (Zornberg et al. 2004). This optimum is not defined by a well-marked peak; over a range of tyre from 20 to 40% the shear strength is close to the optimum.

6.2 Impact tests on confined geo-cells

Impacts on geo-cells refer to both the dynamic response of geo-materials and to geo-material – wire netting cage interaction. Many authors have recently investigated the behavior of geo-materials when submitted to impact by boulders, for instance dealing either with in-situ layers of soil (Pichler et al., 2005) or cushion layers (Montani Stoffel, 1998). Others have studied the confinement effect of an envelop on the characteristics of a geo-material, mainly dealing with cylindrical objects submitted to diametrical or axial static compression (Bathurst & Karpurapu, 1993), (Iizuka et al., 2004). But no study concerns the impact response of cells filled with geo-materials. A series of impact tests was thus preformed varying the fill material (Lambert 2007). The cell response is evaluated through both the resistance it put up to the penetration by the boulder and the force transmitted by the cell to its support. The aim is to identify the conditions for an optimum energy dissipation and impact force attenuation.

The experiments consisted in dropping a 250 kg spherical boulder from a height of 5.5 m onto a cubic cell, 0.5 m in height and filled with different materials. The cell was placed on a heavy pedestal made of concrete. During the impact the lateral faces of the cell were free to deform. Measurements

Table 2. Falling tests conditions and main results.

Filling material	Cell weight (kg)	Impact energy (kJ)	F_{imp} (kN)	F_t (kN)	Att. (−)	t_{imp} (ms)	ΔP (%)
Granulates, #1	215	13,5	127	77	0.6390	≅ 100	–
Granulates, #12	215	13,5	144	43	0.6	≅ 100	46
Sand	202	13,5	81	46	0.6	≅ 80	16
Mixture (30% tyre) #1	190	13,5	48	39	0,8	≅ 65	12
Mixture (30% tyre) #2	190	13,5	74	40	0.5	≅ 65	19

a)   b)

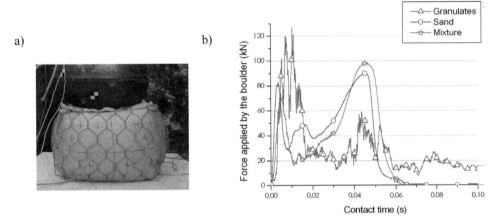

Figure 8. a) 500 mm in height unconfined sand cell, b) Influence of the type of filling material (gravel/sand/tyre-sand mixture) on the force applied by the boulder for boulder falling height of 5.5 m.

made during the impact were the boulder centre acceleration and the force transmitted by the cell to the pedestal, allowing determining the maximum force applied by the boulder, F_{imp} , and the maximum force transmitted to the pedestal, F_t. The attenuation, Att., is here defined as the ratio between these two values. The duration of the impact, t_{imp}, and the cell perimeter variation, ΔP, were also measured. Results clearly exhibit the influence of the filling material (Table 2).

Cells filled with fine materials exhibit rather smooth curves compared to cells filled with granulates (Figure 8). This is mainly due to the coarse nature of the granulates, implying sudden and important collapses due to force chains rupture. Sand and mixtures curves exhibit two main peaks. The second one is due to the solicitation of the envelop: the confining effect increases dramatically with the cell lateral deformation. The first peak corresponds to the response of the fill material to the impact. The response of cells filled with the mixture appears to be variable. It is thought to be due to the difficulty to reproduce exactly the same filling from one test to the other. The attenuation appears to be inappropriate to evaluate the ability of a cell to damp the impact force, as it also takes into account the impact force which depends on the filling material. The transmitted force appears to be significantly reduced using fine materials. Mixtures seem to be slightly more efficient and the impact time duration clearly shows that adding tyres in the sand changes the behaviour of the cell.

7 CONCLUSION

A triaxial test campaign on tyre chip-sand mixtures was carried out varying the tyre content and chips orientation in order to investigate the mechanical behaviour of the mixture: to determine strength parameters, angle of friction and cohesion. In particular the campaign determined the optimum percentage mass of tyres and the optimum unit weight, i.e. which give the maximum, shear strength. The results led to conclusions: the percentage mass of tyres has a great influence on the shear strength of the mixture. Strength increases as the tyre content increases up to an optimum percentage mass of 34%, after which shear strength decreases. This optimum corresponds to an optimum unit weight of 13.5 kN/m^3, the tyre content changes the volumetric variation – axial strain behaviour. It appears that the behaviour of mixtures changes from a sand like behaviour to a tyre only like behaviour at a percentage mass of 34%., the equivalent angle of friction for 100% tyre specimens with no orientation is 25° compared to 41° for the sand.

The Discret Element Method was used to model this particular media and results of the modelling are compared with experimental ones. Good agreement between experimental results and numerical predictions are obtained. This model will allow developing a homogenized constitutive law of these composite materials, witch will be used for composite geo-cells modelling.

Impact tests were preformed varying the fill material of geo-cells to study the capability of composite tyre sand fill. The geo-cell response is evaluated through both the resistance it put up to the penetration by the boulder and the force transmitted by the cell to its support. The aim is to identify the conditions for an optimum energy dissipation and impact force attenuation.

All these developments were used to develop an innovative cellular technique for rockfall protection structures, which consist to build the front face, possibly the core and the back face of the structure with cell assembly. Cells assembly of the core will be performed with tyre sand geo-composite. It is expected that the proposed tyre-sand mixture will have the potential to act as an absorber of energy rather than an energy dissipater.

REFERENCES

AFNOR, N.F. P. 94 074 – Sols: reconnaissance et essais – Essais à l'appareil triaxial de révolution. 1994, AFNOR: Paris. p. 36.
Bertrand D. – Modélisation du comportement mécanique d'une structure cellulaire soumise à une sollicitation dynamique localisée, Application aux structures de protection contre les éboulements rocheux. PhD thesis, Université Joseph Fourier Grenoble 1, Lirigm, Cemagref, Grenoble, 2006, p. 197.
Bertrand, D., Nicot, F., Gotteland, P., and Lambert S., 2005. "Modelling a geo-composite cell using discrete analysis". Computers & Geotechnics, Vol. 32, no 8, pp. 564–577.
Cundall, P.A., and Strack, O.D.L. 1979. A discrete numerical model for granular assemblies. Géotechnique, Vol. 29, pp. 47–65.
Cundall, P.A. and Roger, D.H. 1992. Numerical modelling of discontinua, Engineering computations (9), Pineridge press, 101–113.
Bathurst R.J. and Karpurapu R. – Large-scale triaxial compression testing of geocell-reinforced granular soils. *Geotechnical testing journal*, 1993, Vol. 16, p. 296–303.
Edil T.B. and Bosscher P.J., Engineering properties of tire chips and soils mixtures. Geotechnical testing journal, 1994. **17** (4): p. 453–464.
Foose G.J., Benson C.H., and B.P. J., *Sand reinforced with shredded waste tires.* Journal of geotechnical and geoenvironmental engineering, 1996. **122** (9): pp. 760–767.
Gotteland P., Bertrand D., Lambert S., and Nicot F. – Modelling an unusual geocomposite material barrier against a rockfall impact. *Proc of IACMAG 2005*, Torino (Italy), 2005, pp. 529–536.
Iizuka A., Kawai K., Kim E. R. and Hirata M. – Modeling of the confining effect due to the geosynthetic wrapping of compacted soil specimens. *Geotextiles and geomembranes*, Vol. 22, 2004, pp. 329–358.
Lambert, S. – Modélisation physique du comportement de composants cellulaires, Application a la conception d'ouvrages cellulaire. PhD Thesis in prep, Université Joseph Fourier Grenoble 1, 3S-R, Cemagref, Grenoble, 2007, in prep.
Montani Stoffel S. – Sollicitation dynamique de la couverture des galeries de protection lors de chutes de blocs. PhD Thesis, EPFL, Lausanne, 1998, p. 180.
Pichler B., Hellmich C. Mang H.A. – Impact of rocks onto gravel – Design and evaluation experiments. *International Journal of Impact Engineering*, Vol. 31, 2005, pp. 559–578.

Reddy K.R. and Marella A. Properties of different scrap tire shreds: implications on using as drainage material in landfill cover systems. The 7th International conference on solid waste technology and management. 2001. Philadelphia.

Salot Ch. – Modélisation numérique par éléments discret de matéraux hétérogènes, Application aux mélanges anthropiques pneus – sables. PhD thesis in prep, Université Joseph Fourier Grenoble 1, 3S-R, Grenoble, 2007. in prep

Yang S., Lohnes R.A., and Kjartanson H. Mechanical properties of shredded tires. Geotechical testing journal, 2002. **25** (1). pp. 44–52.

Yoshida H., "Recent studies in rockfall control in Japan", /Joint Japan-Swiss seminar on Impact loads by rockfall and design of protection structures/(Kanazawa) Japan, 1999, pp. 69–78.

Youwai S. and Bergado D.T., Strength and deformation characteristics of shredded rubber tire-sand mixtures. Canadian Geotechnical Journal, 2003. **40**: pp. 254–264.

Wu Y. W., Benda C.C., and Cauley R.F., Triaxial determination of shear strength of tire chips. Journal of geotechnical and geoenvironmental engineering, 1997. **123**(5): p. 479–482.

Zornberg J.G., Cabral A.R. and Viratjandra C. Behaviour of tire shred-sand mixtures. Canadian Geotechnical Journal, 2004. **41**: pp. 227–241.

Scrap Tire Derived Geomaterials – Opportunities and Challenges – Hazarika & Yasuhara (eds)
© 2008 Taylor & Francis Group, London, ISBN 978-0-415-46070-5

In-site experiment about isolation method of ground vibration using by pressed scrap tire

K. Hayakawa, I. Nakaya & M. Asahiro
Ritsumeikan University, Shiga, Japan

T. Kashimoto, M. Moriwaki & H. Koseki
OAK Co., Ltd, Hyougo, Japan

ABSTRACT: The vibration caused by construction work, a plant machinery, a means of transportation in the ground gives a nearby building a difficulty, and an unpleasantness is exerted on the people who live in the building. And it has been being a big environmental problems. It can think about the countermeasure method of such a ground vibration in each process of the vibration source, propagation route and accepted part of the vibration. Isolation wall is set up in the propagation route of this, and isolation countermeasure method to propose here tries to plan a vibration isolation. Some kinds of scrap tires were constructed as isolation wall at experiment site as to the spot, and the verification of the validity as the ground vibration countermeasure method was done as a master from a result of vibration investigation in the surface ground and underground by this report.

1 INTRODUCTION

Research on the effective use of the scrap tire becomes popular year by year, and it proceeds with it in the various industry fields in such cases as the electrical appliances, the building, construction and the car. A scrap tire is added to the usual fuel utilization even of that, and research (Miterai, 2005) about the effective use as a construction material (Morisawa, 2005) is being done. "The building of the circulation type economic society" which is the development subjected of the this country economy is in the background there. Through much is being reused, that most is fuel utilization, and afraid of an influence to give to the environment with a scrap tire in such cases as global warming.

On the hand, the vibration caused by construction work, the plant machinery, a means of transportation in the ground gives a nearby building a difficulty, and an unpleasantness is exerted on the people who live in the building. And it has been being a big environmental problem. It can think about the countermeasure method of such a ground vibration in each process of the vibration source-spread route-accepted part of the vibration. Isolation wall is set up in the spread route of this, and countermeasure method to propose here tries to plan of the vibration isolation (Hayakawa, 1999).

This countermeasure method an epoch-making method with the possibility to use in addition to the effect which reduces the discharge of CO_2 as the atmosphere environmental pollution problem in the ground vibration countermeasure as well. Some kinds of scrap tires were constructed as isolation wall in experiment site as to the spot, and the verification of the validity as the ground vibration countermeasure method was done as a master from a result of vibration investigation in the surface of the earth surface part by this report.

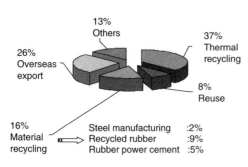

Figure 1. Recycling condition of scrap tire in Japan.

Figure 2. Vibration isolating wall by using scrap tire.

2 THE RECYCLING CONDITION OF THE SCRAP TIRE

The amount of production of the tire is on the increase caused by the development of the automobile industry of our country year by year, and it is 1,030,000 tons in 3,000,000 back of about 1, the weight of the rubber in 2003. But, whereas the amounts of occurrence of the scrap tire increase, too, and they are exceeding 100,000,000 tons in the weight of the rubber in this until 2003 for several years by the continuance. (JATKMA 2000; this 2004) Through 90% is mostly recycled, an influence on global warming by CO_2 that about 40% is thermal recycling, that is, salt peak use and which occurs is regarded as questionable with a scrap tire as shown in the Figure 1 reuse and material recycling (raw material utilization) are delayed in comparison with the Western countries, and it is the subject that the promotion of the reuse in this field is in a hurry in the country.

3 THE USE OF THE SCRAP TIRE AS THE GROUND VIBRATION ISOLATION COUNTERMEASURE (CASE 1)

3.1 *Experiment outline*

Accumulation (Figure 2 reference.) lays it in the pillar line-shaped in the ground, and the counter-measure method of the ground vibration being made the target by this research composes isolation wall with archetype in the pillar-shaped. And, the midair part of the tire is being filled with the tire chip and concrete. Experiment conditions are shown to a Figure 3 and Figure 4 (arrangement pattern of scrap tire isolation wall and the position of measurement of the ground vibration). Arrangement pattern of scrap tire isolation wall is being set at 3 kinds shown in Figure 3. It is width 8 m × depth 6 m laying underground depth 3 m which width 5 m × depth 4 m laying underground which a scrap tire (600 mm) for the car was usually used for, pattern of depth 4 m and pattern for the width 8 m × depth 4 m laying underground depth 2 m or a scrap tire (1000 mm) fro the large car was used for (Hayakawa & Moriwaki, 2005).

The ground condition of experiment site is as shown in the Figure 5. It is silt layer from the surface of the earth side to the establishment depth of isolation wall completely, and weathering gravel of 5–20 m is a little mixed. It is average N value 7 as a result of an examination with Standard Penetration Test to the establishment depth of scrap tire isolation wall, and it is the comparatively soft ground.

As for a source of vibration, traffic from the running vehicles should vibrate, and adopted shovel car in the test vehicles. The vertical component of the vibration acceleration levels were measured by using a vibration level meter as shown in Figure 3. In this case 1 m apart from the vehicles running line to 1 m–10 m and which occurs at the time of heavy equipment running was measured. Scrap tire isolation wall laid in the position of a distance 2.5 m from the vehicles running line. The measurement of the ground vibration was done in every arrangement pattern of each scrap tire isolation wall, and it verified effect on a decrease of the ground vibration and the tendency

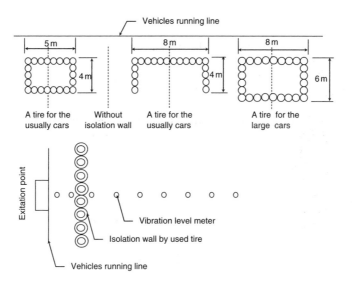

Figure 3. Layout drawing of vibration isolation wall.

Figure 4. Condition of experiment site.

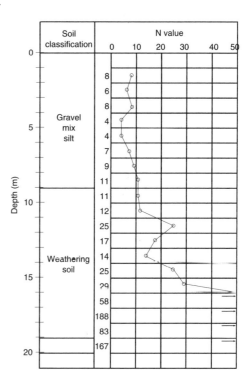

Figure 5. Boring log of the test site.

of the distance decline by comparing a result of measurement at the result of measurement of the ground vibration with a thing after isolation wall execution and the un-execution ground. And, the comparison of effect on a decrease of the ground vibration by look difference of that arrangement pattern was examined by using a result of vibration investigation of every arrangement pattern of the scrap tire about the similar vibration sensing line.

365

3.2 *A results of an experiment of case1 and its considerations*

The result of which effect on a decrease of the ground vibration was compared with in the Figure 6 again is shown about a result of measurement of the vibration acceleration level value at the ground in the Figure 7. The result which the distance decline tendency of the ground vibration was compared with in each experiment pattern. A vertical axis by the Figure 7 of the inside shows it in the difference from the vibration acceleration level value after the scrap tire isolation wall execution and ground. Therefore, minus numerical values is equivalent to the amount of effect on a decrease of the ground vibration more.

When the change tendency of effect on a decrease of the ground vibration along with exciting source distance of the Figure 7 is seen, a big decrease tendency has the ground vibration value to exciting source distance 4.5 m point in arrangement pattern of all scrap tire isolation wall. However, it becomes a gentle decrease tendency in the distance after that, and look difference by arrangement pattern of the scrap tire stops almost being seen in exciting source distance 10 m point. Moreover, an increase in the vibration acceleration level value which it can think about the thing due to the reflection of the wave motion is seen, and this phenomenon is the biggest with a result of the large automobile tire in front of the scrap tire isolation wall. Effect on a decrease of the ground vibration of 5–12 dB was seen in comparison with the un-execution ground. As shown in the Figure 7, effect

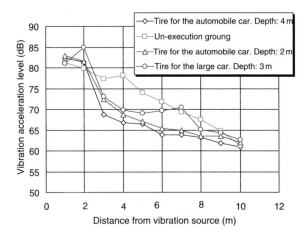

Figure 6.   Relation between the vibration acceleration levels and the distance from vibra.

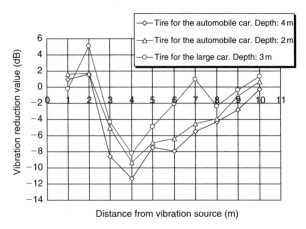

Figure 7.   Vibration reduction effects at each measurement conditions.

366

on a decrease of the biggest ground vibration usually appeared in the thing of the automobile tire (laying underground depth 4 m) even in the inside. The frequency analysis of 1/3 octave band analysis was done about the vibration acceleration level value, and these results were shown as a vibration acceleration spectrum, Difference in level of the vertical axis in the inside of the figure are the thing after the execution of isolation again. It is compared in on wall and a difference in level in each center frequency of 1/3 octave band in un-execution ground respectively. Therefore, when it was above mentioned, it is equivalent in the frequency area of effect on a decrease in vibration which minus value was seen in the same way from. And, the numerical value of the upper left shows distance from a source of vibration. It is compared with the un-execution ground in the comparatively low frequency band of 6.3–8 Hz and under, and the vibration low in weight of 10–15 dB is seen, and that vibration low loss in weight usually occurs in usual tire (laying underground depth 4 m) in (3 m and 4 m point) greatly right after isolation wall according to these in arrangement pattern of all scrap tire isolation wall. When a result of this experiment is examined, it can't think that the diameter (the thickness of isolation wall) of the tire influence effect on a decrease of the ground vibration from on a decrease of the ground vibration having been shown in the automobile tire more usually greatly than a large automobile tire.

4 THE FULL-SIZE EXPERIMENT ABOUT THE ISOLATION EFFECT OF GROUND VIBRATION USED BY A IMPROVEMENT TYPE SCRAP TIRE WALL (CASE 2)

This time, the ground vibration isolation wall which confirms an isolation character is a scrap tire ground wall to propose newly. That structure was composed of the steel pipe pile and the PHC stake in the central part of the scrap tire.

4.1 *An experiment contents*

Figure 8 showed the laying underground conditions of experiment field. An experiment makes a vibration occur by the way that a free fall of heavy weight drops in front of the isolation wall, and the method by running of the heavy equipment and effect on the isolation due to the existence of the isolation wall is confirmed.

Outline: Field dimension; W = 20.0 m L = 19.5 m (two places) isolation wall execution extension L = 10.0 m (two places) isolation wall laying underground depth; D = GL-3.0 mThe composition; outside scrap tire department t = 20 cm × 2 center stake department. 30 cm inside final stage material sand of the dimension ; 0.7 m × length 3.0 m isolation wall of the composition structure isolation wall of the structure; Case 1 scarap tire of the isolation wall, the composition structure Case 2 scrap tire of the steel pipe pile and the PHC stake.

The form of isolation walls are shown in the following. The center form of the form was composed PHC stake wall is shown Figure 9 and by steel pipe wall also is shown in Figure 10.

Figure 8. Condition of measure-ment.

Figure 9. PHC stake wall.

367

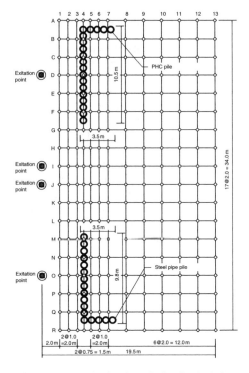

Figure 10.   Steel pipe pile wall.   Figure 11.   Layout drawing of vibration isolation wall.

4.2   A measurement arrangement

Figure 11 showed the arrangement of the points of measurement. The measured points of 13 points of one line hits was arranged as eleven measured line of A to R, and a point of measurement was moved to 234 surface of the earth sides. And, two places of GL-5.5 m were measured as a measurement in the ground before and behind the isolation wall.

4.3   The measurement outline

Six picking out sensors were installed, and the measurement of the vibration level of the measurement of heavy weight drop and a vibration in heavy equipment running. Arrangement was changed and measured in all the point of measurements because the scores which could be measured at the same time were six points.

The arrangement of the sensor was arranged directly to the surface of the earth side. It was inserted into the vinyl chloride palace of the depth 5.5 m that it was dug in advance by using one ground acceleration meter. And measurement in the ground recorded two acceleration wave forms of the front of the wall body and the rear.

Waveform regeneration went for digital wave form data (acceleration wave form). And, Fourer spectrum analysis was performed.

5   EXPERIMENT RESULTS AND ITS CONSIDERATION

5.1   The result of the propagation speed of the wave

The result that the propagation speed of the wave motion in the surface of the earth side of the measurement testing ground place of the elasticity wave speed was measured is shown in the Table 1.

368

Table 1. Spread velocity.

Classification	Spread velocity
Primary wave velocity	436 m/s
Secondary wave velocity	200 m/s

Table 2. Excellence frequency (Hz).

Classification	Exciting source	Excellence frequency
PHC Pile	Heavy weight, without isolation wall	30–40Hz
	Heavy weight, with isolation wall	30–40 Hz
	Heavy equipment, without isolation wall	50–60 Hz
	Heavy equipment, with isolation wall	30–60 Hz
	Vibratory hammer, with isolation wall	50–60 Hz
Steele pipe pile	Heavy weight, without isolation wall	20–40 Hz
	Heavy weight, with isolation wall	20–40 Hz
	Heavy equipment, without isolation wall	30–40 Hz
	Heavy equipment, with isolation wall	40–50Hz
	Vibratory hammer, with isolation wall	40–50 Hz

5.2 Excellence frequency results by using 1/3 octave band analyzer

The result that 1/3 octave band analyzed the vibration which occurred is shown in the following to excitation level set up in the consideration dominant frequency experiment.

Table 2 is the list of the dominant frequency read from data. It is understood that it answers in the number of vibrations of 40 Hz to 60 Hz in the neighborhood, vibration used by vibro-tamper is 20 Hz to 40 Hz by the free fall of the heavy weight, 30 Hz to 60 Hz by the neighborhood, heavy equipment running.

5.3 The vibration decrease effect analysis of the dominant frequency

As for the decrease effect analysis, Figure 12 show the excellent number of vibrations. It is understood that effect on a decrease is big every time the frequency of vibrations increases by the thing that did a vibration acceleration level ratio by the frequency of vibrations of the point in front of the isolation wall from the back of the isolation wall 0.75 m based on the 1.75 m point. And, effect on a natural decline of the ground is shown in the Figure 13.

5.4 The tendency (the surface of the earth side of effect on a decrease by the isolation wall)

The graph which showed it in the Figure 14 showed effects on a decrease of the isolation wall by excitation of heavy weight on the surface of the earth side. For the low loss in weight by the distance, the values about an isolation wall was laid and one's about the natural ground measured line which left the center of the isolation wall 10 m was compared. A decrease in about 8 dB can be confirmed in the rear of the isolation wall.

369

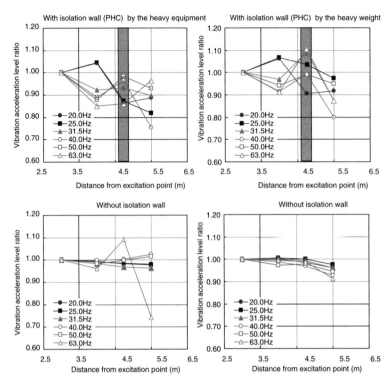

Figure 12.　The vibration decrease effect of the dominant frequency.

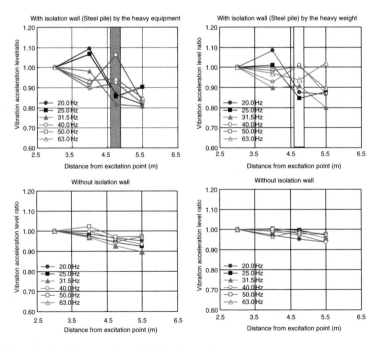

Figure 13.　The vibration decrease effect of the dominant frequency.

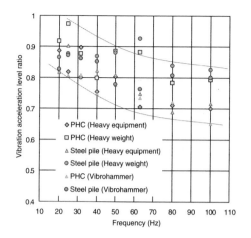
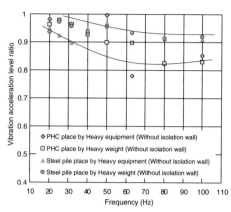

Figure 14. The vibration tendency by isolation wall. Figure 15. The vibration tendency by isolation wall.

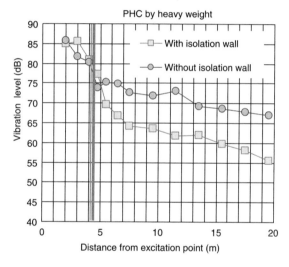

Figure 16. The vibration reduction effect by PHC stake wall.

The graph which showed it in Figure 15 showed effects on a decrease of the isolation wall by heavy weight drop on the surface of the earth side. Low loss in weight by the distance with the natural ground measured line which left the center of the isolation wall that an isolation wall was laid 6 m of the thing and the isolation wall end part was compared. A decrease in about 5 dB can be confirmed.

5.5 *Excitation (comparison with the natural ground of the isolation wall end part) by heavy weight*

The graph which showed it in the Figure 16 (center line date of isolation wall) and Figure 17 (end part date of isolation wall) showed effects on a decrease of the isolation wall by excitation of heavy weight on the surface of the earth side. Low loss in weight by the distance with the natural ground (measured line which left the center of the isolation wall that an isolation wall was laid 6 m) of the thing and the isolation wall end part was compared. A decrease in about 5 dB can be confirmed in the rear of the isolation wall.

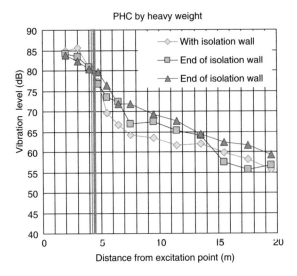

Figure 17.   The vibration reduction effect by PHC stake wall.

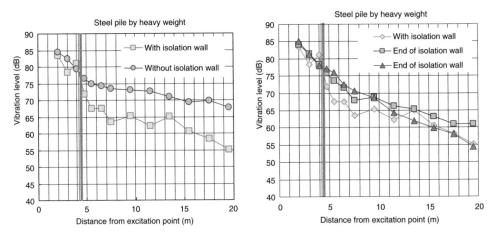

Figure 18.   The vibration reduction effect by steel pipe pile wall.

Figure 19.   The vibration reduction effect by steel pipe pile wall.

5.6   *Excitation (comparison with the natural ground of the isolation wall hair department end part) by the heavy equipment*

The graph which showed it in the Figure 18 showed effects on a base of the isolation wall by excitation of the heavy equipment on the surface earth side. It is compared with the natural ground of the isolation wall end part in the rear of the isolation wall, and a decrease ion about 5 dB can be confirmed.

5.7   *Excitation (comparison with the natural ground of the isolation wall end part) by the heavy equipment*

The graph which showed it in the Figure 19 showed effects on a decrease of the isolation wall by excitation of the heavy equipment on the surface earth side. A decrease in about 5 dB can be confirmed in comparison with the natural ground of the isolation wall end part.

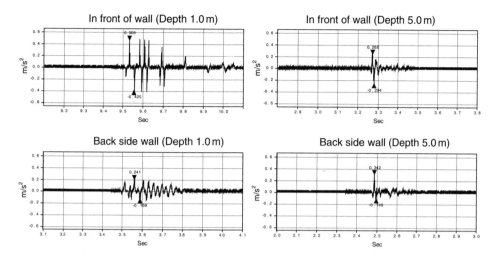

Figure 20.   Vibration acceleration records.

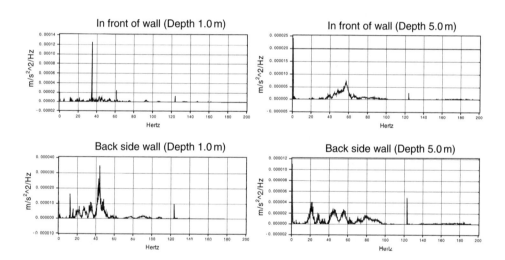

Figure 21.   Spectrum results.

5.8   *Vibration acceleration records and their power spectrum by the excitation of heavy weight*

The vibration acceleration recorded in 1.25 m with the depth 1.0 m and the ground vibro-meter in the depth of 5.0 m before and behind the PHC wick stake are shown in Figure 20 and their power spectrum are show in Figure 21.

5.9   *The influence of the diffracted ray*

Two were common to the isolation wall from contour figure measured on the surface of the earth side, and a diffraction tendency from the end part was compared. And, when an isolation wall is made a type L, it is understood that diffraction phenomenon to that the corner part decrease greatly.

6 CONCLUSIONS

(1) The conclusion which the following could get from this local vibration experiment was expressed. It could get great effect on a decrease in vibration of 5–12 dB, and future validity as isolation countermeasure acting in concern was confirmed by countermeasure method of the ground vibration which a scrap tire was used for effectively.

(2) Effect on a decrease was shown in the ground vibration of the low frequency band 6.3–8 Hz and under as well right after scrap tire isolation wall. However, a search for the durability in the distance of that effect on a decrease is shown as the future subject to use as the countermeasure for the low frequency vibration.

(3) The diameter of the scrap tire became the result which didn't influence effect on a decrease of the ground vibration.

(4) Effect on a decrease in vibration of the isolation wall is shown in the following. It can get about 10% of effect on a decrease in 20 Hz–40 Hz, 80 Hz–100 Hz, and it is here in 40 Hz–60 Hz about 20% of effect on a decease.

(5) When a heart material was made PHC, and it was the case that a steel pipe stake was taken, and effect on a decrease in vibration of isolation wall showed effect on a decrease of about 8 dB together in comparison with the natural ground from the measurement in the vibration acceleration level in the surface of the earth side in the rear of the isolation wall.

(6) It was shown by the vibration acceleration record wave form measured in the ground of the depth 1.0 m and 5.0 m that and the excellence frequency of vibrations before and after the isolation wall changed greatly in the acceleration before and after the isolation wall, it was cut by half.

REFERENCES

Hayakawa, K. (2002). Environmental vibration isolation into the ground, *Journal of INCE/J, Vol. 24, No. 6*: 408–414. (in Japanese)

Hayakawa, K. (1999). The present condition of the environmental vibration problem and subject, the actual condition of the environment vibration in the city stage and countermeasure course textbook, Kansai branch of Japanese Civil Engineering Society: 1–5. (in Japanese)

Hayakawa, K. Kashimoto, T. and Moriwaki, M. (2005). Isolation method of ground vibration by using used tire in considering environmental condition, *BUTURI-TANSA, Vol. 58, No. 4*: 391–396. (in Japanese)

Miterai, Y. et al. (2005). About the effective use as the ground material of the old tire rubber chip and environmental impact, *Proceedings of the 6th Environmental Symposium of the Japanese Geotechnical Society*: 351–358. (in Japanese)

Morisawa, T. et al. (2005). Effect about the tenacity character of the rubber chip mixed solidification management soil, *Proceedings of the 6th Environmental Symposium of the Japanese Geotechnical Society*: 359–364. (in Japanese)

Development of new geomaterials for environmental vibration mitigation

N. Sako
Junior College of Nihon University, Chiba, Japan

T. Adachi
Nihon University, Tokyo, Japan

N. Hikosaka
Nippon Thermonics, Co., Ltd., Kanagawa, Japan

T. Funaki
YACMO, Co., Ltd., Tokyo, Japan

ABSTRACT: In this study, new geomaterials are developed for dealing with the environmental vibration problems from the viewpoint of recycled resources. Two points are aimed at development of new materials. (a) The initial physical properties should be flow body (however, the material must be solidified with the passage of time) (b) The material should have high damping performance. In order to evaluate the mechanical properties of the new materials, cyclic loading tests were conducted using triaxial test equipment. In the case of only adding straight asphalt or emulsified asphalt, the damping ratio was 18–24% even in the elastic range. Field tests were conducted to research the actual vibration mitigation abilities of the aforementioned material using emulsified asphalt as a binder. After the ground was improved, the effectiveness of the vibration mitigation is indicated to be remarkable; however, the effectiveness is dependent on the exciting frequencies.

1 INTRODUCTION

In a modern city, environmental vibrations as represented by high speed trains, massive amount of traffic and factory operations are drawing attention because the residential areas are adjacent to highways, main roads, railways, subways and factories due to the restrictions of land. In Tokyo, residential areas have recently been developing further toward the center area of the city. Increasing condominium construction bordering railways, subways, highways, and main roads will cause people living in the city to call for corrective measures in the near future (Fujimoto, T. et al. 2005).

Additionally, the recycling of construction/demolition (C & D) materials has been promoted from the viewpoint of ecology in the Japanese construction industry (Ministry of Land, Infrastructure and Transport Japan). Yasuhara (2006) introduced various new geomaterials using scrap tire. Recycle technologies on improvement of the ground with a focus on C and D materials have developed in the last few years (Takemiya, H., et al., 2005, Kuno, G., et al. 2006, etc.).

In this study, development of a new geomaterial is attempted in order to take corrective measures against environmental vibration-pollution using recycled resource materials to the existing buildings. In particular, the following two points were aimed at the development of new materials. (a) The initial physical properties should be flow body (however, the material must be solidified with the passage of time). (b) The material should have high damping performance. This paper demonstrates the mitigation effect of the new material from the field tests after confirming its high damping performance in laboratory tests.

Table 1. List of material mixtures making up the new geomaterial.

Specimen	Cutting pieces of scrap tire (%)	Construction/ Demolition soil materials (%)	Straight asphalt (%)	Emulsified asphalt (%)	Cement milk (%)	Water (%)	Mass density (g/cm³)
A	21 (26)	49 (37)	–	–	6.7 (3)	23.4 (34)	1.38
B	40 (46)	40 (29)	20 (25)	–	–	–	1.16
C	37 (46)	38 (29)	–	15 (21)	10 (4)	–	1.10
D	41 (48)	42 (30)	–	17 (22)	–	–	1.01
Rubber	–	–	–	–	–	–	1.54

* The upper value is mass precent the lower is volume percent.

Figure 1. Photo of tire chips.

Figure 2. Photo of specimen D.

2 LABORATORY TEST

2.1 Used materials making up new geomaterial

The new geomaterial consists of aggregate and binder. The cutting pieces of scrap tire (the tire chips) and C & D soil materials are used as the aggregates, and straight asphalt, emulsified asphalt and/or cement milk are used as the binder. Straight asphalt is a material that requires a high heating temperature to solidify, whereas emulsified asphalt is a material that solidifies at room temperature, used herein due to its easy usage in the course of construction. As the cement milk, blast furnace slag cement is used in normal construction.

2.2 Cyclic triaxial test method

The mechanical properties of the new geomaterials are evaluated using by triaxial test equipment. The test specimen is 5 cm in diameter and 10 cm in height. One of the Japanese geotechnical test methods -*Method for cyclic triaxial test to determine deformation properties of geomaterials*- (JGS 2000) applies to the tests conducted in this paper.

After the specimen is set up on the equipment, it is saturated with the aid of CO_2 gas and de-aired water. The tests are conducted in the condition of 1 Hz as a loading frequency, 11 cyclic numbers and 50–100 kPa as the confining stress which correspond to the vertical stress of 5m in thickness of the geomaterials, and are loaded in an undrained condition. In the case of the test specimen with cement milk added, an LDT (Local Deformation Transducer) is used to measure displacement so that bedding error is avoided. The test of the rubber specimen cut off from the construction site using vibration-control material is added as the comparable objection.

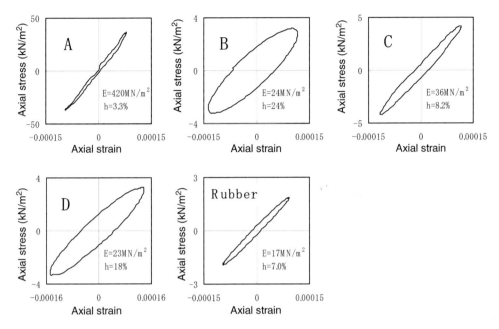

Figure 3. Stress-strain relationship of the specimens, respectively.

Table 1 gives the list of material mixtures making up the new geomaterials, as well as the mass density measured from the specimens. The tire chips used for the geomaterials are shown in Fig. 1, and Fig. 2 shows specimen D.

2.3 Test results

The stress-strain relationships of each specimen are depicted in Fig. 3 as well as the values of Young's modulus and hysteresis damping ratio calculated from the method described in JGS 2000. The specimens treated in this study are not as stiff as soil materials, so we cannot get Young's modulus and hysteresis damping ratio at the 10^{-5} small strain level from the soil test equipment. Therefore, we defined herein that the initial Young's modulus E_0 is the Young's modulus at 10^{-4} small strain level.

The specimen A (only cement milk added)'s Young's modulus is 25 times as large as the rubber's, but the hysteresis damping ratio is half of the rubber's. This shows that the cementation effect by the cement milk occupies the characteristics of the specimen A, and the elasticity of tire chips are not reflected by that.

On the other hands, the Young's modulus of specimens B (only straight asphalt added) and D (only emulsified asphalt added) is the same as rubber's, and the hysteresis damping ratio is 2.5 times or more as large as rubber's even in the elastic range. Therefore, it is considered that much ductility of asphalt and elasticity of tire chips produces a synergy effect, explaining why B and D show high damping performance. The specimen C (both cement milk and emulsified asphalt added)'s hysteresis damping ratio is smaller than that of B and D. This cause is considered to be the cementation effect of the cement milk. Incidentally, the hysteresis damping ratio of alluvial soil in the ground is 2%–3% in the elastic condition.

Fig. 4 depicts Young's modulus and hysteresis damping ratio of each specimen depending on the axial strain. The ordinate in Fig. 4(a) is normalized by E_0. Fig. 4 shows the tendency that the Young's modulus is reduced and hysteresis damping ratio is grown-up with the increasing of the axial strain in the case of A. However in the cases of B and D, Young's modulus is not as reduced as

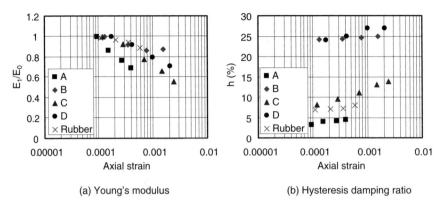

(a) Young's modulus   (b) Hysteresis damping ratio

Figure 4. The Young's modulus and the hysteresis damping ratio depending on the axial strain.

Table 2. List of material mixtures used in the field test.

	Cutting pieces of scrap tire (%)	Construction demolition soil materials (%)	Emulsified asphalt (%)
Mass percent	30	46	24
Volume percent	39	25	36

in the case of A, and the hysteresis damping ratio indicates a level-off trend at the high level. This tendency suggests that B and D will be useful for earthquake hazard mitigation.

3 FIELD TEST

3.1 Information of site

Field tests were carried out to examine the aforementioned new geomaterial's mitigation actual abilities against vibration propagated in the ground. The geomaterial used in the field test was a material mixture with emulsified asphalt as the binder (Tire chips with Emulsified Asphalt -TEA material-) that indicated high damping performance in the laboratory test. Table 2 gives the list of material mixtures used in the field test. A site that was not confirmed an undesirable environmental vibration exercising a harmful influence upon the measurement was adopted for the field test. Fig. 5 illustrates the soil profile at the site.

3.2 Test setup

Fig. 6 illustrates the setup of the field vibration tests. The acceleration pick-ups are arranged straight along the ground (except P2 and P3) from the source loading, and acceleration is measured simultaneously. P2 and P3 are set up to examine how large the amplitude of the diffraction wave is. The source loading is given by a 500 gal exciter-type shaker for the maximum capacity. The vertical loads are imposed directly on the compacted ground surface. The acceleration pick-up and the exciter type shaker are shown in Figs 7 and 8.

The experiment was conducted according to the following procedures:

1) Amplitude of the acceleration before the improvement of the ground with loading on the original condition was investigated.
2) After the trench was dug 0.75 m in width, 3 m in depth and 10 m in length as in Fig. 9, it was filled with TEA material and cured carefully until it is solidified as in Fig. 10.

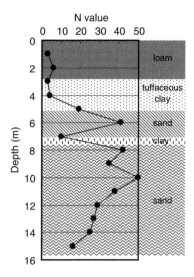

Figure 5.   Soil profile at field test site.

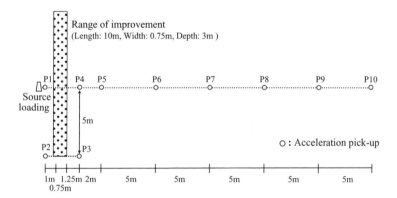

Figure 6.   Field test setup.

Figure 7.   Loading equipment.

Figure 8.   Acceleration pick-up.

3) Amplitude of the acceleration after improved by the TEA material was investigated.

Background vibration was sampled before the measurement, respectively, and the sampled data is considered in the following calculation. Vibration acceleration data obtained from the experiment is processed through the band-pass filter around the objective frequency.

Figure 9. Trench before installation.

Figure 10. After improvement.

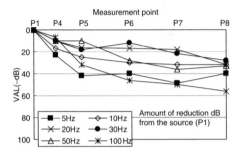

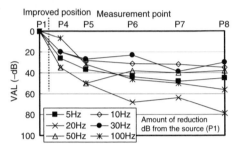

Figure 11. Reduction of VAL on the original condition.

Figure 12. Reduction of VAL after improved condition.

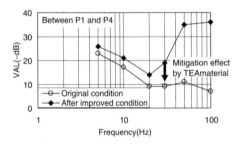

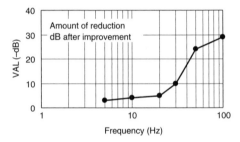

Figure 13. Difference of VAL between P1 and P4.

Figure 14. Mitigation effect of TEA material represented by VAL.

3.3 *Measurement*

The vibration acceleration level (VAL) in the Japanese specification is calculated as follows at each measuring point in order to give an easy understanding of its effect on the environmental vibration assessment.

$$VAL(dB) = 20\log\frac{a}{a_0} \qquad (1)$$

where a = the wave acceleration and a_0 = reference acceleration.

Figures 11 and 12 are the representations of the reduction of VAL on the original condition and after the improved condition, respectively, from the loading source P1. These figures show similar reduction tendencies behind P4 to P10, but show obviously different reduction between P1 and P4.

Then, the difference of reduction VAL between P1 and P4 are computed in Fig.13 in order to investigate the amount of reduction dB. The mitigation effect is appreciated to be remarkably different depending on the exciting frequency from Fig.13 Moreover, the subtraction of reduction dB is computed between the original condition and after the improved condition in Fig.14 to

380

represent the mitigation effect of the TEA material. The reduction is 10 dB–30 dB in the frequency range of 30 Hz-100 Hz and 3 dB–4 dB in the range of 5 Hz–10 Hz in terms of VAL.

There is a possibility that a wave with a low frequency like 5 Hz–10 Hz may propagate diffractively in this experiment, and expanding the width of improvement may be needed to materialize a more effective mitigation against environmental vibration for such low exciting frequencies.

4 CONCLUSION

Developing a new geomaterial is attempted in order to take corrective measures against environmental vibration-pollution using tire chips and the construction/demolition soil materials treated as recycled resource materials.

In the laboratory tests, the mechanical properties of new geomaterials are evaluated using triaxial test equipment. The hysteresis damping ratio of the material using emulsified asphalt as a binder (TEA material) shows great performance, such as 18 % even in the elastic range condition. This value is 2.5 times as large as the rubber's value cut off from the construction site using the vibration-control material.

Field tests were carried out to examine the TEA material's mitigation actual abilities against environmental vibration propagated in the ground. The mitigation effect is appreciated to reduce 10 dB–30 dB in the frequency range of 30 Hz–100 Hz in terms of VAL. However, the effectiveness is depended on the exciting frequency.

In particular, the TEA material has the merit of easy manipulation in the course of construction because its initial property is its flow body (it is solidified at room temperature).

REFERENCES

Fujimoto, T., Notani, M. and Kuroda, S. 2005. Gradual decrease method of a subway vibration to spread in the condominium and the effect, Proceedings of the 40th Japan national conference on geotechnical engineering, pp. 2699–2700 (in Japanese).

Ministry of Land, Infrastructure and Transport Japan. homepage, http://www.mlit.go.jp/sogoseisaku/region/recycle/

Ysuhara, K. 2006. Development of new geomaterials, Tsuti-to-Kiso, The Japanese Geotechnical Society, Vol. 54, No.12, pp. 6–7. (in Japanese).

Takemiya, H., et al 2005. Environmental vibrations, ISEV2005, edited by Takemiya, H., Taylor and Francis

Kuno, G., et al 2006. Recycling technology on improved ground (special section), The foundation engineering and equipment, Sogo doboku kenkyusho, Vol. 34, no. 2 (in Japanese).

JGS (Japanese Geotechnical Society) 2000 Explanation and method of soil test.

Scrap Tire Derived Geomaterials – Opportunities and Challenges – Hazarika & Yasuhara (eds)
© *2008 Taylor & Francis Group, London, ISBN 978-0-415-46070-5*

Pre-cast concrete WIB with tire shreds fill-in for a vibration mitigation measure

H. Takemiya
Okayama University, Okayama, Japan

ABSTRACT: Environmental vibrations are increasingly drawing attention in modern society. Protecting built-in areas from exposures to detrimental effects is desired. Herein, an attempt has been made to develop an effective vibration reduction measure in ground–WIB against the man-made vibrations as induced traffic and machine operation. Theoretically, for effective damping mechanism, the present WIB utilizes the scattering of waves by stiff cell-walls and energy dissipation of fill-in tire shreds. The field test proved the present WIB moiré effective than conventional types of wave barriers.

1   INTRODUCTION

The Environmental vibrations as represented by high-speed train, massive traffic on highway, factory operations and massive events in halls, are increasingly drawing attention in modern society. These vibrations include mostly the low frequency nature from 3 to 8 Hz frequencies that give adverse affects to inhabitants and operation of vibration sensitive equipments in the neighborhood. According to he vibration Regulation Law in Japan (1976)[1], the limit of vibration acceleration level (VAL) is in the range of 60–65 dB during daytime while 55-d0 dB during nighttime. The Shinkansen-induced vibration, on the other hand, is legitimated below 70 dB; otherwise some counter measures should be taken to fulfill the threshold (1976)[2]. Protecting built-in areas from exposures to the detrimental vibration prolusion is strongly desired.

A variety of measures have been attempted to such an environment for avoiding being exposed to vibration sources. They are a row(s) of concrete piles or steel sheet-piles, open trenches of fill-in trenches with concrete or bentonite. These barriers aim to screen the outgoing/incoming vibrations and protect structures and facilities by creating a shadow zone behind the former.

The author and his co-workers have been engaged so far in developing a series of innovative measures called WIBs that can modulate the propagating waves and impeding them across the zone. These include a buried stiff slab construction at a specific depth from the ground surface (1993, 1994, 1996)[3–5], a similar inclusion of used whole tires with treated soil fill-in (2001)[6] and construction of barriers by soil improvement technique (2002)[7], a honeycomb shape stiff barriers with tire shreds fill-in (2004, 2005, 2006)[8–10]. The reduction mechanisms of these are application of wave scattering for damping effect by stiff walls and wave energy absorption by soft fill-in materials. Among them a honeycomb cells with scrapped tires fill-in is demonstrated as a promising practice[10].

Recently, the author has conducted a field test to prove the vibration mitigation effect by the WIB with tire-shreds fill-in in comparison with other types of barriers. In order to give deliberate consideration to the field experimental results, different loading devices are taken from impulse to harmonic types. Because of the size for the proof test, the frequency range is targeted to a relatively low frequency range from 10 Hz to several tens Hz. Along with the improvement of the prediction method of vibration transmission in ground, the effective reduction measures can only be developed[8–9].

2 FIELD TESTS

2.1 *Site condition*

The site for test is an alluvium fan with sandy soils. The water table exists around 2 m deep from the surface. The N values by a bore hall test are listed in Table 1. The predicted shear velocities at layers are indicated in the table. Based on this soil profile, the author computed the wave dispersion characteristics by applying the thin layer method[12]. Figure 1(a) gives the information of the phase velocity vs. frequency and Figure 1(b) the group velocity vs. frequency of resolved modal waves

Table 1. Soil properties at site.

Depth (m)	N-value	Vs (m/s)
0.00–1.00	12	200
1.00–3.15	55	248
3.15–4.15	6	173
4.15–7.40	40	248
7.40–8.00	33	337
8.00–10.00	100	400

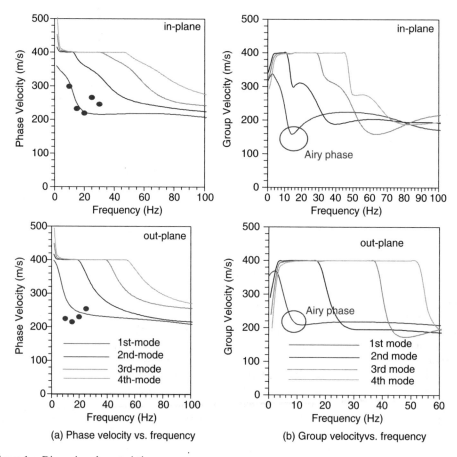

(a) Phase velocity vs. frequency

(b) Group velocityvs. frequency

Figure 1. Dispersion characteristics.

at the site. In these figures, the in-plane motion consists of the P and SV waves combined, whereas the out-of-plane motion consists of the SH wave field. The solid lines are the thin layer method solution and the symbols are field test results from SASW[13,14]. A distinct Airy phase frequency that corresponds to the local minimum of the group velocity is found in the in-plane motion at 14 Hz and slightly observable around 10 Hz in the out-of-plane wave. These specific frequencies are related to the major wave energy transmission so that they suggest that the wave propagation starts appreciable beyond 14 Hz and no propagation below it for the in-plane motion and the same interpretation holds at 10 Hz for the out-of-plane motion. Given the frequency and the wave speed, the corresponding wavelength is determined by a quotient of the latter against the former. Therefore, the maximum wavelength is expected to be 15 m at this site if the shear velocity of the top layer in Table 1 is used. It is around 12 m at 20 Hz and 8 m at 30 Hz. With the above mentioned knowledge, the author schemed the present field test.

2.2 *Loading types*

Different loading equipments are employed for vibration generation: a guided hammer (Photo 1(a) and (b)), a small shaker (Photo 1(c)), and passages of a caterpillar type back-hoe (Photo 1(d)). The guided hamper whose head weight 75 kgf can give an impulse loading in horizontal as well as in vertical directions by changing the set up position of the hammer head. The drop height was kept at 75 cm high for the former case and at horizontal distance 75 cm from the hitting position for the latter case. The shaker of a 30 Kgf weight was adjusted to yield the harmonic driving frequency from several up to 30 Hz. The back-hoe is run by a speed of roughly 4 km/h. This source is oriented to generate vibrations of high frequency of 20 Hz and more.

2.3 *Measurement array*

The measurements are conducted for the vertical and two perpendicular horizontal directions: one is along the measurement line and the other is perpendicular to it. See Photo 2 and Figure 2. Vibration sensors used in the field test are servo-type velocity meters by Tokyo Shokushin Co. Ltd. They are placed on the steel plates positioned at 5 m, 10 m, 15, 25 m distances from the loading position. These sensors have a flat frequency characteristic from DC up to 100 Hz. The vertical component and horizontal component along the measurement line constitute an in-plane motion while the horizontal motion perpendicular to the measurement line an out-of-plane motion. The data acquisition is carried out for the velocity response with time step of 0.005 seconds that gives rise to a Nyquist frequency 100 Hz.

2.4 *Vibration barrier constructions*

A variety of types of vibration barriers are constructed at the site for the purpose to compare the reduction effects with the situation without any measures (Case 1). The author has proposed a

(a) Vertical hammer impulse (b) Horizontal hammer impulse (c) Shaker (d) Caterpillar passage

Photo 1. Different types of loadings.

(a) Vibration measurement array   (b) Three perpendicular sensors on a steel plate stuck in ground

Photo 2.   Measurement array and vibration sensors at measurement points.

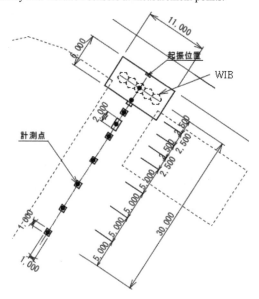

Figure 2.   Measurement array.

honeycomb-shaped WIB comprising cell-walls with the scrapped tire shreds fill-in, as the most effective one among vibration mitigation measures. The theory behind is the wave scattering by multiple stiff thin walls of different angles for the incoming wave and the energy absorption by the soft fill-in of high damping materials. The multiple rows of cells work for increasing wave impediments effect.

In this investigation, because of the most targeted driving frequency being higher than 14 Hz, investigated cases are the following five barriers in order to see the advantage of the above WIB over others stepwise. First, a trench (Case 2) is excavated as seen in Photo 3(a). Secondly, this trench was filled in by tire shreds (Case 3) (Photo 3(b)). The fill-in materials are scrapped tire-shreds of 2 and 4 inch cuts combined in sizes. The scrapped tires are packed in 1 m^3 volume bag each not only for easy handling at work but also for confining scrapped tires to a certain degree. Thirdly, a set of are pre-stressed concrete pieces of double Y-letter cross section are connected which makes a

Tire-shreds

(a) Trench (Case2) (b) Tire-shreds fill-in (Case 3) (c) a zigzag layout of
pre-cast RC (Case 4)

(d) Single-layer honeycomb cells (e) Double-layer honeycomb cells
(Case 5) (Case 6)

Photo 3. Honeycomb cells WIBs.

zigzag layout wall and bags of tire shreds are placed at the both sides (Case 4). Each factory-made concrete piece has the wall height is 120 cm and thickness is 10 cm (Photo 3(c)). The impedance contrast, as defined by the ratio of the product of mass density and wave velocity, of these materials leads to vibration mitigation due to the wave refection and transmission and the angle ups and downs of walls works for the wave scattering and damping consequence. Then, a single layer and a double layer layouts of honeycomb cells (respectively Case 5 and Case 6) composed of the above pre-cast concrete pieces are investigated. The closed section of cells provides the higher structural stiffness and a substantial extent of these cells subdues the waves whose wavelengths are around several times of the honeycomb size in plan view. The side length of the hexagons is 1 m (Photo 3(d) and (e)) For the double layer arrangement of cells, the horizontal to span across them is 6 m, nearly a half of the wavelength for a targeted frequency along the wave propagation direction. The depth is determined as 1.5 m from the top surface layer depth.

The side view of a single WIB in Figure 3(a) and the illustration of the plan views of a zigzag wall, a single layer honeycomb WIB and a double layer honeycomb WIB provides the actual dimensions of the model for the field test.

3 EVALUATION

3.1 *Measured responses*

Representative time histories of ground velocities at measured distances due to impulses of a guided hammer, due to passages of a caterpillar backhoe and due to harmonic driving by a shaker at 20 Hz frequency are shown in Figure 4. The dispersive nature of wave propagation changing the wave form as the distance increases to travel is clearly noted in the responses for the first two loading

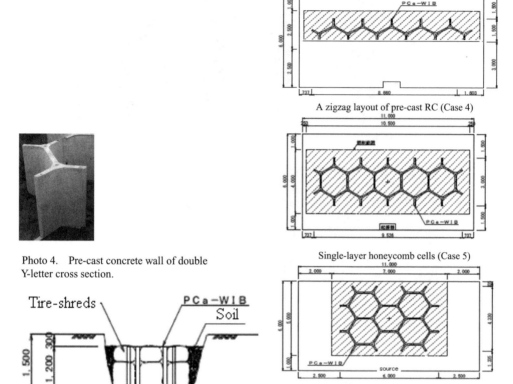

A zigzag layout of pre-cast RC (Case 4)

Single-layer honeycomb cells (Case 5)

Double-layer honeycomb cells (Case 6)

Photo 4.   Pre-cast concrete wall of double Y-letter cross section.

Figure 3(a).   Side view of WIB.   Photo 3(b).   Plan view of WIB.

situations. The results from the SASW method[13,14] for the impact response are plotted in Figure 1 for comparison with the theoretical prediction. These matching confirm the prediction accuracy of the fundamental cut-off frequency from the Airy phase frequency in the group velocity. It is read off as 14 Hz from the figure. The responses due to the shaker are featured by a harmonic response of the given frequency.

The Fourier amplitudes of response due to a hammer impulse indicates a wide spread frequencies near the source location but they tend to show a band limited tendency within 20 through 30 Hz as the observation distance is increased. This fact is a consequence of the ground filtering for vibration transmission. The same trend appears for the passage of a backhoe. The shaker driving possesses a specified 20 Hz frequency only. The consistency in a very harmonic response features guarantees the ground transmission of the harmonic vibrations.

3.2   *Maximum responses*

The maximum values are picked up from the time histories and plotted in Figure 6 to show the attenuation curves with distance.

The case without vibration reduction measures has higher maximum values than those for the cases with employed measures. Some appreciable reductions are noted below 25 Hz. The open and the tire shreds fill-in trenches show smaller response values in front of these barriers. The WIBs of tire shreds fill-in indicates the vibration reduction most in an exponential decay trend.

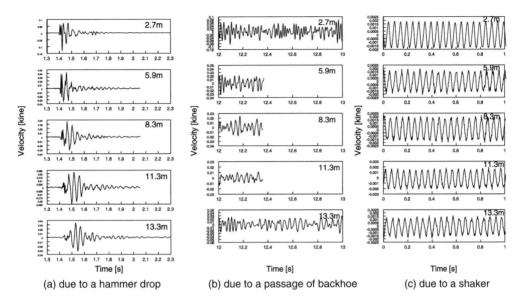

Figure 4.   Ground velocity response time histories, vertical components.

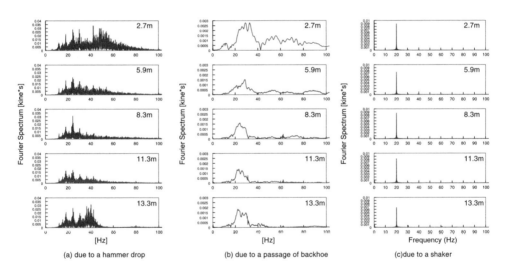

Figure 5.   Fourier amplitudes of the time histories.

Figure 7 is the measured vertical response for the caterpillar-type backhoe passage. The vibration reduction trend is significantly noted in this loading for all the barriers employed. It is also mentioned that a stable reduction is achieved by honeycomb WIBs.

For the harmonic driving by a shaker at 20 Hz, although smooth reduction results along the distance is not achieved, all the barriers worked for the reduction on average.

In order to investigate the change of frequency domain features by installing barriers on field, the Fourier amplitudes are obtained from the Fourier transform of the time histories. An envelope for all the experiment data is taken. The results for the hamper impulse and for a caterpillar-type backhoe passage are depicted in Figure 8. They give predominant peaks, respectively, at certainly determined frequencies. The former is affected by the hammer drop condition and the latter is by the

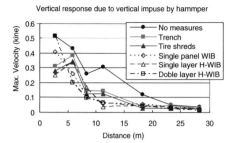

Vertical response due to vertical impuse by hammper

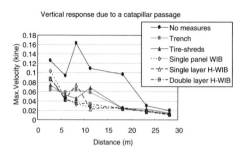

Vertical response due to a catapillar passage

Figure 7. Vibration attenuation characteristic, caterpillar-type backhoe passage.

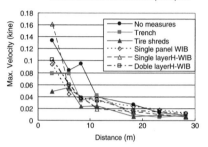

Horizontal response perpendicular to measurement line due to the same directional horizontal hammper impuse

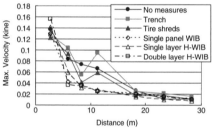

Horizontal response along measurement line due to the same directional hammer impulse

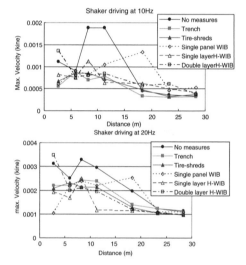

Shaker driving at 10Hz

Shaker driving at 20Hz

Figure 6. Vibration attenuation characteristic, hammer impulse.

Figure 8. Vibration attenuation characteristic, shaker driving.

blade width against the speed. The barrier effect is noted in the higher frequency range than 20 Hz mostly as seen in Figure 9. This leads the above vibration attenuation in terms of the maximum amplitudes.

3.3 *Vibration levels*

In the Japanese regulation regarding the environmental vibration the evaluation is based on the formula below:

$$VAL = 20\,log\left(A\,/\,A_0\right)\left[dB\right], \quad A_0 = 10^{-5}[m\,/\,s^2] \qquad (1)$$

where the A stands for the acceleration in either the maximum or effective rms (root-mean-square of sum of squared) value and A_0 the reference value. Herein, we are interested in the vibration reduction rate by the employed measures, which are given by

$$Reduction\ VAL = 20\,log\left(A_{before}\,/\,A_{after}\right)\left[dB\right] \qquad (2)$$

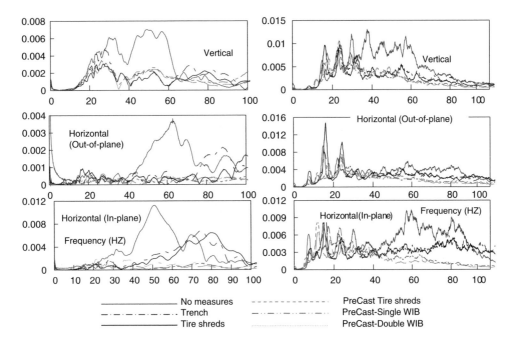

———————— No measures	- - - - - - - - PreCast Tire shreds
—·—·—·—·—·· Trench	—·—··—··—··· PreCast-Single WIB
———————— Tire shreds	············ PreCast-Double WIB

(a) A guided hamper impulse loading at 20Hz   (b) A caterpillar back-hoe passage at 4km/s

Figure 9.   Fourier amplitudes of ground responses at 5.9 m

where A_{before} and A_{after} are the acceleration values before and after the installment of the measures. Since we measured the vibration in terms of velocity response Vs, we may replace Eq.(2) by

$$Reduction\ VAL = 20\ log\left(V_{before}/V_{after}\right)\left[dB\right] \qquad (3)$$

Since the acceleration is obtained by multiplying the velocity by frequency, Eq (3) holds compatibility with Eq.(2) rigorously for the harmonic steady state response. Also, Eq.(3) may be an acceptable good indicator for evaluating the vibration reduction.

Figure 10 indicates such obtained vibration reduction effects in the unit of dB that is averaged over the measurement points for the respective employed barriers. The trench effect of around 4 dB is expected in view of the soil properties, trench size and frequency 20 Hz from the one-dimensional wave theory across it. The reduction more than 10 dB is noted in all the direction by the WIBs, most strikingly for the caterpillar passage loading.

Also considered is the rms vibration value velocity response that is defined as

$$V_{rms} = \sqrt{\frac{1}{2\pi T_d}\int_0^{100} 2V\left(f\right)df} \qquad (4)$$

where the Fourier amplitude $V(f)$ is integrated over the involved frequencies f and T_d is the response duration. The use of this velocity for E.(3) gives alternative evaluation of rms response from the Parseval theory. The result is depicted in Figure 11 for a hammer drop. Almost similar reduction rate is achieved by the barriers with the aforementioned comparison by the maximum values.

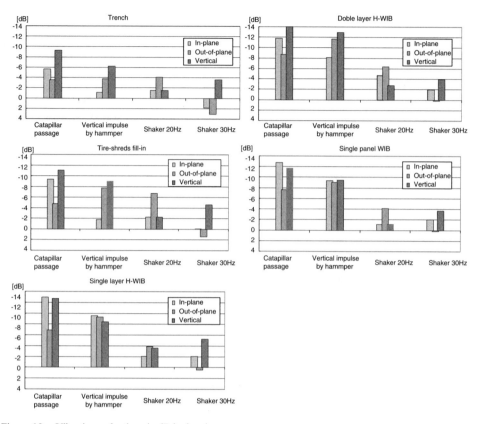

Figure 10. Vibration reductions in dB by barriers

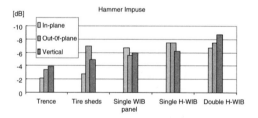

Figure 11. Vibration reductions in dB by barriers, rms.

4 CONCLUSIONS

In this paper, the author has reported the results of the vibration reduction by artificially installed barriers nearby the source location. The considered barriers includes a conventional trench, innovatively designed pre-cast concrete at core in the tire shreds, and single/double layer honeycomb-shaped WIBs. These constructions generates wave scattering and absorb the vibration energy on the way to propagate and work for vibration mitigation.

From the preliminary study on the probable cut-off frequency at the site, different loading are applied by using a guided hammer, a caterpillar-type backhoe and a rammer driving. These devices generate vibrations of higher frequencies than it. Putting this in mind, the barrier effects are compared with those with no measures installed in the attenuation characteristic. Herein, the velocity is preferred for better response indication. Then, it is noted that the honeycomb WIBs are more effective than conventional types of wave barriers, reducing more than several dB or 10 dB in better condition for the targeted frequency range.

ACKNOWLEDGEMENT

The author appreciates the cooperation of Landes Co., Ltd. for the present field test, especially Mr. T. Fujituska in arranging the site and pre-cast concrete fabrication and participation during the field tests. The author's appreciation is extended to the Chugoku Kohsaikai Foundation for its partial support to carry out field tests.

REFERENCES

1) Vibration regulation law, Ministry of Environment, Government of Japan (1976).
2) Environmental Agency, Government of Japan, Recommendation for anti-vibration measures against Shinkansen trains for urgent environmental conservation (1976).
3) Takemiya, H. and Jiang J.Q. Wave Impeding Effect by Buried Block for pile Foundation, *Proc. Japan Society of Civil Engineers*, No. 477/I-25, Structural Eng./Earthquake Eng. JSCE., 1993, 10, 149s–156s.
4) Takemiya H. and Akihiro F. Installation of a wave impeding block (WIB) for dynamic response reduction of soil-structure system, *Proc. Japan Society of Civil Engineers*, No. 489/I-27, 1994. 4, 243–250.
5) Takemiya, H., Goda, K. and Sato, N.: Passive Vibration Control Effect of Wave Impeding Block (WIB), *Proc. Japan Society of Civil Engineers*, 549/I-37, 1996, 221–230.
6) Takemiya, H. and Kii, Y., et al.: On vibration reduction by placing used tires as WIB, *Proc. Annual Conference*, Japan Geotechnical Society, 2001.
7) Takemiya, H., Hashimoto, M., Shiraga, A.: Practices of the WIB columns against road traffic-induced vibrations, *Soil and Foundation*, 9, Japanese Geotechnical Society, 2002, 19–21.
8) Takemiya, H., Field vibration mitigation by honeycomb *WIB* for pile foundations of a high-speed train viaduct, *Soil Dynamics and Earthquake Engineering*, 24, 69–87, 2004.
9) Takemiya, H. and Shimabuku, J.: Honeycomb-WIB for mitigation of traffic-induced ground vibrations when constructed around viaduct foundation, *Proc. of Japan Society of Civil Engineers,* 808/I-74, 2006, 103–112.
10) Takemiya, H. Fujitsuka, T. et al.: Development of a pre-cast-concrete tire shred combined WIB for vibration mitigation, 549, D-07, 1097–1098, Japan Geotechnical Society, Annual Conference, 2006.
11) Takemiya, H., Analyses of wave field from high-speed train on viaduct at shallow/deep soft ground, Journal of Sound and vibration, Special Edition for Euromech 484, 2006 (in print).
12) Kausel, E. An explicit solution for the Green functions for dynamic loads in layered media, Research Report R81-13, MIT (1981).
13) Gucunski, N. and Woods, R.D. "Numerical simulation of the SASW test," Soil Dyn. And Earthq. Eng. 11(1992) 213–227.
14) Stokoe, K H., Wright, S.G., Bay, J.A. and Roesset, J.M. "Characterization of geotechnical sites by SASW method," ISSMFE, TC#10, (1994), 15–25.

Author Index